David SF Portree
5/22/02

AF597450

To David,

a fellow "lunatic,"
who values the
historical aspects of
our quest for the moon,
+ who kindly shared us
the USGS collection.

with warmest regards,

William
Sheehan

Flagstaff, Arizona
may 30, 2012

Epic Moon

A history of lunar exploration in the age of the telescope

William P. Sheehan • Thomas A. Dobbins

Published by:

P.O. Box 35025 • Richmond, VA 23235 • (804) 320-7016 • FAX (804) 272-5920

www.willbell.com

Copyright © 2001 by Willmann-Bell, Inc.
All Rights Reserved

Reproduction or translation of any part of this work beyond that permitted by Sections 107 or 108 of the United States Copyright Act without written permission of the copyright owner is unlawful. Requests for permission or further information should be sent to the Permissions Department, Willmann-Bell, Inc., P.O. Box 35025, Richmond, VA 23235.

Printed in the United States of America

First Printing

Library of Congress Cataloging in Publication Data

Sheehan, William P., 1954-

Epic moon: a history of lunar exploration in the age of the telescope/ William P. Sheehan, Thomas A. Dobbins.

p. cm.

Includes bibliographical references and index.

ISBN 0-943396-70-0

1. Moon--Surface. 2. Moon--Exploration--History. I. Dobbins, Thomas A., 1958- II. Title.

QB591 .S56 2001
523.3--dc21 2001035453

01 02 03 04 05 06 07 08 09 10 9 8 7 6 5 4 3 2 1

Table of Contents

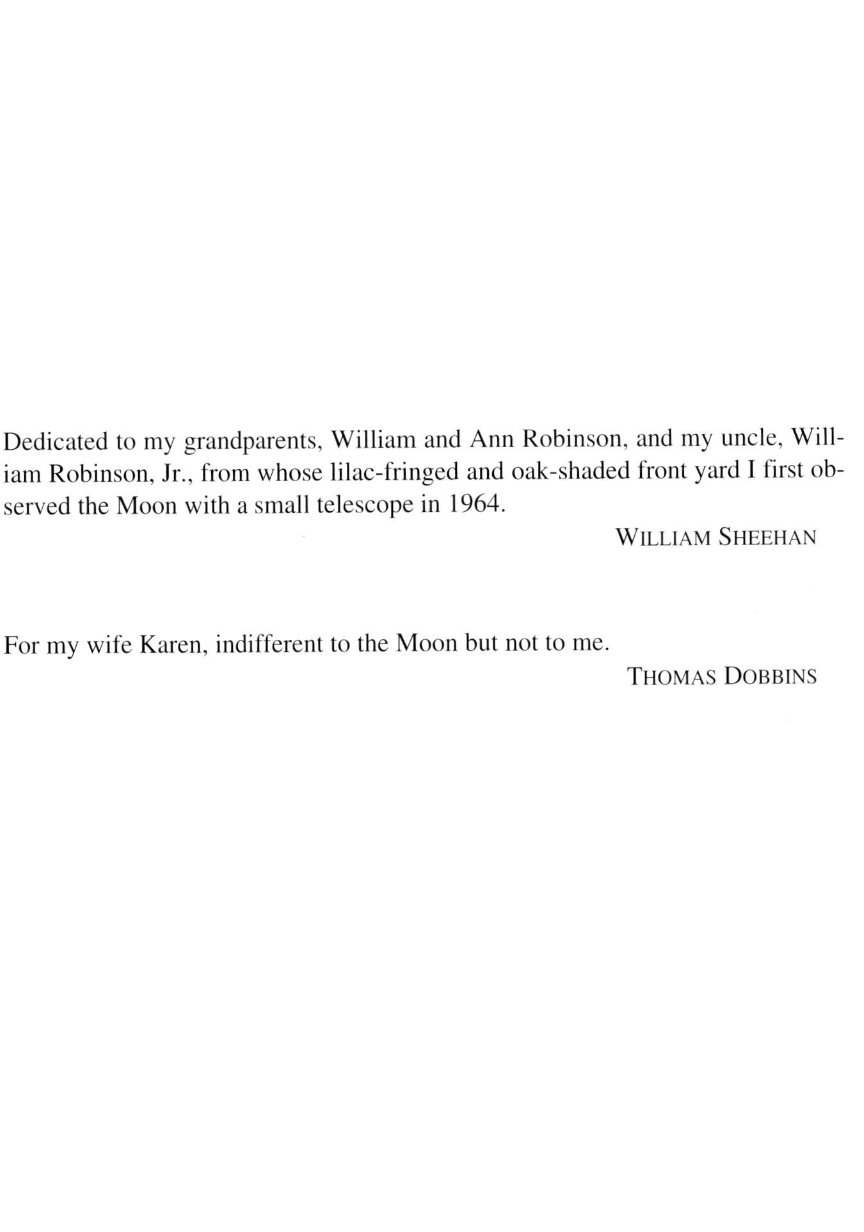

Dedicated to my grandparents, William and Ann Robinson, and my uncle, William Robinson, Jr., from whose lilac-fringed and oak-shaded front yard I first observed the Moon with a small telescope in 1964.

WILLIAM SHEEHAN

For my wife Karen, indifferent to the Moon but not to me.

THOMAS DOBBINS

Acknowledgments

Among the many people who have assisted us in research for this book, provided sage advice, or supplied resource materials and images, we wish to acknowledge the following: Leonard J. Abbey; Leo Aerts; Richard M. Baum; Judith Bausch, librarian of the Yerkes Observatory; the late Wolfgang Bayer; J. Conor Burns; Emmanuel M. Carreira, S.J.; Thomas R. Cave; Amy Chapman; Clark R. Chapman; Michael J. Crowe; Steven J. Dick; Dalvin Donovan; Richard Dreiser, of the photographic services department of Yerkes Observatory; Audouin Dollfus; Ruth S. Freitag of the Library of Congress; the late Dieter Gerdes; Owen Gingerich; David Graham; Walter H. Haas; William K. Hartmann; Alan W. Heath; Harold Hill; Peter Hingley, librarian of the Royal Astronomical Society; the late William Graves Hoyt; George Jones; Imelda Joson of *Sky & Telescope*; the late Alan P. Lenham; Steve Massey; Patrick Moore; José Olivarez of the Chabot Observatory and Science Center; Stephen James O'Meara; Wayne Orchiston; James H. Phillips, M.D.; Leif Robinson; Maurizio di Scullio; the late Eugene M. Shoemaker; Gary Seronik of *Sky & Telescope*; Dorothy Schaumberg of the Mary Lea Shane Archives of the Lick Observatory; Martin Stangl; Chris Szuter; Georges Vicardy; John Westfall; Don E. Wilhelms; Ewen Whitaker; and Charles Wood.

The authors are particularly indebted to John D. Koester for his painstaking review of the manuscript and innumerable valuable suggestions. We also wish to express our gratitude to Perry Remaklus for his unwavering support for this project and his relentless pursuit of perfection on our behalf.

Prologue

The Moon has always been there, floating across our skies, the silent companion of our nights. So elemental are its effects—stirring life on land with its silvery light, life in the sea with the ebb and flow of its tides—that it must already have quickened a strong sense of wonder among our erect stone-age ancestors of more than a million years ago. No doubt they already followed its drift among the stars and noted the cycle of its phases. Unfortunately, they are as enigmatic as the Moon itself. They left little record of themselves, still less of their thoughts.

Even when the dinosaurs roamed and our ancestors were lemur-like mammals, the Moon's death's-head sockets wore the same doleful expression as now. "The Earth is a sepulchre for famous men," said Pericles in his funeral oration; also the sepulchre for dinosaurs, ammonoids, jawed fishes, brachiopods, bryozoans. Before trilobites scuttled across the floors of shallow seas—when the world was the exclusive domain of stromatolite colonies, an anoxic realm of oceans and volcanic eruptions—the Moon, though closer to the Earth, wore an aspect that would have been scarcely distinguishable from what it wears now.

The reality of the Moon is incredibly ancient, its surface a realm of inconceivable changelessness and paralysis. But the appearance is not of a world forever lock-jawed and frozen, but one of changing phases and transmutations. Throughout mankind's long love affair with the Moon, it has evoked liquidity and instability. It has worn a cloudy, watery, changeable mask—a mask, in reality, of the Earth. Even in modern times, as the Moon's surface was scrutinized in great detail with powerful telescopes, the compelling illusion of change—so strongly impressed on the unconscious mind, reinforced through so many lunations over so many centuries—remained irresistible.

In the matter of lunar exploration, the eye arrived before the boot. Those who set out from the eyepieces of telescopes participated in a grand adventure which made them pioneers in their day as much as the astronauts have been in ours. Though they have been largely forgotten, along with the noisy worlds of controversy they once inhabited, they ought to receive their due, for they were, indeed, the first to the Moon.

WILLIAM SHEEHAN
THOMAS DOBBINS

Chapter 1:

The Discovery of a World in the Moone

The Pythagoreans—the followers of Pythagoras of Samos (ca. 565 to 500 B.C.), who after suffering persecution on their native Ionian island escaped to found a school at Croton, in Sicily—are said to have taught that the large naked-eye spots in the Moon known since time immemorial were reflections of the Earth's continents and seas in its mirror-like surface. During the 2nd century A.D., the Greek writer Plutarch, in his dialogue "On the Face in the Orb of the Moon," recalled this doctrine, and attributed it to Clearchus of Soli, who lived around 320 B.C.: "What is called the face consists of... images of the great ocean [the Mediterranean] reflected in the Moon." However, Lamprias, the character who speaks for Plutarch himself, rejected the idea based on what was then known of the terrestrial globe. Ever since the time of Homer, the Greeks had believed the Earth to be circumscribed by the stream of Ocean, a continuous sea, but Plutarch failed to find its image in the Moon:

> In the first place... although the outer ocean is a single thing, a confluent and continuous sea, the dark spots in the moon do not appear as one but as having something like isthmuses between them... [If] the ocean is single, it is not plausible that its reflected image be thus discontinuous.[1]

Plutarch thus concluded that the Moon's surface is no mere reflection but rather a surface like the Earth's, full of mountains and valleys. "It is likely," he wrote,

> ...that the moon... closely resembles in constitution the Earth... It is in fact not incredible or wonderful that the moon... has got open regions of marvelous beauty and mountains flaming bright and has zones of royal purple with gold and silver... bursting forth in abundance on the plains or openly visible on the smooth heights.[2]

Nevertheless, despite Plutarch's objections, the Pythagorean idea that the spots in the Moon were a mirror-image of the Earth long persisted into later times. It even seems to have been used by one Arab mapmaker who, eager to find a shortcut to mapping the unknown coasts of Africa, attempted to trace their outline from the lunar disk.[3]

A Muslim interpreter of Aristotle, ibn-Ahmad ibn-Rushd (1126–1198), known in the West as Averroës, argued that since the Moon and other planets were divine, they shared no substance in common with the Earth. Therefore the spots (*maculae*) in the Moon must be either reflections of the Earth (as the Pythagoreans had believed) or condensations of the lunar matter. Dante employs the second of these ideas in the *Divine Comedy*. As his medieval space-traveler passes through the sphere of the Moon, he describes it as "a cloud... lucid, dense, solid, and polished."

Popularly, the belief in the Moon-as-mirror concept survived almost into modern times. The Holy Roman Emperor Rudolf II, Kepler's patron, accepted it, and even imagined he could see Italy with its two adjacent islands distinctly outlined. And even as late as the nineteenth century this belief had not entirely disappeared; a visitor to Paris from Isfahan, on being shown the Moon through a telescope, told Alexander von Humboldt that in Persia it was thought that "what we see in the Moon is ourselves; it is a map of Earth."[4]

As long as the Moon was regarded as a smooth, homogeneous body, a mirror or a lucid cloud, its surface was of no real interest, and it was not, as historian Scott L. Montgomery argues, regarded as a suitable object for naturalistic representation.[5] Only after Aristotelian views began to be overthrown did the Moon finally become an object worthy of the artist's attentions.[6] In his notebooks, Leonardo da Vinci (1452–1519) entered several sketches of the Moon which show a remarkably human "face."[7] In a note to one of these sketches, Leonardo mentions the reflection theory only to reject it:

> Others say that the surface of the Moon is smooth and polished and that, like a mirror, it reflects the image of our Earth. This view is also false, inasmuch as since the shape of the lands is irregular, when the Moon is in the east it would reflect different spots from those it would show when it is above us or in the west; [but] the spots on the Moon... never vary in the course of its motion.

Leonardo, nevertheless, suspected the lunar spots underwent other changes:

> If you keep the details of the spots of the Moon under observation you will often find great variation in them, and this I myself have proved by drawing them. And this is caused by clouds that rise from the waters in the Moon, which come between the Sun and those waters, and by their shadow deprive these waters of the Sun's rays.[8]

Indeed, as ancient imaginations about a perfect, smooth and homogeneous Moon fell by the wayside, Leonardo's identification of the lunar spots with actual oceans and lands—not merely reflections of the terrestrial globe—acquired new credibility. With the discovery of a New World across a hitherto unnavigated ocean, Europeans began to awaken to the possibility of traveling across another unnavigated ocean to another New World in the Moon.

The first map of the New World in the Moon was drawn up by William Gilbert (1544–1603), the sometime physician to Queen Elizabeth and James I of England who is best remembered today as the author of the treatise *De Magnete*. About 1600, he made an outline map of the Moon from naked-eye observations, but it was never published and Gilbert himself died soon afterwards. However, his manuscript was later studied by eminent scholars such as Francis Bacon (1561–1626) and Thomas Harriot (1560–1621). Like Leonardo, Gilbert identified the dark areas in the Moon as continents and the light areas as seas. The implicit assumption was that water was more reflective than land. He lamented that earlier observers had not provided his generation with maps, for had they done so, the question of changes in the lunar surface features would already have been settled.

Gilbert devised a makeshift set of Latinate names for the lunar features, including, for the large bright area in the middle of the Moon, "Mare Medilunarium," for the main dark areas "Regio Magna Occidentalis" (the modern Mare Serenitatis) and "Regio Magna Orientalis" (the modern Mare Tranquillitatis and Mare Imbrium). None of these coinages is in the least bit captivating, but one feature, near the limb of the Moon (the modern Mare Crisium), was named rather more evocatively—Britannia. Here was an inspiration from Britain's colonial impulse, which had reached virgin territory in the New World in the colony of Virginia (founded by Sir Walter Raleigh in 1584–86) and had become newly confident after the brilliant defeat of the Spanish Armada in 1588. As Montgomery remarks: "'Brittannia' planted a claim on the Moon and implied, historically, that England, the new naval power of Europe, might one day send ships of a different kind to these distant lands and seas."[9] These ships were soon to set out—imaginatively at least—in the works of men who came of age during those years of far-flung aspiration.

Gilbert's map was not actually published until 1651. By then the telescope had been invented. The history of this wondrous instrument can be traced to 1608, when government officials of Zeeland in the Dutch Republic wrote that a spectacle-maker from Middelburg had "a certain device by means of which all things at a very great distance can be seen as if they are nearby."[10] The spectacle maker Hans Lippershey (?–1619) had applied for a patent, and within a span of two weeks, so did two other Dutch claimants, Jacob Metius of Alkmaar (north of Amsterdam) and Zacharias Janssen, another Middelburg spectacle-maker. No patent was granted, however, for the invention was known to many and felt to be too easy to copy, but the States-General did agree to pay Lippershey a handsome retainer in exchange for his exclusive services.

Word of the invention spread rapidly, and by the spring of the following year spyglasses were being offered for sale by spectacle-makers in Paris. By summer the news reached Italy, where it fired the imagination of Galileo Galilei, professor of mathematics at the University of Padua in the Republic of Venice. Galileo at once assembled his own small telescope, magnifying only 3X, by arranging two lenses, one convex and the other concave, in a tube fashioned from a sheet of lead. He soon improved on his work, and before long had a telescope magnifying 10X. Alive to the commerical possibilities that the instrument would confer on the traders in the Rialto, by the end of August Galileo had demonstrated it to the Doge and senators from the highest campaniles in Venice.

Galileo was not the first to grasp the potential of the new instrument for astronomical investigations. In England, Thomas Harriot made the first telescopic sketch of the Moon. Harriot was strongly identified with Sir Walter Raleigh's free-thinking "School of Atheisme" and had served as Raleigh's onetime mathematical tutor and scientific adviser. Another associate was the iconoclastic poet and dramatist Christopher Marlowe (1564–1593), he of the "mightie line." But unlike his colorful poet-companions, Harriot seems to have been personally reserved, a quiet perfectionist whose knowledge of mathematics was extensive but who published little during his lifetime.[11] In 1585, when he was in his early twenties, he accompanied Raleigh as surveyor and cartographer on an expedition to Virginia. Though he planned an encyclopedic account of the New World, he actually produced only "A brief and true report of the new-found land of Virginia", which became widely known through Richard Hakluyt's *Principal Navigations* (1598–1600), a collection of the voyages made by English adventurers and dreamers such as Raleigh—men in whom "the desire to see new worlds is interwoven with the lust for gold and the hope of conquest."[12]

But these had been the adventures of a younger man. Harriot was already fifty when he first turned a telescope toward the five-day old Moon, on the evening of July 26, 1609 (OS; August 5 NS). Though magnifying 6X, his "trunke," as he called it, must have been of very poor optical quality, since it showed very little detail.[13] For that matter, Harriot seems himself to have been singularly insusceptible to romantic possibilities. (One can only wonder what might have happened if Marlowe or Raleigh had been the first to view the Moon with such optical means!). Certainly he had very little insight into the true nature of what he was seeing; about as little as was shown on a slightly later occasion by his friend Sir William Lower, who used one of Harriot's telescopes to make his own observations of the Moon, from Kidwelly in Wales. Lower wrote memorably to Harriot on February 6, 1610 (OS):

> According as you wished I have observed the moone in all his changes. In the new I discover manifestlie the earthshine, a little before the dichotomie that spot which represents unto me the man in the moone (but without a head)... A little after neare the brimme of the gibbous parts like starres, much brighter then the rest and the whole brimme along, lookes like unto the description of coasts, in the dutch bookes of voyages. In the full she appeares like a tarte that my cooke made me the last weeke. Here a vaine of bright stuffe, and there of darke, and so confusedlie al over.[14]

Though Harriot was first to peruse the Moon with a telescope, it was Galileo (Figure 1.1) who boldly annexed its landscapes for the human imagination. In November 1609, he equipped himself with a telescope magnifying 20X—easily the most powerful instrument in existence at the time. He pointed it toward the Moon on November 30, 1609, four lunations after Harriot had done so. He observed intensively until December 19, making at least half a dozen lunar drawings (Figure 1.2). Soon afterwards he summarized his observations in a little book, *Sidereus Nuncius* (The Starry Messenger), one of the classics of observational astronomy, written at white-heat and published early in 1610:

Figure 1.1 Galileo Galilei, the "father of modern astronomy." Galileo constructed telescopes not only for his personal use and research, but also for well-paying customers. For centuries many mathematicians and astronomers had doubled as artisans, making and selling astrolabes, sundials, and surveying instruments, as well as casting the occasional horoscope. Courtesy Yerkes Observatory.

> portion, while on the other hand some dark patches invade the illuminated part. Moreover a great quantity of small blackish spots, entirely separated from the dark regions, are scattered almost all over the area illuminated by the sun with the exception only of that part which is occupied by the large and ancient spots... There is a similar sight on Earth about sunrise, when we behold the valleys not yet flooded with light though the mountains surrounding them are already ablaze with glowing splendor.[15]

Galileo realized that if there were lands and seas on the Moon (and he did not actually affirm that there were), the dark areas (the "ancient spots") must be the seas. He explained:

> I have never doubted that if our globe were seen from afar when flooded with sunlight, the land regions would appear brighter and the watery regions daker.[16]

Even before the invention of the telescope, the German astronomer Johannes Kepler (1571–1630) had concluded that the Moon "is a body of such a kind as this our earth, comprising one globe out of water and land."[17] But Kepler had erroneously interpreted the dark areas on the Moon as lands and the light spots as water, a conclusion he based on an observation that he made from the top of Mt. Schöckel in Styria; the river below had appeared brilliant when all the surrounding land had appeared dark. However, this effect was caused by the reflection off the water of the bright sky; when viewed from directly overhead, the river would have appeared darker than the surrounding land.

Now, reading Galileo, Kepler was immediately converted, and in a rapidly composed *Conversation with Galileo's Starry Messenger* declared without further hesitation: "The dark spots are seas, the light areas are lands."[18] Others too read Galileo as a revelation. Instead of fumbling over confusing telescopic images as he had done earlier, Lower, with new confidence, conveyed to Harriot the excitement of a whole new world being made accessible for the first time to exploration:

Figure 1.2 A pirated edition of Galileo's *Sidereus Nuncius* was hastily issued in Frankfurt in 1610. This German edition was illustrated with woodcuts that were far less skillfully rendered than the engravings of the original edition. With steadily decreasing clarity and aesthetic appeal, these crude woodcuts served to illustrate many subsequent editions of the book, leading more than one unwary scholar to unfairly deprecate Galileo's artistic talents. Courtesy E. M. Carreira, S.J.

> I am sure with more ease and safetie to him self & more pleasure to me… [I]n the moone I had foremerlie observed a strange spottedenesse al over, but had no conceite that anie parte thereof mighte be shadowes.[19]

Harriot's map of the Moon from the summer of 1610 (Figure 1.3) shows an equally striking improvement over his rough impression of a year earlier, no doubt attributable in large measure to the fact that he was by now using far more powerful telescopes that magnified 10X and 20X. However, the principal factor was no doubt the influence of Galileo's book.

Indeed, though Galileo himself had easily the best telescopes available at the time, we must not, as historian Albert Van Helden has pointed out, "allow this instrumental superiority to obscure Galileo's uncanny talent as an observer."[20] The contrast between Galileo's decisive results and Harriot and Lower's unpolished early attempts indicate just how great was Galileo's genius as an observer—the brain-directed power of his sight. Galileo was certain that the Moon was "sprinkled over with prominences and depressions," and was actually measuring the heights of lunar peaks—they proved to be about four Italian miles in height—before his contemporaries even realized that some of the features they were seeing were shadows cast by mountains.

The quality of Galileo's representations of the lunar surface far surpassed those of his contemporaries. On the whole, his engravings must be regarded as remarkably accurate. Better yet are his sepia ink-wash images (Figure 1.4), which are highly suggestive of direct telescopic impressions and even seem to record the increasing assurance of his views. (It has been argued that under the circumstances Galileo would have been too excited to make ink washes and that his brushwork would have been, in the words of art historian Samuel Edgerton, "composed not *en plein air* but in the studio."[21] However, given the orientation of the ink washes on the sheet—the upper middle one was first, then the one to its right; then Galileo turned the paper around and started with the one below it—it seems almost certain

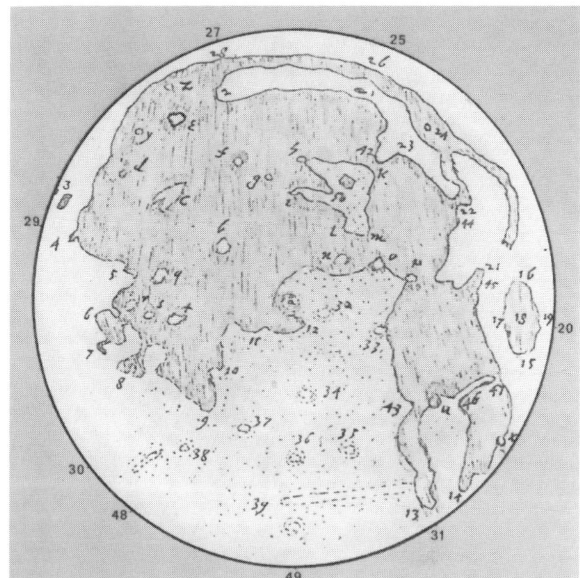

Figure 1.3 The first telescopic British map of the Moon, Thomas Harriot's chart of 1610 was a pen-and-ink drawing measuring only six inches in diameter. Harriott used a combination of numbers and letters to designate the most prominent lunar features. Courtesy Ewen Whitaker.

they were, indeed, composed *en plein air.*)

The poet and clergyman John Donne (1571?–1631), though opposed to Copernicanism, wrote in 1611 of "Galileo the Florentine… who by this time hath thoroughly instructed himself of all the hills, woods, and cities in the new world, the Moone."[22] Actually, what Donne wrote was truer of Kepler than Galileo. Captivated by Galileo's discoveries, the German astronomer had boldly speculated in his *Conversation*:

> I cannot help wondering about the meaning of that large circular cavity in what I usually call the left corner of the mouth [of the face in the Moon]. Is it a work of nature, or of art? Suppose there are living beings on the Moon… It surely follows that the character of the inhabitants must agree with their dwelling place. Since the Moon has much bigger mountains and valleys than our earth, they no doubt have more massive bodies, and construct gigantic projects. During their day, as long as fifteen of our days, they feel insufferable heat. Perhaps, lacking stone for building shelters against the Sun, they instead make do with the sticky soil. They dig up huge fields, and as they carry out the Earth, they heap it in a circle, in order to draw out the moisture from below. So they hide themselves in the shadows of their excavated mounds, and as the Sun moves across the sky, shift their positions so as to remain ever in the shadows. They make, then, as it were, an underground city, and live in caves in the circular embankment.[23]

Galileo's own views were always more cautious and reserved. In 1613, he published some comments on lunar conditions in his *Letters on Sunspots,* a work in which he was mainly intent on making his position clear about the spots he observed on the Sun with his telescope. The sunspots had been independently discovered by Galileo's rival, Christoph Scheiner (1575–1650) in Germany (Figure

Figure 1.4 Galileo's attractive sepia ink-wash rendering of the Moon, which were almost certainly completed outside at the telescope. Courtesy Ewen Whitaker.

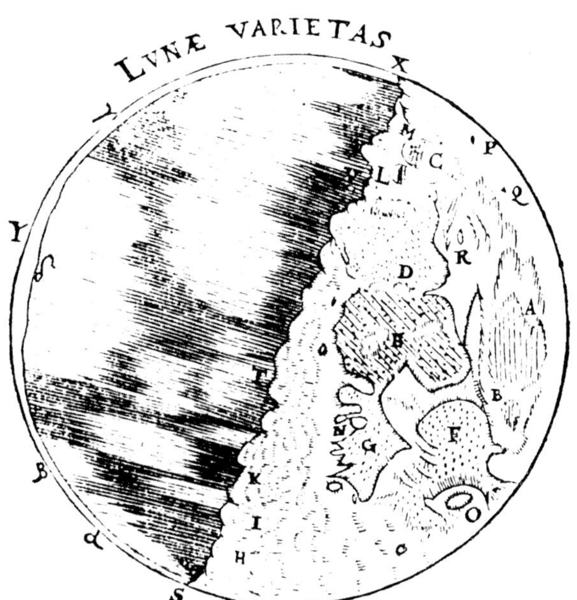

Figure 1.5 This crude chart of the Moon at First Quarter, drawn by Galileo's rival, the German Jesuit Christoph Scheiner, was published in Georg Locher's *Disquisitiones Mathematicae* of 1614. This chart bears careful comparison with Harriot's map of 1610. Only the outlines of the maria and a handful of bright features and craters are rather confusingly portrayed. Courtesy E. M. Carreira, S.J.

1.5). A Jesuit, Scheiner wished to avoid attributing blemishes to the Sun itself (which would, *contra* Aristotle, make it a less than perfect body). He suggested these features were actually small planets seen in transit across the solar disk. But Galileo disagreed. He regarded the spots as something actually on the surface of the Sun, "vapors, or exhalations… or clouds or fumes," since they could be shown to rotate with it. And yet at the same time he was careful not to get mixed up in Kepler's enthusiasms. "I agree with Apelles [Scheiner]", he wrote, "in regarding as false and damnable the view of those who put inhabitants on Jupiter, Venus, Saturn, and the Moon, meaning by inhabitants, animals like ours, and men in particular."[24]

Again in 1616, Galileo told Giacomo Muti, "I do not believe that the body of the Moon is composed of earth and water."[25] And by the time he came to write his celebrated *Dialogue Concerning the Two Chief World Systems*, published in 1632, he had found no reason to change his mind:

> If there were in nature only one way for two surfaces to be illuminated by the Sun so that one appears lighter than the other, and that this were by having one made of land and the other of water, it would be necessary to say that the Moon's surface was partly terrene and partly aqueous. But because there are more ways known to us that could produce the same effect, and perhaps others that we do not know of, I shall not make bold to affirm one rather than another to exist on the Moon.[26]

Nonetheless, in contrast to what he said and wrote about the Moon, Galileo's drawings do tend to emphasize analogies with terrestrial landscapes rather than differences. Montgomery has underscored "the irregularity of the terminator to which he gave an excessively scalloped appearance," and his suppression of the "prominent 'explosive' appearance of several craters (Tycho, Copernicus, Kepler)." Such depictions suggest that "Galileo… comprehending the Moon in terms of terrestrial features, chose at some level to make its surface look more Earth-like than it was, removing the most alien features and contouring others in accord with certain conventions of geographic representation for maps."[27]

Others, inevitably, rushed in where even Galileo feared to tread. Though the ancient idea of the face in the Moon as a literal reflection of the Earth was no longer acceptable, the idea reappeared in modified form. It proved only too tempting to fashion the telescopic image of the Moon into another Earth and to imagine on the Moon (and later on the other planets) counterparts of the terrestrial oceans and lands.[28]

In his posthumously published *Somnium*, Kepler suggested that to the inhabitants of Levania [the Moon] their world would appear to be just as motionless as the Earth seems to ourselves. He was even at some pains to distinguish the differences in the lifestyles of what he called Subvolvans and Privolvans, those living on the near- and far-sides of the Moon from Volva [the Earth]. In 1623 the Scottish poet William Drummond wrote: "Some affirme there is another world of men and sensitive creatures, with cities and palaces in the Moone." Admittedly, Drummond himself mentioned this belief only to counter it: "Thus Sciences… have become Opiniones, nay Errores, and leade the Imagination in a thousand Labyrinthes."[29]

In only a few years the mood would change again. In 1638, John Wilkins, a 24-year old English divine and amateur *virtuoso* (as the scientists of the day were called), wrote an influential little book in defense of Copernicanism, *Discovery of a World in the Moone*, in which he endeavored "to prove, that 'tis probable there may be another habitable World in that Planet." Wilkins took as one of his basic tenets that "those spots and brighter parts which by our sight may be distinguished in the Moone, doe shew the difference betweixt the Sea and Land in that other World."[30] And he concluded: "Tis probable there may be inhabitants in this other World, but of what kinde they are is uncertaine."[31]

That same year there appeared a work of a very different stamp—Bishop Francis Godwin's posthumously published *The Man in the Moone: or, A Discourse of a Voyage Thither by Domingo Gonsales, The Speedy Messenger*. Godwin's work is an imaginative flight of fancy, purporting to tell the adventures of a Spaniard who is transported to the Moon on a "flying chariot" drawn upward by a flock of geese. Yet Godwin takes his inspiration equally from Hakluyt's *Voyages* and Galileo's *Sidereal Messenger,* noting that the knowledge of the inhabited world in the Moon is "properly reserv'd for this our discovering age: In which our *Galilaeusses*, can by advantage of their spectacles gaze the Sunne into spots, & descry mountains in the *Moon*."[32] The idea of the Moon as another world, even as another Earth—with the presumption that it would show changes to the carefully perceptive observer—was clearly gaining ground.

References

1. Plutarch, "De Facie quae in Orbe Lunae Apparet" [Concerning the Face which appears in the Orb of the Moon], *Plutarch's Moralia*, edited with an English translation by Harold Cherniss and William C. Helmbold (Cambridge, Massachusetts: Loeb Classical Library, 1957), p. 43.
2. Ibid., p. 141.
3. As suggested by Philip Stooke, "The Mirror in the Moon," *Sky & Telescope*, 91, 3 (1996), 96–98.
4. Alexander von Humboldt, *Kosmos* (Stuttgart, 1850), vol. 3, p. 544.
5. Scott L. Montgomery, "Expanding the Earth: Seeing and Naming the Skies—The Case of the Moon," in *The Scientific Voice* (New York: Guilford Press, in press), p.201.
6. S. L. Montgomery, "The First Naturalistic Drawings of the Moon: Jan Van Eyck and the Art of Observation," *Journal for the History of Astronomy,* 25 (1994), 317–320.
7. G. Reaves and C. Pedretti, "Leonardo da Vinci's Drawings of the Surface Features of the Moon," *Journal for the History of Astronomy*, 28 (1987), 55–58.
8. Leonardo da Vinci, *The Notebooks of Leonardo da Vinci*, compiled and edited by Jean Paul Richter (New York: Dover, 1970), vol. 2, p. 167. It is not entirely clear what variations Leonardo refers to. He may have noticed some of the changes produced by the Moon's librations.
9. Montgomery, "The Case of the Moon," p. 211.
10. For a complete account of the early history of the invention of the telescope, see Albert Van Helden, *The Invention of the Telescope*, American Philosophical Society, *Transactions*, 67, part 4 (1977), pp. 5–16.
11. See J. W. Shirley, *Thomas Harriot: A Biography* (Oxford: Clarendon Press, 1983).
12. *Norton Anthology of English Literature* (New York: W. W. Norton, 3rd edition, 1974), vol. 1, pp. 1077–1078.
13. For Harriot and Lower, see S. P. Rigaud, *Miscellaneous Works and Correspondence of the Rev. James Bradley*, Supplement (Oxford: Oxford University Press, 1833) and Terrie F. Bloom, "Borrowed Perceptions: Harriot's Maps of the Moon," *Journal for the History of Astronomy*, 9 (1978), 117–122.
14. Sir William Lower to Thomas Harriot, February 6, 1610 (OS); quoted in Ewen Whitaker, "Selenography in the Seventeenth Century," in *Planetary Astronomy from the Renaissance to the Rise of Astrophysics,* ed. R. Taton and C. Wilson, The General History of Astronomy, vol. 2, part A (Cambridge: Cambridge University Press, 1989), p. 120.
15. Galileo Galilei, "The Starry Messenger," in *Discoveries and Opinions of Galileo*, trans. Stillman Drake (New York: Doubleday, 1957), p. 32.
16. Ibid., pp. 33–34.
17. For a discussion of Kepler's early views, see Steven J. Dick, *Plurality of Worlds: The Origins of the Extraterrestrial Life Debate from Democritus to Kant* (Cambridge, England: Cambridge University Press, 1982), pp. 70ff.
18. Johannes Kepler, *Kepler's Conversation with Galileo's Starry Messenger*, trans. Edward Rosen (New York: Johnson Reprint Co., 1965), pp. 33–34.
19. Whitaker, "Selenography," pp. 120–121.
20. Galileo Galilei, *Sidereus Nuncius, or The Sidereal Messenger*, translated with an introduction by Albert Van Helden (Chicago: University of Chicago Press, 1989), p. 23.
21. S. Y. Edgerton, *The Heritage of Giotto's Geometry: Art and Science on the Eve of the Scientific Revolution* (Ithaca, New York: Cornell University Press, 1991).
22. Quoted in Dick, *Plurality of Worlds*, p. 84.
23. Johannes Kepler, *Kepler's Conversation with Galileo's Sidereal Messenger,* trans. E. Rosen (New York: Johnson Reprint Corp., 1965), pp. 27–28.
24. Galileo, *Discoveries and Opinions of Galileo*, ed. and trans. Stillman Drake (New York: Doubleday, 1957), p. 137.
25. Galileo to Giacomo Muti, February 28, 1616.

26. Galileo, *Dialogue Concerning the Two Chief World Systems: Ptolemaic and Copernican*, trans. Stillman Drake (Berkeley: University of California Press, 1962), p. 99.
27. Montgomery, "Seeing and Naming the Skies," p. 225.
28. The telescope refocused the ancient "many-worlds controversy" on the Moon, as Steven J. Dick has noted. Dick, *Plurality of Worlds*, p. 74.
29. John Wilkins, *The Discovery of a World in the Moone; or, A Discourse tending to prove that 'tis probable there may be another habitable world in that planet* (Delmar, New York: Scholars' Facsimiles and Reprints, 1973), p. 95.
30. Ibid., p. 187.
31. William Drummond, *The Poetical Works of William Drummond of Hawthornden*, vol. 2, ed. L. E. Kastner (Manchester, 1913), p. 78.
32. Grant McColley, *The Man in the Moone* and *Nuncius Inanimatus* by Francis Godwin (1638), *Smith College Studies in Modern Languages*, 19, 1 (1937), p. 2. The poet John Milton echoes Godwin's "descry mountains in the Moon." In *Paradise Lost*, he mentions Galileo who views

    At Ev'ning, from the top of Fiesole,
    Or in Valdarno, to descry new Lands,
    Rivers, or Mountains in her spotty Globe.

    Milton was mistaken about the scene of Galileo's observations—they were made at Padua, not at Fiesole or Valdarno, near Florence. Incidentally, Milton met Galileo in 1638 at Arcetri, where he was living under house arrest.

Chapter 2:

Longitudes and Long Telescopes

In the same year Harriot and Galileo began observing the Moon with their small telescopes, Kepler published his classic findings, based on the Danish astronomer Tycho Brahe's data, about the true shape of the orbit of Mars. It was, he showed, an ellipse, with the Sun at one focus. Gone at once were the intricate wheels-on-wheels of epicycles that had cluttered astronomical science since the time of Ptolemy.

Kepler, naturally, attempted a similar solution for the Moon, placing it too in an elliptical orbit, of which the Earth occupied one focus. It was an important step, though it did not completely solve the problem, since, as we now know, the Keplerian ellipse the Moon describes around the Earth is actually distorted in various ways, and turns and wobbles in a fantastically complicated fashion. Kepler made various attempts to solve the problem, but in the end he was unsuccessful.[1] The best model of the Moon's motion was long that put forward by a young Englishman, Jeremiah Horrocks (1619–1641). Horrocks made the "impudent star" of the Moon move in an ellipse, where the line of apsides—the line between the Earth and the perigee or apogee of the Moon's orbit—precessed slowly; at the same time, the lunar orbit librated and oscillated slightly. It was a marvelous piece of workmanship, and would tax the resources of Isaac Newton and his theory of gravitation to better it. One can only guess that had he lived, Horrocks would have accomplished great things—he did become the first person to observe a transit of Venus across the face of the Sun, the feat for which he is best remembered today—but tragically he died at the age of only twenty-two.[2]

The problem of the Moon's motion that so exercised the minds of Kepler and Horrocks was not as academic as might be imagined. It was tied in with the problem of finding longitude at sea, the great scientific project of the seventeenth and eighteenthth centuries, forming the underpinning of the ability of sea-faring nation-states like England and Spain, with their colonies and rich trade routes, to project imperial power around the globe.

In order to navigate safely, one must be able to determine accurately one's position on the surface of the Earth. Latitude is easily determined, at least when the sky is clear. In the northern hemisphere, one simply measures the height that the Pole Star stands above the horizon by night or the altitude of the Sun by day.

On the other hand, finding a ship's longitude—the difference between the meridian the ship happens to lie on and a standard meridian such as Greenwich or Paris—is quite another matter. It would be easy to determine with the aid of a clock that accurately kept Greenwich or Paris time, since all one would need to do would be to determine the local noon (the moment the Sun stands highest above the horizon) and note the difference from Greenwich or Paris time. From this, the difference in longitude follows.

To illustrate: there are 360 degrees of longitude in the circumference of the globe. Since there are 24 hours in a day, the sun will appear to traverse 360 degrees/24 = 15 degrees every hour. Suppose that a navigator finds that local noon at his ship's location occurs when a clock reading Greenwich time is reading five o'clock. This five hour difference in time corresponds to a difference of 5 x 15 degrees = 75 degrees of longitude.

Unfortunately, during the seventeenth century no clock in existence was able to keep time with sufficient accuracy during a sea voyage lasting many months, and hence even the best terrestrial maps of the seventeenth century invariably contain gross distortions of longitude but not of latitude. In this regard, it is notable that even Thomas Harriot's crude lunar chart of 1610 depicts the outlines of the lunar "seas" far more faithfully than did contemporary maps of their Earthly counterparts.

In principle, the Moon can serve as a clock, because it appears to move against the background of stars at a rate of some 13 degrees per day. On average, moonrise occurs about 50 minutes later from one night to the next. By accurately measuring the Moon's position relative to stellar background, one can determine the local time, provided—and here's the rub—one has an accurate star chart and ephemerides of the Moon's motion worked out in advance for the standard meridian.

This method of longitude-finding, which became known as "the lunar-distance method," or more simply as the method of "lunars," was first proposed by the German astronomers Johann Müller ("Regiomontanus," 1436–1476) and Johann Werner (1468–1528), and had been crudely applied by the explorer Amerigo Vespucci in determining the difference in longitude between South America and Spain. However, as straightforward as it is in principle, it is exceedingly difficult to apply in practice. The ephemerides must be accurate—even those worked out from Horrocks's theory, while the best of the seventeenth century, were not nearly good enough to give a position at sea accurate to within two or three hundred miles. Moreover, only a small error in the measured position of the Moon taken at sea will result in a comparatively large error in the calculation of terrestrial longitude. As Sir John Herschel was later to put it, "A Lunar at sea seems rather a bungling business."[3]

As early as 1598 Philip III of Spain had offered the substantial prize of 6,000 ducats for anyone able to present a practical solution to the problem. Apart from the lunar-distance method, various other approaches seemed promising. Galileo, for instance, proposed that sailors make careful records of the positions of his newly discovered satellites of Jupiter. By comparing the observed positions with ephemerides worked out in advance, one could derive the difference in time and hence work out the longitude of the ship. As early as 1612, he sent the details to Philip III. His negotiations with Spain continued for another two decades, during which time he worked on the problem sporadically, even devising a special helmet with a telescope affixed for making observations of the satellites of Jupiter. (He actually tested the device aboard a ship in the harbor of Livorno. Unfortunately, it proved difficult for him to sight on the planet and to keep it in view—even his heartbeat was enough to jiggle the image.) In the end, the plan came to naught, as the Spanish authorities decided it would be too difficult to make the requisite observations. When Galileo died in 1642, his dream of guiding the movements of navies around the globe with his astronomical tables was no closer to fulfillment.

Others were no less captivated by the same dream. But they were immediately beset with difficulties. Even apart from the almost insurmountable mathematical problem of unraveling the complicated motion of the Moon, there was the difficult practical problem of measuring the Moon's position with sufficient accuracy.

Pierre Gassendi (1592–1655), canon of Digny in Provence, believed the most workable method would involve exact timings of the moments when lunar mountains and other features passed into or emerged from the Earth's shadow during lunar eclipses. He and a colleague, Nicolas Claude Fabri de Peiresc (1580–1637), made a first attempt to apply this method during the lunar eclipse of January 1628. By comparing their timings from Aix-en-Provence with those made by astronomers at Paris, they did succeed in approximately working out the difference in longitude. However, they realized that in order to improve their timings, a more reliable map of the Moon was required.

Thus the seventeenth century's great race to map the Moon—the equivalent of our own century's "Moon Race"—began. By 1636, Gassendi and Peiresc had enlisted the services of a well-known artist, Claud Mellan, to produce images of the Moon (Figure 2.1) which might be "remembered for all time." Using a telescope supplied by Galileo, Mellan began a study of the Moon that resulted in three engravings. These were, Montgomery notes, "such a startling leap over the stylized, manneristic drawings of the 1620s and early 30s—even over Galileo's own pictures—that we are left almost stunned."[4] Unfortunately, with Peiresc's death in 1637 the project was abandoned, and Mellan's superb drawings were never combined into an actual map.

Gassendi continued an occasional correspondence about lunar mapping matters with the Spaniard Juan Caramuel y Lobkowitz (1606–1682), the archbishop of Otranto, who aspired to win the still unclaimed prize offered by Philip III. Lobkowitz offered a new scheme of nomenclature: "All our friends will be there," he assured Gassendi; "yourself, Peiresc, Mersenne, and Naude."[5] However, the first lunar map worthy of the name (Figure 2.2) was produced neither by Gassendi nor by Lobkowitz but by a Belgian engineer and mathematician, Michel Florent van Langren (1600–1675), better known as Langrenus. He too was in on the great chase for Philip III's ducats. The best way to determine terrestrial longitudes, he

Figure 2.1 The French artist Claud Mellan's remarkably detailed, accurate, and appealing depiction of the third quarter Moon. The improvement over the earlier renderings of Galileo, Harriot, and Scheiner is astounding. Mellan's feat is all the more impressive when one considers that he used a telescope of the Galilean design that employed a simple biconcave lens for an eyepiece. This optical configuration affords a very narrow field of view, even at low powers. While modern telescopes magnifying 30X have a field of view three or four times the apparent diameter of the Moon, at the same power a Galilean telescope takes in less than one quarter of the Moon's diameter. Today the Galilean design survives only in the form of cheap opera glasses magnifying only 2 or 3X. Courtesy E. M. Carreira, S.J.

Figure 2.2 The Langrenus map of 1645. In addition to his system of nomenclature, which was soon abandoned, this map introduced the convention of representing all the craters as if they were being illuminated by a late morning Sun, an innovation that would be imitated on many subsequent maps. Of the 325 names that Langrenus gave to lunar features, only a handful have been retained. Courtesy E. M. Carreira, S.J.

decided, would be to use sunrise or sunset on various peaks on the Moon's terminator as a time signal that would be visible over a wide expanse of the Earth. By comparing the local time of such an event with a standard time in tables calculated for the purpose, it ought to be possible to work out the longitude at any time the Moon was visible at land or sea. Again it was obvious that the first step must be to draw up a reasonably accurate map of the lunar surface, and this is precisely what Langrenus set out to do.

Unfortunately, he was distracted with a number of more down-to-earth projects, including designing canals and drawing up plans for fortifying the ports at Mardyck and Ostende. Not until 1625 did he even get around to presenting his ideas about the Moon to Isabelle, the princess of the Spanish Netherlands and Philip III's sister. She was won over completely. Through her Langrenus obtained an introduction to the new king, Philip IV, who was so impressed that he appointed Langrenus "Royal Cosmographer."

After Isabelle's death in 1634, Langrenus suspended work for several more years. Only in 1645, when it appeared that he was about to be forestalled by a rival, did he publish his *Philippian Full Moon, adorned with proper names*—actually, a highly politicized version of the Moon reflecting the sharp divisions of Europe on

Figure 2.3 The wealthy brewer and amateur astronomer Johannes Hevelius, the seventeenth century's leading lunar observer. From Arthur Berry's *A Short History of Astronomy* (1898).

religious and political grounds. The whole continent was being convulsed by the Thirty Years' War, in which the Spanish Habsburgs struggled to join their empire with that of the Austrian Habsburgs in Germany. (In the main, allegiances were drawn between Protestant and Catholic, with one important exception—Bourbon France, despite being Catholic, joined the fighting on the Protestant side in order to thwart Spain.) Not surprisingly, Langrenus chose names that honor the Catholic Habsburgs, the Bourbons, and the popes. There is an "Oceanus Philippicus," after Philip IV, and a "Mare Eugenianum" for his queen. There was also a Mare Austriacum and a Mare Borbonicum. Yet so ephemeral were the shifting alliances of the day that most of his designations have vanished without a trace. Among the few that have survived is Langrenus's own name.

Langrenus's scheme was decisively rejected by Johannes Hevelius (1611–1687) (Figure 2.3), a wealthy Protestant brewer who held the post of Rathsherr (city councillor) at the Baltic port of Danzig (now Gdansk, Poland). One of several brewers who have made significant contributions to the science of astronomy, Hevelius was the rival whose entry into the field Langrenus feared. In striking contrast to the obsequious "Royal Cosmographer," who was emasculatingly and parasitically dependent on royal patronage, Hevelius was more fortunate, having the independence that went along with the considerable fortune he had inherited from his father.

In the early 1640s Hevelius established a private observatory, the Sternenberg, that was at once the envy of the crowned heads of Europe. Its centerpiece was a telescope that measured 150 feet (64 meters) in length (Figure 2.4). This colossal dimension resulted from a heroic effort to overcome the limitations of the primitive optics of the day, and here we must turn, briefly, to the difficulties faced by the pioneer instrument-makers of the seventeenth century.

Figure 2.4 Hevelius' Sternenberg Observatory, located in the heart of the Baltic port of Danzig, was easily the finest in the world in its day. From T. E. R. Phillips and W. H. Steavenson, *Splendour of the Heavens* (1925).

The simple single lenses of seventeenth century telescopes suffered from a near-fatal optical defect known as *chromatic aberration*. A cross section through such a lens resembles two narrow prisms placed base to base, each dispersing white light into its constituent colors. The various spectral colors are brought to focus at differing distances along the optical axis of the lens, blue falling shorter than red. As a result, the image of a star or planet is surrounded by an obtrusive purple halo composed of defocused blue and red light.

To make matters worse, even light of a single color cannot be sharply focused because rays of light that pass through the periphery of the lens are bent more steeply than those that pass through nearer the center, reaching focus sooner to produce a blurring of the image called *spherical aberration*. Lensmakers soon realized that both the chromatic and spherical aberrations could be greatly reduced by making very thin lenses with very shallow curves, but at a price—such lenses have very long focal lengths relative to their diameters.

This remedy was pushed to desperate extremes by Hevelius, who mounted his objective at one end of a T-shaped wooden spar composed of four sections of 40-foot beams joined end to end, stiffened by a series of diaphragms spaced at short intervals and by a network of cords whose tension could be adjusted by a team of assistants to prevent the whole affair from sagging under its own weight. The telescope was slung from a mast that towered to a height of 90 feet (27 meters), and was raised or lowered by means of cranks, ropes, and pulleys.

It must have been exasperating to use, for the slightest breath of wind caused the instrument to quiver like a reed, and at certain wind speeds it must have resonated like the rigging of a sailing ship. Despite his seemingly boundless patience and enthusiasm, Hevelius employed this monster only occasionally.

In 1647, Hevelius published at his own expense a lavish book on the Moon,

Selenographia. It contains a suitably baroque frontispiece in which the allegorical figure Contemplation, armed with a telescope, is depicted dispelling the clouds of ignorance as Hevelius's images of the Moon and the Sun peek through breaks in the clouds. A banner containing the title is borne aloft by two figures on pedestals—the Arab Alhazen and Galileo himself, who holds a telescope in one hand while the fingers of the other point to Hevelius's name; Hevelius was obviously not given to false modesty. In all, *Selenographia* contains forty copper-plate engravings cut with his own hand, showing the Moon in its various phases, as well as several versions of his celebrated lunar map (Figure 2.5). These were the product of four years' persistent observation with rather modest telescopes, the largest of 12-foot (3.6 meters) focal length and magnifying only about 50X. This is mute testimony to the failure of far larger and more unwieldy instruments to provide results commensurate with their size.

Hevelius regarded himself as the "muse of the Moon," and his efforts were duly hailed by several of his contemporaries as eclipsing even those of Galileo himself. Hevelius certainly carried the analogy of the Earth to the Moon much farther than Galileo had dared. He imagined in the Moon a distorted map of the classical world, and designated its seas accordingly: Mediterranean, Adriatic, Black, and Caspian. The mirror-image world of the Pythagoreans had thus, apparently, received another incarnation in Hevelius's map.

In the midst of the Mediterranean appears Sicily, drawn as "a large volcanic island," whose center Hevelius named "Mt. Aetna" (the feature corresponding to the crater known today as Copernicus; Figure 2.6). Indeed, as Montgomery points out, "the manner in which this is drawn approximates the general morphology of a strombolian-type volcano, with radiating ridges resulting from lava flows." Hevelius also says that his *Mons Porphyrites* (the crater Aristarchus in modern nomenclature) is "without doubt" a volcano in the midst of continual eruption, "similar in its red color to those known to us as Aetna, Heckla, Vesuvius, etc."[6] Hevelius thus clearly foreshadowed what would become the reigning vision of the Moon as a volcanic landscape—a vision which would maintain a compelling hold on lunar observers for the next three centuries.

Hevelius introduced his names for the lunar features partly, he said, "because he knew not with what else to compare them," partly also to avoid introducing names which would arouse human vanity and petty jealousies (as those of Langrenus, by implication, had done). On the other hand, it would be a mistake to suppose that his use of geographical terms was merely conventional. He took it for granted that the Moon was Earthlike—and inhabited. He even coined the name "Selenites" for the lunar inhabitants.

The copper plate that bore the image of Hevelius's lunar map was melted down after his death to make a teapot, while that depicting the full Moon survived into the twentieth century as a serving tray. His system of names, though widely employed for a century in Northern Europe, in predominantly Catholic Southern Europe fared little better than his copper plates. With only a handful of exceptions (the *Alps* and *Apennines*, *Promontorium Acherusia*, *Promontorium Aenarium*, and

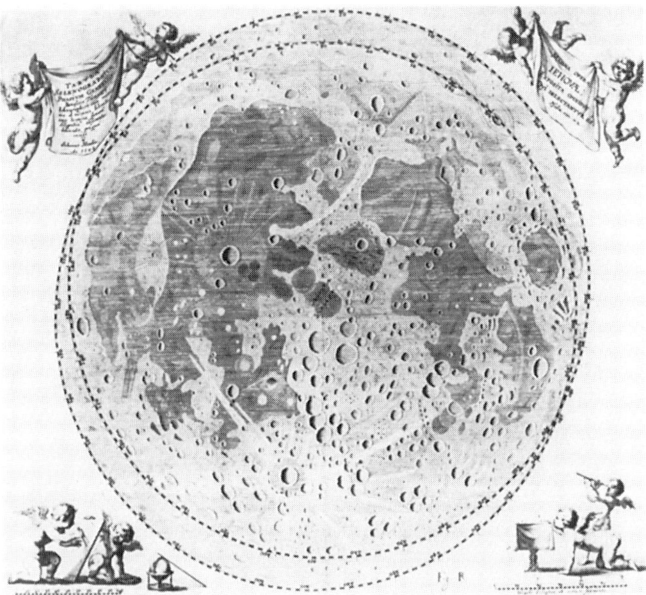

Figure 2.5 The Full Moon from Hevelius' *Selenographia* of 1647. The double outline is an attempt to represent the areas of the Moon revealed by extremes of libration. Courtesy E. M. Carreira, S.J.

Figure 2.6 A section of another version of Hevelius' 1647 map, centered on the crater now designated as "Copernicus." Many of the ray systems were depicted as mountain chains. The halo of rays surrounding the crater was given the name "Sicilia" by Hevelius, while the central crater itself was called "Mount Etna." This was probably the first—and certainly not the last—explicit comparison of a lunar formation to a terrestrial volcano. Courtesy Martin Stangl.

Cape Agarum), his nomenclature has been relegated to the slag heap of history. It was immediately challenged by an entirely different scheme, introduced by Giovanni Battista Riccioli (1598–1671), a Jesuit professor of philosophy, theology, and mathematics at the University of Bologna.

In 1651 Riccioli published a profusely illustrated, comprehensive treatise on theoretical and practical astronomy. The scope of the work is underscored by its magesterial title: *Almagestum Novum*. Though he praised the elegance of the Copernican system, Father Riccioli did not accept that the Earth went around the Sun. Instead, he adopted a compromise system rather like that worked out by Tycho Brahe: Mercury, Venus, and Mars moved around the Sun, but the Sun, Jupiter, and Saturn moved around the Earth.

Apart from its lunar map, the contents of the *Almagestum* are little remembered today. Even the map itself was largely based on Hevelius's images (as a side by side comparison shows; Figure 2.7), though supplemented by observations made by Riccioli's Jesuit colleague, Francesco Maria Grimaldi (1618–1663), professor of mathematics at Bologna and the scion of an illustrious family which gave eighteen bishops and seven cardinals to the Church. Though Riccioli denied that there could be either water or inhabitants on the Moon (he engraved on his chart: *Nec Homines Lunam incolunt, Nec Animae in Lunam migrant*; "neither do humans inhabit the Moon, nor do spirits migrate there"), he did not depart from the tradition of calling dark areas *maria*—seas. However, he devised for them a fresh set of fanciful, poetic Latin names: Mare Imbrium ("Sea of Rains"), Oceanus Procellarum ("Ocean of Storms"), Mare Tranquillitatis ("Sea of Calm"), Mare Serenitatis ("Sea of Peace"), Lacus Somniorum ("Lake of Dreams"), Sinus Iridum ("Bay of Rainbows").

Riccioli's greatest innovation—apparently first suggested to him by Grimaldi, though as earlier noted, similar schemes had already been floated by Lobkowitz and Gassendi—was to name the lunar craters after philosophers, scientists, mathematicians, and other celebrities, both real and mythical. He gave the names of the ancients to most of the features in the northern part of the Moon, while his contemporaries generally were honored in the south. Some of the choices are rather surprising. For instance, there is a Julius Caesar on the Moon, in recognition of his role in the reform of the calendar. Other selections clearly bear out Riccioli's prejudices. Thus the most prominent crater of all honors Tycho, and Ptolemy too receives a splendid feature. But the proponents of the heliocentric system—Aristarchus, Copernicus, and Kepler—were all flung into the Ocean of Storms. Several of Riccioli's Jesuit colleagues were honored, including Fathers Clavius and Scheiner, and he reserved choice features for himself and Grimaldi on the western limb.

With Riccioli's nomenclature the Moon became, as the Abbé Moreux later called it, "the cemetery of astronomers" and "the pantheon of savants."[7] Not all of Riccioli's choices were just; the crater named for Galileo, in particular, is of a size that is by no means worthy of his contributions. There is even some truth to the remark of W. H. Pickering that as a general rule the larger the crater, the smaller

Figure 2.7 The famous Moon map of Giovanni Riccioli, which appeared in the *Almagestum Novum* of 1651. It is often said that the map was prepared from observational materials provided by Riccioli's older colleague Francesco Grimaldi, but a comparison with the map of Hevelius (**Figure 2.5**) reveals it to have nearly the identity of a tracing. The majority of the names that Riccioli assigned to the lunar formations have been retained right up to the present day, including a dark-floored crater near the western limb (astronautical convention) which Riccioli named for himself. Courtesy Ewen Whitaker.

the contribution to the study of the Moon by the person after whom it was named. But despite its shortcomings, Riccioli's scheme was certainly more convenient than those proposed by Langrenus and Hevelius. It would be adopted by Schroeter in the eighteenth century, and has been employed by all astronomers since. It has now been in use far too long to be changed.

References

1. The best account of Kepler's lunar theory and other details of Keplerian astronomy is Bruce Stephenson, *Kepler's Physical Astronomy* (Princeton: Princeton University Press, 1987).
2. On Horrocks (or Horrox), see: Arundell Blount Whatton, *Memoir of the Life and Labours of the Rev. Jeremiah Horrox* (London: Wertheim, Mackinstosh, and Hunt, 1839), J. E. Bailey, *The Palatine Note-books*, abridged in *The Observatory*, 6 (1883), 318–328, and Allan Chapman, "Jeremiah Horrocks, the transit of Venus, and the 'New Astronomy' in seventeenth century England," *Quarterly Journal of the Royal Astronomical Society*, 31 (1990), 333–357.
3. Sir John Herschel, December 20, 1833; in David S. Evans, et. al., editors, *Herschel at the Cape: Diaries and Correspondence of Sir John Herschel, 1834–1838* (Austin and London: University of Texas Press, 1969), p. 16.

4. Montgomery, "The Case of the Moon," p. 258.
5. P. Humbert, "La premiére carte de la lune," *Revue des questions scientifiques*, 20 (1931), 183–204:200.
6. Montgomery, "The Case of the Moon," p. 261.
7. The Abbé Th. Moreux, *A Day in the Moon* (London: Hutchinson and Co., 1913), pp. 57ff. The triumph of Riccioli's names over those of Hevelius is attested to by the various editions of Bernard le Bovier de Fontenelle's popular *Conversations on the Plurality of Worlds*. In the first edition, in 1686, Fontenelle writes of the dark regions (presumed seas) in the Moon: "All these different parts are… thoroughly recognized. There are a Caspian Sea, the Porphyrian Mountains, the Black Lake; in short, the description of the Moon is so exact that a learned man who found himself there nowadays could no more go astray than I could in Paris." In the 1708 edition, Fontenelle changes this to: "One place is called Copernicus, another Archimedes, another Galileo, there is a Promontary of Dreams, a Sea of Rains, a Sea of Nectar, a Sea of Crises." See Fontenelle, *Conversations on the Plurality of Worlds*, trans. H. A. Hargreaves (Berkeley: University of California Press, 1990), pp. 29–30 and note.

Chapter 3:

The Compelling Spell of Change

Though the complete triumph of Riccioli's map did not take place for more than a century, it was already preferred by many of the leading astronomers of the day, including Robert Hooke (1635–1703) and Christopher Wren (1632–1723) of England's Royal Society. Wren was to achieve fame as the architect of St. Paul's Cathedral, but early in his career he was Savilian professor of astronomy at Oxford. In 1661, he prepared a globe of the Moon for the Royal Society, on which "the hills, eminences and cavities [were] molded in solid work."[1]

Hooke was a versatile genius, a pioneer equally in the use of the microscope and the telescope. In his *Micrographia* of 1665, he included a detailed drawing of the lunar formation that Riccioli had named Hipparchus (Figure 3.1). It had, he noted, "much the figure of a Pear." He went so far as to suggest that it seemed to be a "fruitful place, that is, to have its surface all covered over with some kinds of vegetable substances."[2]

More important were Hooke's pioneering speculations on the origin of the lunar craters. The craters, he found, resembled the structures formed by dropping round bullets onto moistened pipeclay. Hard-pressed to imagine a source of such bodies striking the Moon, he had another inspiration. He heated a thick slurry of water and powdered alabaster in a pot, forming rising bubbles. When these bubbles burst, they left circular depressions, leading him to speculate:

> These pits in the Moon seem to have been generated much after the same manner that the holes in Alabaster, and the *Vulcans* of the Earth are made. For first, it is not improbable, but that the substance of the Moon may be very much like that of our Earth, that is, may consist of an earthy, sandy, or rocky substance, in several of its superficial parts, which parts being agitated, undermin'd, or heav'd up, by eruptions of vapours, may naturally be thrown into the same kind of figured holes, as the small dust, or powder of the body of Alabaster. Next, it is not improbable but that there may be generated, within the body of the Moon, divers such kind of internal fires and heats, as may produce such Exhalations.[3]

Other lunar observations were made by Christiaan Huygens (1629–1695) of Holland (Figure 3.2), one of the leading *virtuosi* of the age. He was a skillful observer who in 1655, using a telescope built with his own hands, discovered the

Figure 3.1 (Left) Robert Hooke's 1651 drawing of the 138-km diameter crater Hipparchus, as seen through a "good long telescope." The drawing appeared in Hooke's *Micrographia* ("Small Drawings"), published in 1665, which also contained studies of snowflake crystals and depictions of the cellular structure of cork through a microscope. Courtesy Ewen Whitaker. (Right) For comparison, an oblique photograph of Hipparchus taken from lunar orbit with the Metric Mapping Camera by the Apollo 16 astronauts in 1972. Courtesy NASA.

largest satellite of Saturn (now known as Titan). He also made a critical series of observations which led him to the correct solution of the strange appearances of Saturn first noted by Galileo. The globe of the planet, Galileo had found, sometimes appeared "solitary," while at other times the planet seemed to have appendages, in the form of handles or ears. Galileo himself never grasped the solution, and later astronomers were equally baffled. Even Hevelius could do no better than to suggest that the triple-form "arises only from hallucinated vision."[4] Equipped with a superior telescope, Huygens was the first to realize the correct explanation: the planet is surrounded by a thin flat ring, sometimes appearing edgewise to the Earth, at other times tilted at an angle to our line of sight.

In 1656, Huygens built a clock regulated by swinging pendulums. Two years later, in his treatise *Horologium*, he suggested that the device might be made perfect enough to keep accurate time at sea. Unfortunately, he found that his pendulum clocks, though capable of keeping time admirably in calm seas, were easily thrown off by rough weather.

Ten years later Huygens went to Paris at the invitation of Louis XIV, who wanted to add to the renown of the recently founded Académie des Sciences. There he continued to tinker with his clocks, and applied for a patent for the spiral balance-spring which promised more reliable timekeeping on long voyages. (His claim was immediately attacked by Hooke, who accused Huygens of stealing the invention from him).

Although he was Protestant, Huygens remained in France for fifteen years as

Figure 3.2 Christiaan Huygens, best remembered today not for his lunar work but for his discovery of Saturn's satellite Titan and his recognition that Saturn is surrounded by a ring. In addition, he developed the pendulum clock and was a leading proponent of the wave theory of light (*contra* Isaac Newton, who advanced a corpuscular theory). Courtesy Yerkes Observatory.

the most prestigious member of the Académie, but after the publication of the Edict of Nantes, which opened the way for the persecution of Protestants in an increasingly militant Catholic France, he returned to Holland, settling on a country estate at Hofwjick near The Hague. There, aided by his brother Constantyn, he ground lenses and constructed ever larger telescopes.

To overcome many of the problems associated with the massive wooden beams and spars of instruments like those built by Hevelius, the brothers Huygens developed a design that came to be known as the "aerial telescope" (Figure 3.3). They mounted their objective lens on a ball-and-socket joint that could be moved up and down a tall pole. The instrument was aimed by pulling taut on a wire that connected the objective and the eyepiece assembly, which the observer held in his hands while resting his arms on a moveable wooden support. These gossamer, tubeless instruments were all but immune to buffeting by breezes, but protracted work must have sorely tested the observer's patience. When studying an object near the celestial equator, the observer had to move the eyepiece assembly at a uniform rate of several inches per minute to compensate for the Earth's rotation, which certainly required a very steady hand. On very dark nights when the objective lens was difficult to see, just pointing the telescope was a time-consuming task requiring an assistant equipped with a lantern.

Despite these difficulties, the brothers persisted. Their efforts culminated in instruments with focal lengths of 123 feet (37 meters) and 210 feet (63 meters). With these telescopes Christiaan Huygens succeeded in recording a number of lunar features that would not be recognized again for another century—the Straight Wall (Figure 3.4), the Hyginus Cleft, and, named after its re-discoverer, Schroeter's Valley.

A more systematic lunar observer was the Italian Giovanni Domenico Cassini (1625-1712) (Figure 3.5). Born at Perinaldo near Nice in 1629, he gained experience as a young astronomer at the private observatory of the Marquis Malvasia

Figure 3.3 An observer at the eyepiece of one of Huygens' smaller aerial telescopes. From A. von Schweiger-Lerchenfeld's *Atlas der Himmelskunde* (1898).

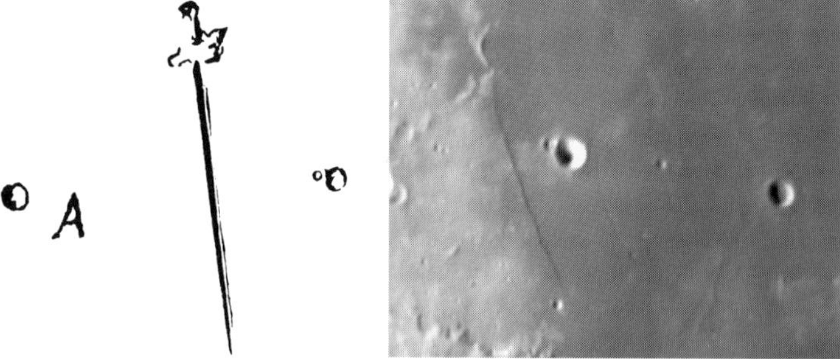

Figure 3.4 (Left) Huygens' discovery drawing of the Straight Wall, the appearance of which he compared to a sword. This notebook sketch remained unpublished and unknown for nearly two centuries after the observation was made. Courtesy Ewen Whitaker. (Right) The Straight Wall under morning illumination. Despite the simplicity of Huygens' sketch, in its essential features it compares favorably with this CCD image by T. A. Dobbins.

Figure 3.5 Giovanni Domenico Cassini, with Huygens, the greatest observational astronomer of the second half of the seventeenth century. Courtesy Yerkes Observatory.

at Bologna, where he determined the rotation periods of Mars and Jupiter and, in 1668, published new tables of the motions of the satellites of Jupiter. The accuracy of these tables and the possibility they might at long last yield the solution to the longitude problem brought Cassini to the attention of Louis XIV, who invited him to France to take charge of the newly founded Paris Observatory; Cassini accepted, an arrived in Paris the following year.

His priorities included coordinating a project in which the four Galilean satellites of Jupiter were used to work out the difference in longitude between Paris and Tycho Brahe's famous observatory on the Danish island of Hven. Cooperative observations of the satellites by astronomers throughout Europe soon led to such sweeping revisions of the maps in use at the time that Louis XIV complained that he was losing more territory to astronomers than to all his enemies combined. While making timings of the exact moments when the satellites became immersed in the planet's shadow cone, the Danish astronomer Olaus Roemer (1644–1710) made the important discovery that the eclipses occurred earlier than predicted when the planet was close to the Earth and later than predicted when it was far away, a fact he attributed correctly to the finite velocity of light.[5]

Cassini hedged his bets. Despite his long-standing interest in the Jovian-satellite method of determining terrestrial longitudes, he also paid a great deal of attention to the method of "lunars." As early as 1671, he used a 17-foot Campani telescope that he had brought with him from Italy to commence an ambitious survey of the Moon. His drawings are preserved scrapbook-style in a large volume in the Paris Observatory, and include admirably detailed studies of small areas of the lunar surface. One of the most interesting shows the Promontorium Heraclides on the western side of Sinus Iridum. Cassini represented the formation in the form of a "Lady's Head" (Figure 3.6).

Another feature to which he paid more than passing attention was "a kind of whitish cloud," first noted near the crater Gauricus on October 21, 1671. A few days later he found only a few remnants of the "cloud"; but he recovered it, in exactly the same place, on November 12, 1671. There are then no further references to this particular feature until October 18, 1673, when he recorded a "new large crater" between Pitatus and Walter, in the very spot where the whitish cloud had appeared two years previously.[6]

In fact, this first candidate for a lunar change was a weak one. Whenever this region is near the terminator (as had been the case on October 18, 1673) a crater is distinctly visible. Under high angles of illumination a few days before or after Full Moon the crater vanishes, leaving only the "Cassini Bright Spot"—actually a particularly conspicuous patch in one of the rays extending from Tycho (Figure 3.7)—in its place.

Eventually Cassini's drawings of the Moon were used as the basis of a large map, engraved not by Cassini himself but an assistant, Jean Patigny, in 1679 (Figure 3.8). It is usually known as the Cassini "large map," to distinguish it from the smaller, more widely known version, published in the *Memoirs* of the Académie des Sciences in 1692 (Figure 3.9). On the whole, its accuracy is disappointing. The English portrait artist John Russell, who made a series of beautiful crayon drawings of the Moon at Oxford a century later, wrote:

> ... perhaps it was too hastily concluded that the large dark parts upon the Moon's Face, were Seas; I am apprehensive that, if the Engraver had been faithfull to his trust, this must have led that great Astronomer Cassini to represent these parts of one almost uniformly smooth, and unvaried effect, which upon a strict inspection will appear to be full of parts as various and nearly as multitudinous, as that portion of the Moon, which has generally been considered to be Land.[7]

The Cassini map, notes lunar historian Ewen Whitaker,[8] is not only wanting in accuracy but aesthetically inferior to another small map (Figure 3.10) published about the same time by Philippe de la Hire (1640–1718), a professor of mathematics in Paris. De la Hire at one time resolved to produce an elaborate lunar atlas, but like so many lunar mapping projects his was never completed.

It is important to remember that all these efforts to map the Moon were not ends in themselves, but were made with the express purpose of serving as the basis of terrestrial longitude determinations. The main features on de la Hire's small map are, for instance, numbered in the order in which they would be covered by the umbra of the Earth's shadow during a lunar eclipse. Cassini's method followed that of Langrenus and depended on careful observations of the times of passage of lunar features into or out of shadow at the terminator. However, in order to be practical, a precise knowledge is needed of the coordinates of the various features, which in turn requires detailed knowledge of the so-called librations of the Moon. This is a complicated subject; nevertheless, some attempt had to be made to come to terms with it, since the Moon's librations enter inextricably into all observations of the lunar features, most especially those in the neighborhood of the limbs.

Since the Moon, at least to a first approximation, always keeps the same face beholden to the Earth, it is obvious that it must turn once on its axis in nearly the same period as that in which it revolves. In 1632, Galileo had proposed that the centers of the Earth and Moon were joined by a fixed line, which always passed

Figure 3.6 (Top)The lunar drawings of Cassini that ultimately served as the basis of his map are preserved scrapbook-style in a large book at the Paris Observatory. The example shown here is his drawing of the feature Heracleides Promontorium, which looked, to Cassini's eye, for all the world like a Lady's Head. It illustrates the eye's tendency to exaggerate regularities in low-resolution views of natural features and calls to mind the widely publicized case the Viking Orbiter spacecraft's image of the "Face on Mars." Courtesy Ewen Whitaker. (Bottom) For comparison, a photograph of Sinus Iridum and the Heraclides Promontorium by E. M. Carreira, S.J., with the 16-inch Zeiss refractor of the Vatican Observatory's Castel Gandolfo station near Rome.

Figure 3.7 The so-called Cassini Bright Spot, one of the earliest examples of a lunar feature suspected of regular changes. Surprisingly, it is included in the definitive catalog of transient lunar phenomena, even though it is actually nothing more than a bright patch on one of the rays flung from the crater Tycho. Photograph by E. M. Carreira, S.J., with the 16-inch Zeiss refractor of the Vatican Observatory's Castel Gandolfo station near Rome.

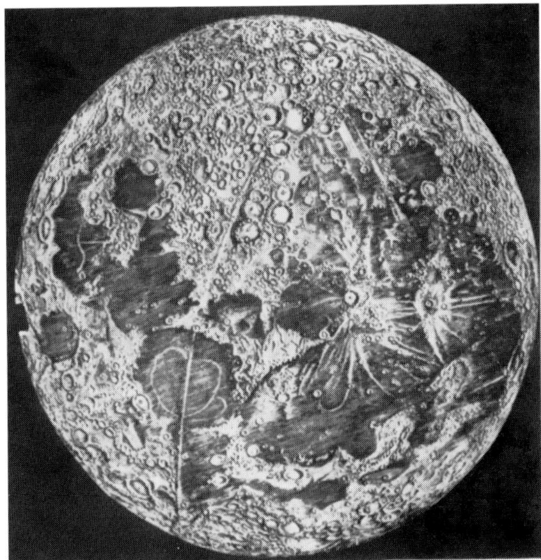

Figure 3.8 The Cassini large map of 1679, once thought to have been engraved by Claude Mellan. It is, however, clearly not up to Mellan's usual standard, and it has now been demonstrated that the engraver was actually an inferior talent, Jean Patigny. Patigny, with Sebastien Leclerc, assisted Cassini in making 60 drawings of the lunar surface using aerial telescopes built around optics crafted by the Roman artisan Giuseppi Campani. Despite the impressive scale of this map, careful scrutiny reveals that it is by no means a marvel of accuracy. Many of its smaller details are entirely spurious. Courtesy E. M. Carreira, S.J.

Figure 3.9 Cassini's small lunar map is often referred to simply as "Cassini's Map of the Moon" since it was widely circulated in the *Memoirs* of the Royal Academy of Sciences. Like Galileo's engravings for *Sidereus Nuncius*, it was so often copied during the century following its first appearance in 1692 that it came to exist only in highly degraded form. Courtesy E. M. Carreira,S.J.

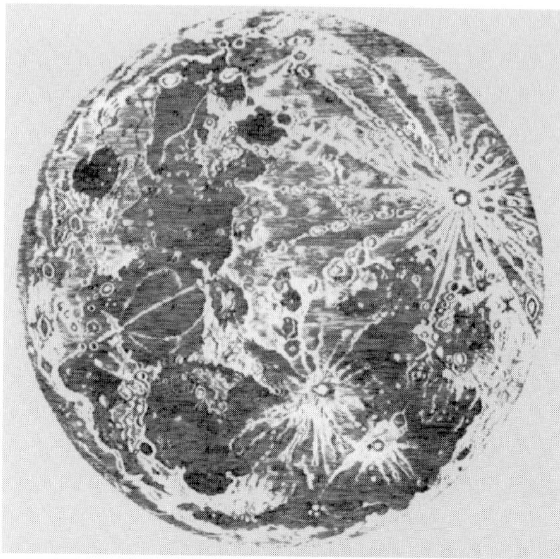

Figure 3.10 Map of the Moon by Philippe de la Hire. Artistically far superior to the Cassini maps, it does share a number of features with them, notably the prominent "phi" (ϕ) marking in Mare Serenitatis. Courtesy Ewen Whitaker.

through a given point on the lunar surface. From inspecting features near the limb like Mare Crisium, he knew that the Moon appears to sway gently from side to side, the effect that we now refer to as libration in longitude. At first he supposed that this effect was attributable to the effect of parallax due to the different positions of the observer on the rotating Earth, but by 1637 he detected a corresponding libration in latitude—an apparent nodding motion of the Moon, with a period of about a month. His initial idea of any rigid line between the Earth and Moon had to be abandoned.

One result of these apparent swaying and nodding motions is that we are actually able to glimpse somewhat more than half of the Moon's surface. Hevelius included the limb regions revealed by libration on his map, and tried to account for the libration in longitude. He assumed that the Moon moved in an elliptical orbit, with the Earth occupying one focus. However, instead of pointing toward the Earth, the center of the Moon's disk always pointed toward the center of this ellipse.

In his *Principia* of 1687, Isaac Newton showed that the libration in longitude is due to the fact that the Moon's rate of spin on its axis is uniform, but its speed along its elliptical orbit varies in accordance with Kepler's second law, moving faster when it is close to the Earth and slower when it is farther away. Constant rotation and variable orbital velocity cause the Moon to show more of its western hemisphere at times, more of its eastern at others, so the Moon appears to oscillate back and forth over the course of a month as seen from Earth. (There is also a diurnal effect due to the rotation of the Earth, as Galileo had supposed, but this is far smaller in magnitude and is most pronounced when the Moon is on the horizon).

Newton was not as successful in accounting for the libration in latitude, since he accepted Hevelius's statement that the Moon's axis was perpendicular to the ecliptic (the plane of the Earth's orbit around the Sun). Finally in 1693, after twenty years of study, Cassini revealed the exact laws of the Moon's rotation:

1. The inclination of the Moon's axis to the ecliptic remains constant. (Cassini put its value at about 2.5°, but its actual value is now known to be 1°32′9″.)
2. Three planes—the Moon's orbit, the Moon's equator, and the ecliptic—all intersect in a single line, the line of the nodes of the lunar orbit.
3. The Moon's period of axial rotation is equal to its period of revolution.

Because of libration, the apparent and mean centers of the Moon, as seen from the center of the Earth, rarely coincide. When the Moon's orbital motion is running ahead of its rotation, a portion of the western surface is carried out of sight even as the corresponding portion of the eastern surface is unmasked. When it is running behind its rotation, the reverse is true. In addition, the combined effects of the inclination of the lunar equator and of the 5° inclination of the plane of the lunar orbit to the ecliptic produce librations in latitude that bring extremities of the Moon's north and south polar regions alternately into and out of view. In all, the effects of the various librations combine to allow earthbound observers to see some 59 percent of the surface of the Moon; the other 41 percent would remain inaccessible until the dawn of the Space Age.

In order to describe precisely the circumstances of any particular observation of the Moon, it is necessary to know the prevailing conditions of libration. For this purpose selenographical coordinates were later introduced (as soon as micrometric measurements of the lunar features took the place of mere eye-estimates and drawings). The distance between the mean center of the Moon and the point where the apparent center intersects the equator gives the libration in longitude; the distance of the apparent center from the lunar equator gives the libration in latitude. These coordinates are known, respectively, as the selenographical longitude and latitude.

The maximum libration in longitude shifts formations to or from the mean center of the Moon by as much as 7°53′ in either direction, while the libration in latitude varies by as much as 6°50′ from the apparent equator. (Not even these figures represent the most extreme conditions, since they do not take into account the parallactic librations.)

In addition to the selenographical coordinates, it is necessary to know the location of the terminator—the dividing line between day and night on the Moon—at the time of the observation. Because it lacks any appreciable atmosphere, a fact that was already apparent in the seventeenth century (Huygens had declared categorically that "the Moon has no water or air"), its shadows are always of a profound depth and blackness. Hence the apparent coldness and harshness of the lunar landscape. Near the terminator, the length of the shadows accentuates the apparent relief of the various features; low swells take on the appearance of imposing masses, long shadows gradually taper into spires to give the impression they are being cast by needle-sharp crags. The dominant forms look like "the mouths of craters, or at least of deep calderas with very prominent walls."[9] On the termi-

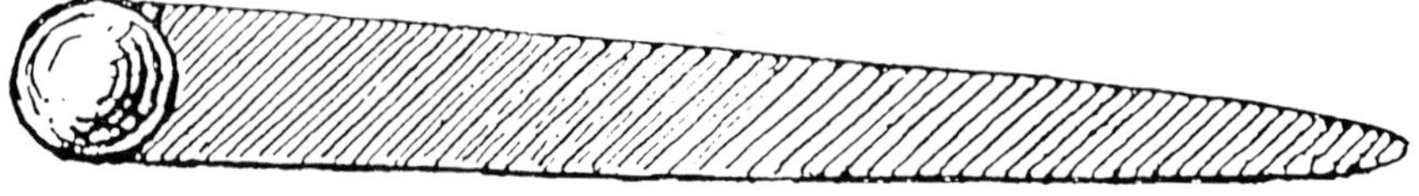

Figure 3.11 The shadow cast by a pea under a glancing illumination, illustrating how the incidence of lighting on a lunar formation is literally capable of making a mountain out of a molehill. From Philipp Fauth, *The Moon in Modern Astronomy* (1907).

nator these grand formations appear exactly like the dark, brooding orifices of great volcanoes. Irresistible as such impressions are they are mere illusions, like the legs of the looming giants that frighten small children who do not recognize their own shadows.

In fact, even the acuteness of a shadow-peak is no indication of actual steepness of the form giving rise to it, for under a glancing angle of illumination even a round pea will cast a long and ominous shadow (Figure 3.11).

Here, then, we have the actual conditions of observation that confront and conspire to baffle the lunar observer. The effect of the librations is to shift formations now toward, now away from, the apparent center of the Moon. As a result, features in the extreme regions near the limbs or the poles remain inaccessible to observation for prolonged periods of time. Indeed, they are presented most favorably only when the same conditions of libration repeat (i.e., whenever the line of the nodes returns to the same position relative to the Earth) after an interval of some 18 years, the Saros cycle. The perplexing effects of the changes in apparent form, position, and dimension of the lunar formations are in turn compounded by the continually varying conditions of illumination over the course of every lunation.

Under the circumstances, it is hardly surprising that so many careful students of the Moon should have been mesmerized by the impression of change in its features. For the prospect of such change already hinted at by naked-eye observers such as Leonardo becomes infinitely more compelling once the Moon is inspected by means of optical aid. According to the skillful nineteenth-century observer, the Reverend T. W. Webb:

> As the angle of illumination increases, a fresh aspect of things creeps in, and extends itself successively over the whole disc, and in its progress the inexperienced observer will find himself astonished at the change, and frequently bewildered in the attempt to trace out the landmarks of the surface. Objects recently well recognised under the relief of light and shade will become confused by a novel effect of local illumination, and the eye will wander over a wilderness of streaks and specks of light, and spots and clouds of darkness, where it may sometimes catch the whole, sometimes a portion, sometimes nothing of many a familiar feature; while unknown configurations will stand boldly out defying scrutiny, and keeping their post immovably till the decreasing angle of illumination warns them to withdraw. Nothing can be more perplexing than this optical metamorphosis, so complete in parts as utterly to efface well-defined objects; so capricious as in some instances to obliterate one, and leave unaffected the other, of two similar and adjacent forms.[10]

"A change," noted Agnes M. Clerke (1842–1907), the eminent nineteenth-century historian of astronomy, "always seems to the inquistive intellect of man like a breach in the defences of Nature's secrets, through which it may hope to make its way to the citadel." Certainly the wish is often father to the thought. "What is desirable," Clerke added, "easily becomes credible;... thus statements and rumours of lunar convulsions acquired credence, only to be rejected on closer acquaintance":

> The subject is one as to which illusion is peculiarly easy. Our view of the moon's surface is a bird's-eye view. Its conformation reveals itself indirectly through irregularities in the distribution of light and darkness. The forms of its elevations and depressions can be inferred only from the shapes of the black, unmitigated shadows cast by them. But these shapes are in a state of perpetual and bewildering fluctuation, partly through changes in the angle of illumination, partly through changes in our point of view, caused by what are called the moon's "librations." The result is, that no single observation can be exactly repeated by the same observer, since identical conditions recur only after the lapse of a great number of years.[11]

Separating the wheat from the chaff—the reality of change from the mere appearance—would become the project of generations of astronomers from Cassini's time down to our own century. The "error of the Moon" would make men mad—or at least tinged with lunacy—as many a moonstruck astronomer would expend the effort of a long career in pursuit of changes (or the shadow of changes) on this world of illusion where "all did not alter that alteration finds." Hinted at already in the seventeenth century, here is the theme which would become the great obsession of lunar science: the changes of the Moon.

References

1. Whitaker, "Selenography in the Seventeenth Century," p. 140.
2. Robert Hooke, *Micrographia, or some physical descriptions of minute bodies made by magnifying glass with observations and inquiries thereon* (London, 1665).
3. Quoted in J. Green, "Hookes and Spurrs in Selenology," *Annals of the New York Academy of Sciences,* 123 (1965), 373–402:382–383.
4. Albert Van Helden, "'Annulo Cingitur': The Solution of the Problem of Saturn," *Journal for the History of Astronomy*, 5 (1974), 155–157.
5. Roemer and Cassini observed not only the eclipses but also the transits of the satellites and their shadows across the disk. In the course of this work Cassini noticed various spots on the disk, including one (also observed by Robert Hooke) which may have been the now-famous Great Red Spot. He found that equatorial spots rotated slightly faster than those in higher latitudes, and derived rotation periods that are in tolerable agreement with modern ones. In December 1690, he followed the development of a particularly remarkable spot which appeared abruptly as a small, sharp black dot, then over the next eighteen days became crescent-shaped, finally stretching into a series of narrow elongated patches. G. D. Cassini, "Nouvelles d'escouvertes dans le globe de Jupiter," MSS in Paris Observatory. For a discussion, see Thomas Hockey, *Galileo's Planet: Observing Jupiter before Photography* (Bristol and Philadelphia: Institute of Physics Press, 1999), p. 32.
6. Joseph Ashbrook, "The 'Long Night' of Selenography," in *The Astronomical Scrapbook: Skywatchers, Pioneers, and Seekers in Astronomy* (Cambridge, Massachusetts: Sky Publishing Corp., 1984), p. 242.

7. E. J. Stone, "Note on a Crayon Drawing of the Moon by John Russell, R.A., at the Radcliffe Observatory, Oxford," *Monthly Notices of the Royal Astronomical Society,* 56 (1896), 92.
8. Whitaker, "Selenography," p. 140.
9. P. Fauth, *The Moon in Modern Astronomy*, trans. by Joseph McCabe (New York: D. Van Nostrand Company, 1909), p. 64.
10. Rev. T. W. Webb, *Celestial Objects for Common Telescopes* (New York: Dover reprint of 5th ed., 1917), vol. 1, p. 88.
11. Agnes M. Clerke, *A Popular History of Astronomy in the Nineteenth Century* (London: Adam & Charles Black, 3rd ed., 1893), p. 327.

Chapter 4:
The Long Night of Selenography

As the great century of astronomical discovery drew to a close, two books with particular relevance to the study of the Moon appeared. The first was *Entretiens sur la pluralité des Mondes* (Conversations on the Plurality of Worlds) by Bernard le Bovier de Fontenelle (1657–1757), published in 1686 when the author was twenty-nine. Written as a dialogue between Fontenelle and a beautiful, intelligent, and (alas) imaginary countess at her country seat, the conversations take place over the course of five evenings, of which the second and third are devoted to the Moon. Fontenelle is well aware that the Moon has no water, and is apparently without clouds.[1] Though he acknowledges that this would argue against the existence of lunar inhabitants, he speculates that there might be other "exhalations and vapors," enveloping the Moon in a thin atmosphere of its own; but as soon as this is allowed, he concludes, the "inhabitants are reborn, and we have the foundation necessary for their subsistence."[2] Fontenelle recalls Kepler: without clouds, the Sun's heat must beat down fiercely on the Moon. "In fact it may be because of this that Nature has sunk those things like pits in the Moon, which are huge enough to be seen by our telescopes… How do we know that the Moon's inhabitants, distressed by the Sun's perpetual strength, don't take refuge in these great pits?"[3]

Fontenelle's *Conversations* was a tremendous success; it went through numerous editions despite—or perhaps because of—being placed on the Vatican's *Index of Forbidden Books* the year after it appeared. Partly out of a spirit of emulation, Christiaan Huygens, who was as near to the end of his brilliant career as Fontenelle was near the beginning of his, penned a similar work, *Cosmotheoros, or Conjectures Concerning the Inhabitants of the Planets.*[4] Published in 1698, three years after Huygens' death, it failed to become as popular as Fontenelle's work, though in some ways it is more interesting since it is informed by Huygens' work as an observational astronomer, "most evident," notes historian Steven J. Dick, "in his discussion of the Moon."[5] Huygens' observations with "a good long Telescope" had shown him that so-called seas were full of little round cavities, thereby proving they were not of uniform level and they did not contain water or any other liquid. No rivers had ever been observed there, no clouds. That being the case, Huygens could not imagine "how any Plants or Animals, whose whole nourishment comes from liquid Bodies, can thrive in a dry, waterless, parch'd Soil."[6]

In part because of the lack of information forthcoming about the possibility of lunar inhabitants, selenographical studies fell into neglect in the new century. Admittedly this was only one of the factors. Another was the fact that the long telescopes employed by Huygens and Cassini were terribly cumbersome and difficult to use—"optical dinosauri, with small heads on large bodies."[7] After his death, Huygens' telescopes were bequeathed to the Royal Society of London, but they were scarcely used. Cassini, it is said (probably with little exaggeration), usually dispensed with the mountings of his long telescopes and simply fixed the objectives to the roof of the Paris Observatory, then waited patiently, eyepiece in hand, for the celestial body he wished to observe to transit the narrow field of view.[8]

Needless to say, such circumstances did not encourage long sessions at the eyepiece. Little wonder the results were so meager over the next three quarters of a century, a period Joseph Ashbrook has aptly described as "the 'long night' of selenography." [9] Yet hope springs eternal. In particular, the fond hope of Kepler, Wilkins, and Hevelius that the Moon might prove to be inhabited did not fade easily. In opposition to Huygens's somber if informed opinions, Fontenelle's Marquise expressed a far more romantic view: "I confess my weakness," she said, "I'm not capable of such perfect impartiality; I need to believe... Let's preserve [the inhabitants of the Moon] if possible. I've taken a liking to them that I'd be sorry to lose."[10] The pursuit of changes on the Moon—of life, perhaps of traces of the Selenites themselves—would continue only when more manageable telescopes became available.

Huygens died in 1695; Cassini not until 1712. For the next several decades there would be few able students of the Moon. One who does deserve mention was the priest who gave Cassini the last rites of the Roman Catholic Church, Francesco Bianchini (1662–1729). He was librarian to Cardinal Ottobini and, when the latter was elected Pope Alexander III, became his domestic prelate. In the summer of 1725, with the aid of a long Campani telescope erected on the Palatine Hill at Rome, Bianchini observed a reddish ray of sunlight stretching like a comet across the darkened floor of the crater Plato (Figure 4.1). He wondered whether it might be a shaft of light admitted through a breach in the wall of the crater, or perhaps a ray refracted, and hence reddened, by "some denser medium existing in the form of an atmosphere round the Moon's globe."[11] In retrospect, the ruddy hue was no doubt spurious—an artifact of the chromatic aberration of the simple lens of his telescope.

The next day, as he was hurriedly taking measurements of the ground-plan of the ruins of the palace built on the Palatine by the Roman emperor Augustus, recently discovered and cleared of rubble in the Farnese Gardens, Bianchini stumbled into a deep hole and broke his leg. His return to his long telescope was long delayed—not until September 22, 1727 did he again study Plato and its surroundings, when he made the discovery of the long "gash" in the lunar Alps now known as the Alpine Valley.

Figure 4.1 The Italian astronomer Francesco Bianchini's unwieldy 75-foot Campani telescope. Campani's lenses became highly prized after G. D. Cassini used a 17-foot Campani telescope to discover a satellite of Saturn and to accurately measure the rotation periods of Mars and Jupiter. The inset is Bianchini's attractive depiction of the lunar Alps and the crater Plato. From *Hesperi et Phosphori Nova Phenomena sive Observationes Planetam Veneris* (1728). Courtesy E. M. Carreira, S.J.

Bianchini was a rarity among astronomers of the day in having any interest in the surface detail on the Moon, and even he devoted less attention to the Moon than to the New World he thought he was discovering in Venus. (The continents and seas he thought he found on our sister planet have, alas, proved to be entirely illusory.) For the most part, interest in the Moon, during the first three quarters of the eighteenth century, was stimulated less by curiosity about the conditions of its surface or its possible inhabitants than by the still burning practical question of determining terrestrial longitude.

Methods of determining longitude from telescopic observations remained tantalizing, though in the end they foundered on the complex problem of the Moon's motion. After Horrocks matters had reached an apparent impasse, since even his theory was not accurate enough to allow the determination of longitude to within less than a few hundred miles. The next advance awaited the development of Isaac Newton's majestic theory of gravitation. It has been said that "altogether the most important circumstance in what may be called the history of the Moon, is the part which she has played in assisting the progress of modern exact astronomy. It is not saying too much to assert that if the Earth had had no satellite the law of gravitation would never have been discovered."[12] There is some truth in this; were it not for the Moon, its disclosure would certainly have been more difficult and longer delayed.

In his immortal *Principia*, published in 1687, Newton demonstrated how the inverse-square law of force held for bodies moving in any of the conic sections (ellipse, parabola, hyperbola); various theorems related to the attraction of the Sun, Moon, and planets; investigations of the shape of the Earth (oblate, i.e., flattened at the poles); and explanations of the tides, which are caused by the pull of the Moon on the oceans (with a lesser contribution from the Sun) and of the precession of the equinoxes. This last one was the grand phenomenon clearly described by Hipparchus two thousand years earlier, which had utterly baffled astronomers until Newton was able to show how it arose from the pull of the Moon on the equatorial bulge of the oblate earth.

And yet even Newton found it difficult to improve on Horrocks' theory of the motions of the Moon—it was, he once said, "the one problem that made my head ache." Basically, the Moon's orbit around the Earth is a Keplerian ellipse. Newton conceived of the Moon as being drawn from this idealized curve by the disturbing force of the Sun, a force always directed toward the Sun as it constantly changes its position relative to the Earth and the Moon. He could explain some aspects of the Moon's irregular motions. However, he was baffled by one particular feature. The line between the Earth and the points in its orbit where the Moon is closest or farthest away (perigee and apogee) is known as the line of apsides. This line advances in the same direction as the Moon's motion at an average rate of 3°23′ per month, or 40°41′ per year (again, this is an average; the actual rate is highly irregular, and makes up what Newtonian historian Nicholas Kollerstrom has described as the "nightmare of the rocking apse").

Qualitatively, Newton understood the cause of the overall advance very well. In the undisturbed ellipse that the Moon would follow if the Earth alone acted upon it, the orientation of the Moon's orbit would remain constant and fixed in space. The solar force destabilizes the situation, however. Averaged over the entire orbit of the Moon, it decreases (by 1⁄357) the central pull of the Earth on that body. Thus the actual orbit of the Moon is slightly less curved than the undisturbed ellipse, bringing the Moon to each successive perigee slightly later than it otherwise would. The line of apsides twists in the same direction as the Moon's motion causing it to trace out an intricate rosette over successive orbits. Finally, after 8.9 years, the line of apsides returns to its original position. Unfortunately, when Newton attempted to carry out this analysis quantitatively, he was sharply disappointed; the value he found for the advance of the line of apsides came out to just half the observed value.

As late as April 1695, Newton was still confiding to the Astronomer Royal, John Flamsteed (1646–1719), to whom he appealed for observations, that he hoped to finish the Moon's theory within a few months. His frustration was clearly mounting, however, and he added that "when I have done it once I would have done with it forever."[13] In fact, he never did conquer the Everest of lunar theory. He attempted to try to place most of the blame on Flamsteed, whom he accused, quite unjustifiably as it turns out, of not furnishing him all the observations he needed. Instead of producing a pure derivation of the advance of the line of apsides from the law of gravitation, he resorted to tinkering with Horrocks' model of a rotating Keplerian ellipse of variable eccentricity.

A version of Newton's jerry-rigged Horrocksian lunar theory was published in 1702, and contained an appended note to the reader, probably by Edmond Halley:

> The Irregularity of the Moon's Motion hath been all along the just Complaint of Astronomers; and indeed I have always look'd upon it as a great Misfortune that a Planet

> so near to us as the Moon is, and which might be so wonderfully useful to us by her Motion, as well as her Light and Attraction (by which our Tides are chiefly occasioned) should have her Orbit so unaccountably various, that it is in a manner vain to depend on any Calculation of an Eclipse, a Transit, or an Appulse of her, tho never so accurately made. Whereas could her place be but truly calculated, the Longitude of Places would be found everywhere at Land with great facility, and might be nearly guess'd at Sea without the help of a Telescope, which cannot be there used.[14]

Though Halley computed tables using Newton's theory which he claimed were accurate within two or three minutes, Flamsteed (somewhat gleefully, no doubt) found them off by as much as four times that amount. This made Newton's theory no better than Horrocks'. (If we repeat the calculations today using a modern computer, we can say that—if one gets Newton's sixth equation the right way around—the accuracy is generally within four to five minutes, with a maximum deviation of seven to eight minutes). Unfortunately, this was obviously still far too inaccurate to be of practical use for determining longitude at sea.

Moreover, the problem seemed even more acute in 1707 after a much-publicized naval disaster off the Scilly Isles. The British Admiralty, the merchants of London, and the ships' captains themselves all cried out for something to be done. A petition was presented to Parliament in May 1714, asking that an award be offered for a satisfactory solution. After a hearing that Newton himself attended, Parliament passed an act "for Providing a Publick Reward for such Person or Persons as shall Discover the Longitude at Sea," and offering a substantial monetary award of up to £20,000, the exact amount to depend on the accuracy of the method. A Board of Longitude was appointed to sit in judgment on the proposals which, given the sheer size of the award, were expected to be numerous. Newton was one of the members, but personally saw little point in the Board continuing to meet until his own lunar theory had been perfected. When he died in 1727, the prize had not yet been claimed.

Newton's greatest successors—the Swiss Leonard Euler (1707–1783) and the Frenchmen Alexis Claude Clairaut (1713–1765) and Jean-le-Rond d'Alembert (1717–1783)—made further improvements to lunar theory. By 1749, Clairaut succeeded in deriving the correct motion of the apsides within the grand gravitational framework, and soon afterwards d'Alembert arrived at the same result and was actually the first to publish. All three astronomers taxed themselves in working out new tables. Even so, the best of their tables remained out by as much as five minutes; thus, they were not sufficiently accurate to be of practical use in determining longitude at sea—such an error in the Moon's position would have led to errors of 2½° on the Earth's surface, corresponding to 250 nautical miles.

The next step forward did not come from an improved theoretical understanding of the Moon's motion but from a more skillful discussion of the observations. This was the achievement of the German cartographer Tobias Mayer (1723–1762). The son of a Württemberg wheelwright, Mayer received little formal education, but did manage to teach himself mathematics and mechanical drawing. As a young man he first found employment with the copper-engraving and map-mak-

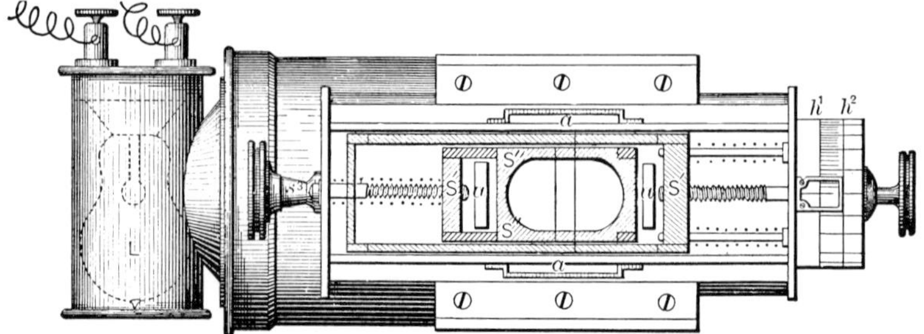

Figure 4.2 Micrometer. By the end of the nineteenth century, micrometers featured electrically illuminated wires, but their principle of operation remained much the same as in Tobias Mayer's day. From David Peck Todd, *A New Astronomy* (1897).

ing firm of Johann Andreas Pfeffel in Augsburg. In 1746 he was hired by the Homann Cartographic Bureau in Nürnberg to put the finishing touches on a large atlas of the Earth. While engaged in this project he realized the utter unreliability of much of the longitude data with which he had to work.[15]

A very old method for determining the difference in longitudes between two cities had been to make simultaneous timings of the beginning of a lunar eclipse. Hevelius had suggested that the accuracy of this method could be much improved if instead of timing the beginning of the eclipse, observers noted the exact moment at which individual features such as craters and peaks entered into the Earth's shadow, but he himself had never attempted to put the method into practice. Mayer, while grasping the method's promise, realized that it could never be successfully applied unless observers were equipped with a better lunar map. It was long overdue. Nearly a century had passed since the heyday of Hevelius, Langrenus, and Riccioli.

Between February and August 1748, Mayer made a series of lunar observations with a very modest telescope of only 1.4 inches (35mm) aperture, using a micrometer (Figure 4.2) to fix the positions of two dozen points on the lunar surface and interpolating the positions of other points from those actually measured. For the first time in the history of selenography, selenographic coordinates were used to plot the positions of the various lunar features. Using these measures as the basis, Mayer produced an artistic map (Figure 4.3), on a scale of 7.5 inches (190 mm) to the Moon's diameter. He also drew up an 18-inch (457mm) map, and began to work on sections which he hoped eventually to assemble into a large lunar globe (Figure 4.4).

None of Mayer's maps was published in his lifetime, in large part because he was increasingly immersed in the calculation of new tables of the Moon's motions. He had begun this research when, in 1751, he received an appointment as professor of mathematics and political economy at the Georg-August University in Göttingen (he was not obliged to lecture on the latter subject, in which he professed no special knowledge). When the Göttingen Observatory opened in the summer of 1753, he commenced a series of observations of the Moon's positions relative to

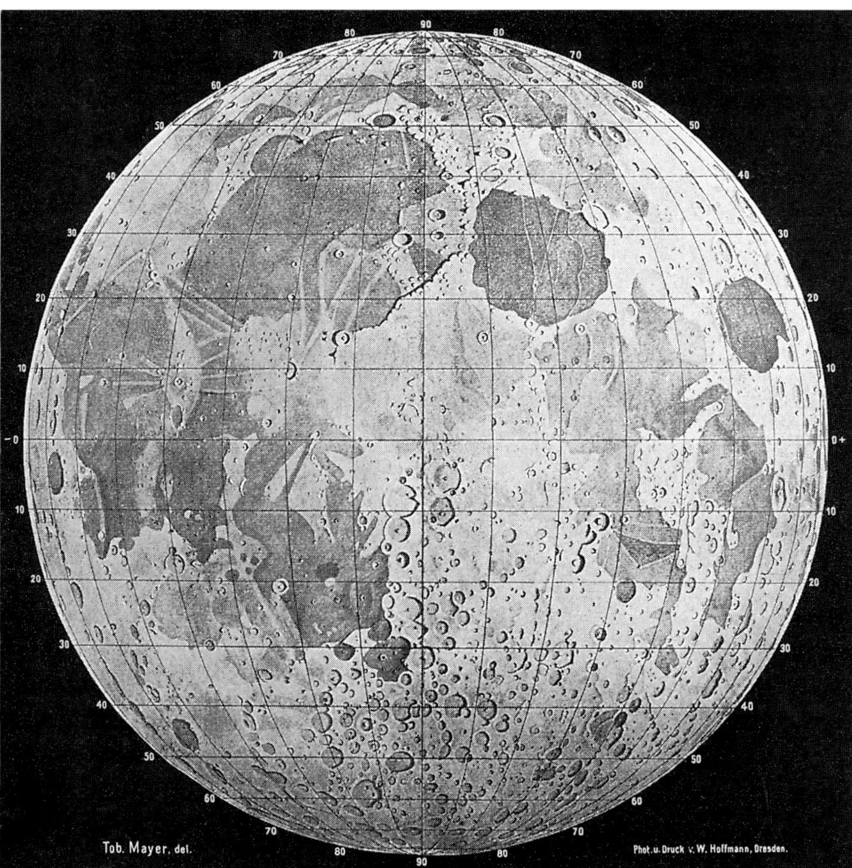

Figure 4.3 Tobias Mayer's Moon map, completed in 1749 and published in 1775 by Georg Lichtenberg, who added the lines of latitude and longitude. Mayer's was the first lunar map to be based on painstaking micrometric measures of the positions of the various features, and as a result was far more accurate than any of its predecessors. This version was published by Klinkerfaess in 1881. Courtesy Dieter Gerdes.

the stars, using a micrometer of his own design to measure the distances of these stars near the limb of the Moon. Almost at once he realized that much of the apparent discrepancy in the tables was due not to still-unsolved problems of lunar theory, as had been thought, but to inaccuracies in the charted positions of these stars. A notable by-product of this research was his conclusion that the Moon is for all intents and purposes airless, since he observed the stars to snap out instantaneously when they were covered by the limb of the Moon, with no preliminary dimming or displacement by refraction that would betray the presence of even a tenuous lunar atmosphere.[16] However, this conclusion was immediately attacked by Euler, who from other considerations calculated that the Moon had an atmosphere with a density of about 1/200 the Earth's. On the other hand, it won influential support from the Slovene Jesuit astronomer Roger Boscovich (1711–1787).[17]

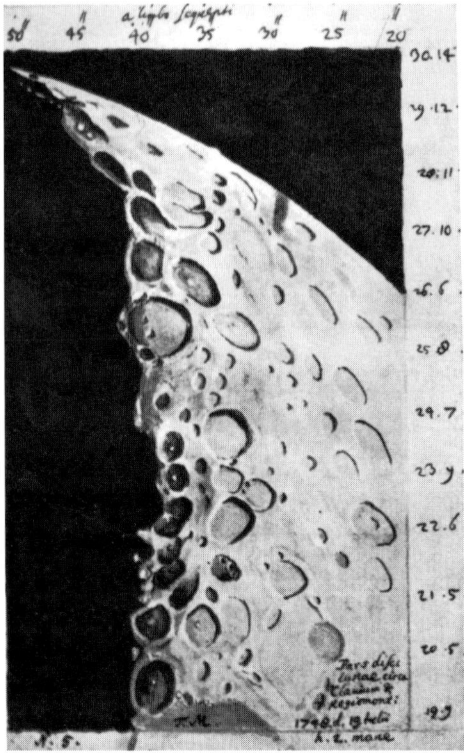

Figure 4.4 One of 40 detailed drawings of various areas of the Moon at different phases that Mayer made with the intent of producing a lunar globe. Although six of the twelve gores for the globe were completed, the project was ultimately abandoned because the publisher, the Cosmographical Society of Nuremberg, chronically flirted with bankruptcy. Courtesy John D. Koester.

By incorporating these insights Mayer believed that he had at last produced tables accurate to a mere two minutes, and he now submitted his methods to the British Admiralty. Unfortunately, a test at sea was delayed by the Seven Years' War which had broken out between France and Britain, and Mayer died in 1762 at the tragically early age of thirty-nine without receiving any of the prize money. Moreover, none of his lunar maps was published during his lifetime.

A rather different approach to the problem of longitude at sea, and in the end a more practical one, came in the form of an exquisite timepiece fashioned by the unschooled son of a joiner, the English clockmaker John Harrison (1693–1776), who disdainfully referred to the learned university astronomers as "lunar men." His fourth marine chronometer, the culmination of thirty years of frustration and dogged labor, pioneered the use of jeweled bearings and proved capable of keeping time with remarkable accuracy even on a pitching ship on rough seas. When the British Admiralty finally made its recommendations in 1765, it granted £10,000 to Harrison, but awarded only £3,000 to Mayer's widow.

Though overshadowed by the success of Harrison's marine chronometer, Mayer's work was the culmination of a grand tradition begun by Langrenus and extended by Hevelius and Riccioli. His map, sanctioned by the British Admiralty and finally published by his friend Georg Christoph Lichtenberg in 1775, harks back to earlier efforts of selenographers to extend to the Moon the empire of Spain or the dominion of the Roman Church—they were all set-pieces in the great era of navigation and empire, when the Moon was seen mainly as an extension of terrestrial conquests. Not because of the intrinsic interest of its features was it mapped by these pioneer cartographers, but to provide advantages to sea-farers on Earth.

So the metaphor of exploration and conquest which had thrilled the first observers with their optic tubes had been largely co-opted to serve earthly needs. The pale enchantress of the night had been rationalized and subdued to the most practical of purposes. The "world in the Moon" of Galileo and the other early telescopists had been reduced to mere matter in motion by Newton's laws; the satellite redefined as a mere time-piece, servant of navigators and image of terrestrial affairs.

Mayer's work marked both an end and a beginning. It would await the advent of new and better telescopes—and, no less important, the stirrings of the more romantic imagination of one of the greatest astronomical observers who ever lived—to bring the curtain down on the "long night of selenography."

References

1. Fontenelle, *Conversations on the Plurality of Worlds*, trans. Hargreaves, p. 38.
2. Ibid., p. 39.
3. Ibid., p. 43.
4. The title of the first English translation, which appeared in Glasgow in 1702. The full title of Huygens' work was *Kosmotheoros, sive De Terris Coelestibus, earumque ornatu, Conjecturae* (The Hague, 1698).
5. Steven J. Dick, *Plurality of Worlds*, p. 131.
6. Ibid.
7. Zdenek Kopal, "Topography of the Moon," in *Physics and Astronomy of the Moon*, ed. Z. Kopal (New York: Academic Press, 1962), p. 238.
8. A. Danjon and A. Couder, *Lunettes et Télescopes* (Paris, 1935), p. 651.
9. Ashbrook, *The Astronomical Scrapbook*, p. 236.
10. Fontenelle, *Conversations*, p. 38.
11. Francesco Bianchini, *Observations Concerning the Planet Venus*, trans. by Sally Beaumont, assisted by Peter Fay (Berlin: Springer-Verlag, 1996), p.23.
12. Richard Anthony Proctor, *The Moon: Her Motions, Aspect, Scenery, and Physical Condition* (New York: Appleton, 2nd ed., 1878), p. 41.
13. Newton to Flamsteed, April 23, 1695; *The Correspondence of Isaac Newton,* ed. H. W. Turnbull (vols. 1–3), J. F. Scott (vol. 4) and A. R. Hall and L. Tilling (vols. 5–7), (Cambridge: Cambridge University Press, 1959–1977), IV, 106.
14. I. Newton, *A New and Most Accurate Theory of the Moon's Motion* (London, 1702), p. iii–iv.
15. For Tobias Mayer, see Eric G. Forbes, "Who Discovered Longitude at Sea?" *Sky and Telescope*, 41 (1971), 4–6; and for a more technical account, Forbes, "Tobias Mayer's Contributions to the Development of Lunar Theory," *Journal for the History of Astronomy*, 1 (1970), 144–154. For a popular account of the longitude question, it is impossible to do better than Dava Sobel, *Longitude* (New York: Walker and Co., 1995).
16. Tobias Mayer, "Tobias Mayers Beweis das der Mond keinen Luftkreis habe," *Kosmographische Sammlungen auf d. J. 1748* (Nürnberg, 1750), 397–419.
17. L. Euler, *Sur l'atmosphere de la Lune* (Berlin, 1748); R. Boscovich, *De Lunae Atmosphera* (Rome, 1753). For a general discussion, see Ashbrook, "Roger Boscovich and the Moon's Atmosphere," in *The Astronomical Scrapbook,* pp. 244–247.

Chapter 5:
Eruptions in the Moon

Friedrich Wilhelm Herschel (Figure 5.1) (or William Herschel, as he later naturalized his name) was one of the greatest observational astronomers of all time.[1] He is known as "the father of stellar astronomy" who cataloged thousands of star clusters and nebulae, discovered that many double stars are actually binary systems gravitationally bound to one another, and made an early attempt to map the shape of the Galaxy. On the whole, he paid relatively little attention to the Moon. His program was far more wide-ranging and ambitious, aiming at nothing less than revealing the "construction of the whole heavens." Fascinating as it may be, the Moon is admittedly but a minor detail of the grand scheme and could hardly be expected to enjoy high priority in the mind of a man like Herschel.

Or so it is generally thought. However, what has been said holds true only for the mature Herschel. During his early years he frequently cast more than a casual eye upon the Moon.

Herschel was born in Hanover (or Hannover) in 1738, the son of an oboist in the Hanoverian Guard. At first young Wilhelm attempted to follow in his father's footsteps. However, in the spring of 1757, after a disastrous campaign of the Guard against the French during the Seven Years' War during which Herschel came under fire at the battle of Hastenbeck, he obtained his discharge and decided to emigrate to England.

At the time England and Hanover enjoyed close ties. In order to assure the Protestant succession, a Hanoverian Elector, George Louis (George I), had ascended to the British throne in 1714. This event inaugurated the era of "personal union" between the two countries, which would last until the accession of Victoria in 1837. Though they were absentee Electors, both George I and his son George II continued to take a strong interest in their hereditary domains. Indeed, it was George II who had founded the Georg-August Universität in Göttingen (in 1737), the scene of Tobias Mayer's Moon-mapping exploits. Only George III thought of himself as a true Englishman born and bred, and during his long life he never set foot in Hanover.

Naturally the court soon teemed with Germans, and many more, like Herschel, came to England to try to make their fortunes. Herschel struggled at first, but by 1766 had established himself as a successful musician in the fashionable resort city of Bath. His principal position was as organist in the new Octagon

Figure 5.1 William Herschel, the father of stellar astronomy," was also an avid lunar observer, especially early in his career. Courtesy Yerkes Observatory.

Chapel, but he also gave private music lessons and was a prolific composer of symphonies and concerti.

Though for the next ten years he would earn his living as a musician, Herschel's fascination with astronomy was growing into an all-consuming passion. Following a frustrating interlude with a succession of cumbersome refractors, he embarked upon the painstaking work of casting and grinding mirrors for small reflectors. As the invention of a method of chemically depositing a layer of metallic silver onto a glass substrate lay almost a century in the future, Herschel's mirrors were made from speculum metal, a hard, brittle bronze alloy composed of 68 parts copper and 32 parts tin, whitened with a small admixture of arsenic or antimony. Speculum metal mirrors slowly tarnished in the open air and had to be periodically repolished and refigured. In the damp climate of England, a surface would often go bad in little more than a month, and routinely became completely unusable in a mere three or four months. Only those capable of performing the difficult task of figuring could keep a reflector in service, so the successful user of such an instrument had to be at once an observer and a master optician.

After over one hundred failures, Herschel finally succeeded in making a serviceable mirror and, having patiently mastered the techniques, was soon able to keep relays of specula on hand. His first recorded observation was a sketch of the Orion Nebula made on March 4, 1774. He devoted much time in learning "how to see," which he considered "in some respects an art which must be learnt," as he later told his friend Dr. William Watson of the Bath Philosophical Society:

> To make a person see with... power is nearly the same as if I were asked to make him play one of Handel's fugues upon the organ. Many a night have I been practicing to see, and it would be strange if one did not acquire a certain dexterity by such constant practice.[2]

Naturally, the Moon came up early for its turn in the course of Herschel's telescopic studies. One of his first recorded lunar observations is dated May 28, 1776. That night he was testing a new reflector with a focal length of ten feet (Figure 5.2), when he found himself

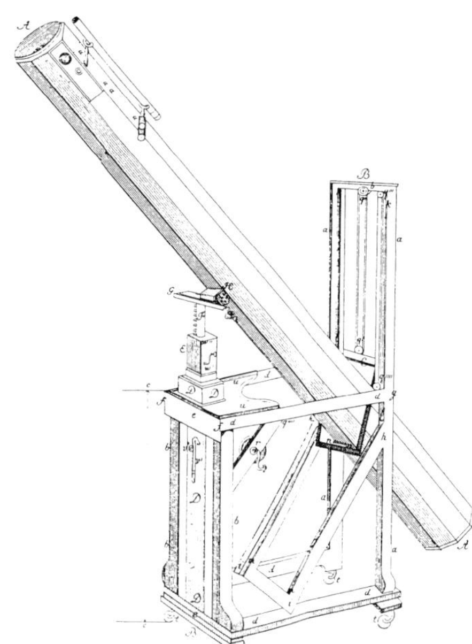

Figure 5.2 A contemporary engraving of a 7-foot Herschel reflector. Most of his lunar observations were maid with instruments of this manageable size. Following his discovery of the planet Uranus, he was inundated with orders for similar instruments. Courtesy Dieter Gerdes.

> struck with the appearance of something I had never seen before, which I ascribed to the power and distinctness of my instrument, but which may perhaps be an optical fallacy… I believed [myself] to perceive something which I immediately took to be *growing substances*. I will not call them trees, as from their size they can hardly come under that denomination.[3]

The region which had seized Herschel's attention was the Mare Humorum, near the crater Gassendi (Figure 5.3). Herschel was well aware that even the tallest trees on Earth would be invisible from such a great distance. Indeed, he calculated that "the borders of forests, to be visible, would require Trees at least 4, 5 or 6 times the height of ours."[4] Nevertheless, at the moment he could think of no better explanation:

> It has hitherto been supposed that those seas, as they are called, consisted of a different kind of soil which reflected light less copiously than the Hills and Mountains. I conclude them to be Woods or Forests… These Forests… cast shadows all along the side opposite the Sun. Now I think if we admit this dark coloured ground to be forests it will solve all these appearances at once; which in other Hypothesis will be found very difficult.[5]

Under higher lighting on the following night, the aspect of the region was markedly changed—so typical of the Moon—and Herschel could "perceive none of those appearances of the border of Woods I saw yesterday."[6]

Herschel's next journal entry about the Moon, dated July 30, 1776, is even more remarkable and shows how was his progress in "learning how to see." This entry deserves to be quoted at length.

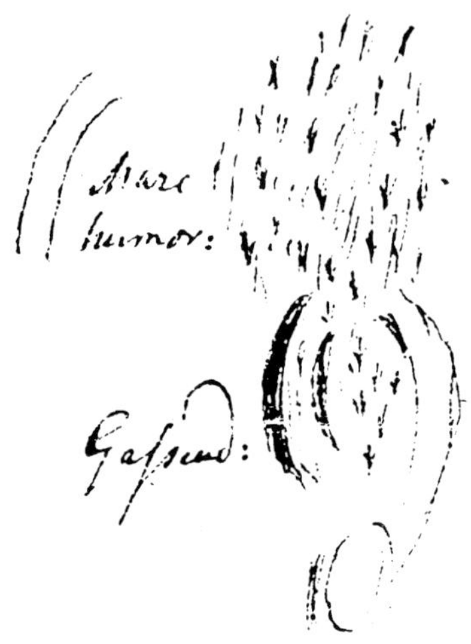

Figure 5.3 The victim of a beginner's giddy enthusiasm, Herschel entered into his observing log for May 28, 1776 this crude sketch of a "wood" on Mare Humorum, close by the crater Gassendi. Courtesy Peter Hingley, librarian, Royal Astronomical Society.

> Perhaps conclusions from the analogy of things may be exceedingly different from the truth, but... we have no other way to come at knowledge... For instance, seeing that our Earth is inhabited and comparing the Moon with this planet: finding that in such a satellite there is a provision of light and heat: also, in all appearance a soil proper for habitation full as good as ours, if not perhaps better—who can say that it is not extremely probable, nay beyond doubt, that there must be inhabitants on the Moon of some kind or other? Moreover it is perhaps not altogether so certain that the Moon is out of the reach of observation in this respect. I hope, and am convinced, that some time or other very evident signs of life will be discovered on the Moon.
>
> When we call the Earth by way of distinction a planet and the Moon a satellite or attendant, we should consider whether we do not perhaps, in a certain sense, mistake the matter... The Earth acts [the] part of a Carriage, a heavenly waggon to carry about the more delicate Moon, to whom it is destined to give a glorious light in the absence of the Sun; whereas we as it were travel on foot and have but a small lamp to give us light in our dark nights and that too, often enough extinguished by clouds. For my part, were I to chuse between the Earth and Moon I should not hesitate a moment to fix upon the moon for my habitation.[7]

Herschel spells out in considerable detail his views about lunar conditions. He refers to the craters, or "circuses" (by which term he meant not the Barnum and Bailey variety, but circular places like London's celebrated Picadilly Circus, where several streets come together; an imposing structure of the kind was completed in Bath itself in 1756, which remains one of the prominent landmarks of the city to this day). These circuses were evidently the dwelling-places of lunar inhabitants, and he ponders whether "perhaps... on the Moon every town is one very large Circus." But if this is admitted, a very large area of research is opened up:

> Ought we not to watch the erection of any new small Circus as Lunarians may the Building of a new Town on the Earth. Our telescopes will do this... Now if we could discover any new erection it is evident an exact list of those Towns that are already built will be necessary. But this is no easy undertaking to make out, and will require

the observation of many a careful Astronomer and the most capital Instruments that can be had. However this is what I will begin."[8]

Like so many astronomers before and since, Herschel was "moonstruck." In his lunar writings we find a wilder, more speculative observer than the sober man portrayed in the usual accounts that place emphasis on his later stellar work. Historian Michael J. Crowe has suggested that Herschel's expectation that signs of lunar life might lie within the reach of direct observation stimulated his quest for ever larger and more powerful telescopes—the very telescopes that he later applied so successfully to bold and wide-ranging investigations of the stellar system and nebulae.[9]

This may be overstating the case, but it is obvious enough that Herschel was briefly enamored of the idea of carrying out an ambitious program of lunar research. In his notebook, under the headings "Metropolises, Cities, Villages" (which he later amended to "Large places, Middling places, Small places"), he prepared to catalog the lunar "circuses."[10] On June 17, 1779, he recorded "a Cut or Canal that seems evidently to be the effect of Art rather than of Nature." A month later, on finding a "new" spot on the Mare Crisium, he declared: "I find it is a city."[11]

Many of his observations during 1780 were devoted to measuring the heights of lunar mountains. He found that most did not exceed an altitude of half a mile. These results were summarized in a paper, "Astronomical Observations Relating to the Mountains of the Moon," which was communicated to the prestigious Royal Society of London by Dr. Watson and read at its meeting in May 1780. It is seldom remembered that Herschel made his scientific debut before the Royal Society as an observer of the Moon.

He introduced his treatise with this comment:

> I grant that there are more necessary and more useful objects of inquiry in the science of astronomy [than measuring the heights of lunar mountains]; but when we consider that the knowledge of the construction of the Moon leads us insensibly to several consequences... such as the great probability, not to say almost absolute certainty, of her being inhabited, we shall soon agree, that these researches are far from being trifling."[12]

His methods for measuring the heights of lunar mountains, but more especially the passage just quoted, aroused the interest, as well as perhaps the skepticism, of Nevil Maskelyne (1732–1811), the Astronomer Royal at the Greenwich Observatory, who asked Herschel to elaborate on his views of lunar life.

In his reply, Herschel implored Maskelyne to "promise not to call me a Lunatic," and admitted that his belief in an inhabited Moon might be due to "a certain Enthusiasm which an observer, but young in the Science of Astronomy can hardly divest himself of when he sees such wonders."[13] But then he shed his inhibitions and proceeded to quote to Maskelyne extended passages from his earlier cited notebook entry of July 30, 1776.

Herschel's life changed forever when he discovered, using one of his seven-

Figure 5.4 A 1794 engraving of Herschel's favorite telescope, the 20-foot reflector. Completed in 1785, this instrument was later refurbished by Herschel's son John and taken by him to a site near Cape Town, where he used it to carry out a systematic survey of the southern skies. Courtesy Peter Hingley, librarian, Royal Astronomical Society.

foot reflectors, the planet he called *Georgium Sidus*, the Star of the Georges—later to become known as Uranus—on March 13, 1781. Henceforth he became a leading figure in British science and received a pension from George III that allowed him to give up music and devote all his time to telescope making and astronomical research.[14] He made ever larger instruments for his personal use (Figures 5.4 and 5.5), while his optics enjoyed a reputation for quality and commanded very high prices.[15]

Gradually Herschel seems to have become less fervent in his views about the Moon, since he never published any of his supposed observational evidence for the existence of lunar life. It remained buried in his notebooks, unpublished until the present day.[16] One gathers from his reticence that he may have become skeptical about most of it.

Although Herschel's evidence for lunar life never entered into the realm of scientific discourse, another suggestion—also based on terrestrial analogies—certainly did. In 1787 he published reports of volcanic eruptions on the Moon, which had an electrifying effect on selenographical studies.

Lunar volcanoes were very much on Herschel's mind as early as September 1780, when he made the following entry in his observing book:

> What I have just now seen on the Moon is one of the most extraordinary Phenomena I ever saw: A well-defined line, like the trace of a fluid matter that had run down along the side of [a] Hill. It struck me immediately that it might be the Lava of a Vulcano... [It] is indeed very extraordinary that I should have overlooked the appearance [hitherto]... if it had been there. For which reason I am convinced that it happened since last month.[17]

There were further developments on May 4, 1783. Observing with friends,

Figure 5.5 Herschel's 40-foot telescope, the world's largest until the mid-nineteenth century, was erected at Slough, near Windsor, in 1789. The performance of this instrument was not proportional to its size, and Herschel generally preferred to use his more manageable instruments of 7 to 20 feet focus. He cautioned that "a 40-foot telescope should be used only for examining objects that other instruments will not reach." From Edwin Dunkin, *The Midnight Sky* (1879).

Dr. James Lind and his wife, Herschel perceived in the dark part of the crescent Moon that was feebly illuminated by earthshine "a luminous spot" that he estimated to be of about the same brightness as a star of the fourth magnitude. Mrs. Lind was called to the telescope, whereupon she "immediately saw it, tho' no person had mentioned it, and compared it to a star."[18] Less than two weeks later, on May 13, Herschel saw in the same location on the Moon "two conical mountains, which I suppose to have been thrown up in the last eruption… They are situated just by a third much larger which I have often seen before and remarked, tho' the two small ones were never before perceived in that place nor expressed in a drawing I had made of the spot."[19] In his observing book, he recorded also what he took to be flows of "Lava," and "an opening… as if a channel were dug ready for the Lava to run in."[20]

There can be little doubt that Herschel's sudden interest in lunar volcanism owed much to the contemporary fascination with terrestrial eruptions, the result of the highly publicized accounts of Vesuvius by Sir William Hamilton (1730–1803), the English Envoy Extraordinary to the Court of Naples who is remembered today chiefly on account of his wife's scandalous affair with Britain's greatest naval hero, Admiral Horatio Nelson. Between 1767 and 1771, Hamilton recounted his ascents of the volcano and his eyewitness observations of its recent eruptions in a series of reports that appeared in the *Philosophical Transactions of the Royal Society of London.* Hamilton was astonished by the sheer power of the natural forces he saw at work as Vesuvius grew hundreds of feet before his very eyes. Rejecting the prevailing view that mountains were "consumed" by volcanoes, he suggested instead that "mountains are produced by volcanoes, and not volcanoes by mountains."[21] He followed up these reports with a book, *Campi Phlegraei*, in 1776.[22] Here he included superb engravings by an Italian artist, Pietro Fabris, showing not only Vesuvius itself but the broad expanse of volcanic basins which constitute the Campi Phlegraei ("Fields of Fire") proper; the latter, as astronomical historian Roderick Home has noted, "much more closely resembles a typical lunar landscape than do the immediate surroundings of Vesuvius."[23]

The striking resemblance between the Campi Phlegrai as depicted by Fabris and the lunar landscape as seen through a small telescope had impressed at least one observer even before Herschel. In 1778 Franz Aepinus (1724–1802) of Rostock, a councillor at the St. Petersburg court of Catherine the Great, examined the Moon for the first time with a small refractor shortly after reading Hamilton's book. He soon surmised that the craters of the Moon were produced by the eruption of ash from vents, which on falling back to the ground produced the surrounding ring-wall. This is the first clear statement of the volcanic theory of the origin of the lunar craters which was soon to become so influential.[24]

Until 1787 Herschel himself had remained circumspect about these supposed lunar volcanoes. His attitude changed, however, after he witnessed what he took to be an even more dramatic "eruption." He was ready at long last to break his silence with a paper in the *Philosophical Transactions of the Royal Society*, claiming that on the evening of April 19, he had found

> three volcanoes in different places of the dark part of the Moon. Two of them are either already or nearly extinct, or otherwise in a state of approaching eruption; which, perhaps, may be decided next lunation. The third shows an actual eruption of fire, or luminous matter."

The next night, he had found that

> the volcano burns with greater violence. It is of an irregular round figure, and very sharply defined on the edges. The other two volcanoes are much farther towards the centre of the Moon, and resemble large, pretty faint nebulae, that are gradually much brighter in the middle; but no well defined luminous spot can be discerned in them... The appearance of what I have called the actual fire or eruption of a volcano exactly resembled a small piece of burning charcoal, when it is covered by a very thin coat of white ashes, which frequently adhere to it when it has been some time ignited.[25]

Such reports from the greatest observer of the day, the celebrated discoverer of the *Georgium Sidus*, created a sensation, and telescopes around the world were soon pointed at the Moon. But after an initial flurry of excitement, a mood of skepticism set in. As early as May 1788, the eminent French astronomer Joseph Jérôme de Lalande (1732–1807) wrote politely to Herschel: "The volcano in the Moon has been visible these last few days; but there are astronomers who are inclined to believe that Mount Aristarchus, which is naturally very brilliant, might very well reflect the light of the Earth in such a manner as to produce this bright appearance across the pale light of the Moon."[26]

Similarly, the pair of lesser "volcanoes" located nearer to the center of the Moon can only have been the imposing craters Copernicus and Kepler. Each is surrounded by a halo of bright rays that take on a diffuse, nebulous appearance when illuminated by the wan blue light reflected by the Earth.

Herschel seems to have acquiesced to this embarrassingly mundane explanation. At any rate, he said no more publicly on the matter. In retrospect, it is hard to imagine how he could have made so fundamental an error. It was clearly a case of what Francis Bacon called *anticipatio mentis* (mental preconception). One can only say that "seeing by analogy, or with the eye of reason," which had served Herschel so well on many other occasions, led him in this one instance down a primrose path.[27]

References

1. Herschel's papers are collected in *The Scientific Papers of Sir William Herschel*, J. L. E. Dreyer, editor, two volumes (London: The Royal Society and the Royal Astronomical Society, 1912). For Herschel's lunar work, a detailed discussion is provided by Michael J. Crowe, *The Extraterrestrial Life Debate 1750–1900* (Cambridge: Cambridge University Press, 1986). Also very helpful was the senior thesis of Crowe's student J. Conor Burns, "Barking at the Moon" (unpublished; West Bend, Indiana: University of Notre Dame, 1991).
2. William Herschel to Dr. William Watson, January 9, 1782; in *Scientific Papers*, 1, xxxiii.
3. Microfilm (reel 17) of the Royal Astronomical Society Herschel MSS, W. 3/1.4, p. 1. Referred to henceforth as *Herschel MSS*.
4. Ibid.,, p. 4.
5. Ibid., pp. 2–3.
6. Ibid., *d*, p. 4.
7. Ibid., *d*, pp. 7–8.
8. Ibid., pp. 8–10.
9. Crowe, *Extraterrestrial Life Debate*, p. 65.
10. *Herschel MSS*, p. 17.
11. Ibid.
12. W. Herschel, *Scientific Papers*, 1, 5.
13. Letter to the Rev. Dr. Maskelyne, June 12, 1780; in ibid., I, xc.
14. Simon Schaffer, "Uranus and the Establishment of Herschel's Astronomy," *Journal for the History of Astronomy*, 12 (1981), 11–26.
15. There can be little doubt that Herchel's specula were well worth the princely sums that he charged. In 1924 several of his mirrors, which had sealed been in soldered tins to prevent tarnishing several years before his death in 1822, were evaluated by W. H. Steavenson, F.R.A.S. Applying the Foucault knife-edge test after a lapse of more than a century in storage, Steavenson found that "on the whole the figure of all the mirrors was up to a very high standard," comparable in quality and performance to optics competently made with the benefit of modern methods of figuring and testing. See *Amateur Telescope Making*, A. Ingalls, ed. (New York: Scientific American, 1945), p. 332.
16. Nevertheless, Herschel's belief that the Moon was inhabited was unwavering. Indeed, he was equally certain of the existence of inhabitants within the Sun, a belief that only a few years earlier had been alleged as proof of insanity in an English courtroom! See Ashbrook, "William Herschel and the Sun," *The Astronomical Scrapbook,* p. 348.
17. *Herschel MSS*, p. 56.
18. W. Herschel, *Scientific Papers*, I, xxxviii.
19. Ibid.

20. *Herschel MSS*, p. 70.
21. *Philosophical Transactions of the Royal Society of London*, 54 (1769), 18–22:21.
22. William Hamilton, *Campi Phlegraei* (Naples, 1776).
23. Roderick W. Home, "The Origin of the Lunar Craters: An Eighteenth Century View," *Journal for the History of Astronomy*, 3 (1972), 1–10:3.
24. Franz Aepinus, "Über den Bau der Mondfläche, und den Vulcanischen Ursprung ihrer ungleicheiten," *Schriften der Berlinschen Gesellschaft Naturforschender Freunde*, 2 (1781), 1–40. In his *Selenographia*, Hevelius had named several lunar craters after terrestrial volcanoes.
25. W. Herschel, *Scientific Papers*, 1, pp. 315–316.
26. J. J. Lalande to W. Herschel, May 21, 1788, in Lubbock, *Herschel Chronicle,* p. 218.
27. On the other hand, one must admit that at times other astronomers have fallen into the same trap. To give but one example, the skilful German astronomer Wilhelm Tempel made the following observations at the Marseilles Observatory, June 10, 1866: "I have recently considered the Moon with attention... and found a remarkable light-appearance upon the dark side... which bright part agreed with the position of Aristarchus... I am aware that in astronomical observation one must be extremely on one's guard against optical deceptions. But this light-apperance has certainly not deceived me; for the light was not a faint white light, like that shown by other craters in the first days after the new Moon upon her dark side, but it was star-like, diffused, in colour reddish-yellow, certainly different from all the other visible bright spots. Of course I am far from surmising a still active volcanic outbreak, as such an outbreak supposes water and an atmosphere, both of which are universally allowed not to exist in the Moon—so that the crater-forming process can only be thought of as a dry, chemical, although warm one." *Astronomical Register*, 5 (1867), pp. 218–220.

Chapter 6:
A Compulsion to Observe

William Herschel's failure to understand the Moon may help to create a more sympathetic frame of mind from which to regard the work of his much-maligned contemporary, Johann Hieronymous Schroeter (Figure 6.1), the true father of modern selenography.[1] Though Herschel's more extreme notions about the Moon had no direct influence on Schroeter—for the simple reason that they remained unpublished—their overall views of lunar conditions were remarkably alike. Independently, Schroeter struck out along the same lines of investigation that Herschel had confided only to his observing notebook in the late 1770s, and went much farther in actually carrying it out. Indeed, compared to Schroeter, and despite the importance he attached to the subject at various times, Herschel can only be regarded as a dilettante of the Moon.

Schroeter (or Schröter, the form of the name that is often used, although it is without authority in contemporary documents) was born on August 30, 1745 in Luther's old city, Erfurt. A delicate infant, he was named for the local physician who helped deliver him, Johann Hieronymus Kniphof. Schroeter's father was an attorney, though not a very prosperous one. His main interests seem to have been in mechanical work and music. He suffered from poor health and died when Schroeter was not yet twelve. After distinguishing himself in the local schools, Schroeter matriculated at the University of Gera in 1761 as a student of theology, and was already a keen student of astronomy, sometimes carrying a small telescope to the top of one of the church towers in order to make observations.

In 1764 Schroeter abandoned theological studies in order to follow his father's footsteps into the law, enrolling *ob paupertaten gratis* at the University in Göttingen. When he arrived, the old city's tranquil academic groves were still haunted by the shade of Tobias Mayer. One of Schroeter's first acquaintances was young Georg Christoph Lichtenberg (1742–1799), who later became well known for his experiments with electricity. At Mayer's death, Lichtenberg had been placed in charge of all his effects, including his unpublished lunar maps, and Schroeter may well have learned something of the work of the dead master from him. However, a greater influence on him was Abraham Gotthelf Kastner (1719–1800), an inspiring lecturer in mathematical physics and astronomy.

After completing his studies in Göttingen, Schroeter's budding scientific career entered a dormant period that would last for several years while he concentrated on building a prosperous legal practice. He toiled at the law for ten years

Figure 6.1 A contemorary portrait of Johann Hieronymous Schroeter, the Geman magistrate who emulated Herschel in enthusiasm and perserverence. Courtesy Dieter Gerdes.

and, in 1777, was appointed secretary of the Royal Chamber of King George III in Hanover. There he became acquainted with two of William Herschel's younger brothers, Johann Alexander and Dietrich. The Herschels shared Schroeter's love of music and reawakened his interest in astronomy. Through Dietrich he acquired a 2.25-inch (57-mm) aperture refractor by the London optician Peter Dollond, and made his first observations with it in 1779.

Schroeter's life took a decisive turn in 1781, with William Herschel's discovery of the "Georgian." In a spirit of emulation, Schroeter resolved to devote himself to astronomy. Resigning his position in Hanover, he applied for that of *Oberamtmann* (chief magistrate) of Lilienthal, a sleepy little hamlet near Bremen. His application was approved, and on May 2, 1782, he moved into the Amthaus in Lilienthal. Since his official duties were light, he was able to devote most of his time to astronomy, and he set up the Dollond refractor in an observatory he called "Urania Temple" in the Amthaus garden.

The small refractor failed to satisfy Schroeter's ambitions, so he again made use of his contacts with the Herschel family in Hanover to acquire from William mirrors of 4.75 and 6.5 inches (120 and 165 mm) in diameter. For the latter, he paid Herschel 600 Reichsthalers, nearly half his annual salary. He constructed two Newtonian reflectors around these optics which would serve as his main instruments for several years. The larger, a seven-foot reflector, was a virtual twin to the instrument Herschel used to discover Uranus, and in 1786 was the largest working telescope in Germany. (Incidentally, in keeping with the convention of the time, telescopes were referred to by their focal lengths, not by their apertures as would be the case today; thus the Dollond refractor was his "three-foot," the Herschel reflectors his "four-foot" and "seven-foot" telescopes).

Now Schroeter could do much more than simply amuse himself. He embarked on a program of serious astronomical work, which included observing sunspots and a series of "extremely dark small spots" on Jupiter, from which he independently discovered that features in the planet's equatorial zone rotate more swiftly than those at higher latitudes.[2]

In winter 1787–88, Schroeter turned much of his attention toward the Moon, in large part due to his excitement over Herschel's reports of volcanoes. He announced his observing plans: "As a main occupation the observation and delineation of lunar features under all conditions of illumination, for the *topographiam lunarem*."[3] He confided to the Bremen physician and amateur astronomer Heinrich Wilhelm Matthäus Olbers (1758–1840) that he hoped that this program of research would lead to the completion of a lunar map 46.5 inches in diameter, based partly on the 7½-inch chart of Tobias Mayer which Lichtenberg had published in 1775. Schroeter soon had second thoughts, and decided that his proposal was too dauntingly ambitious for one observer. Instead of mapping the entire Moon, he would concentrate on making "special charts" of selected regions.

It was a prudent decision, for the Moon is a notoriously difficult object to draw. Herschel had quickly despaired of doing so, and made only a few small sketches because, as he confided in his notebook:

> By inspecting those maps I have already seen of the Moon, I find how exceedingly difficult it would be to give any delineations true enough to depend upon, as not one of them, in the least agrees upon reality. I shall therefore depend chiefly upon verbal descriptions... My figures will serve well enough to explain my meaning; more they are not meant to do.[4]

Herschel had supplied one of his reflectors to the noted portrait-artist John Russell (1745–1806), who in 1785 commenced a series of sketches in preparation for a large color drawing of the gibbous Moon. And yet even Russell, a professional artist, would spend ten years in making one finished drawing, which was donated by his daughter to the Radcliffe Observatory, Oxford, after his death. But Schroeter was no Russell. Though he had an indisputably keen eye and good instruments, no one has ever claimed he possessed outstanding skill as a draughtsman. On the other hand, his drawings, though sorely wanting from an aesthetic standpoint, are at least much more accurate than is sometimes appreciated.[5]

For purposes of measurement, Schroeter employed an ingenious device of his own invention, which he called his "projection machine."[6] (Figure 6.2) This consisted of a white screen divided into a grid of half-inch squares, supported by a bar at a right angle to the tube of his reflector. From a distance of 32½ inches, it gave a scale of 20 seconds of arc to the half-inch. The telescopic image seen by one eye had to be superimposed onto the image of the screen simultaneously viewed by the other eye—not by any means an easy thing to do! Schroeter was one of only a handful of observers who ever managed to use such a contrivance successfully. In his hands it did prove capable of fairly accurate measures, and he was able to faithfully render the proportions of many lunar formations.

Where Cassini (or his engraver Patigny) had represented the dark grey areas of the Moon as completely smooth and uniform, Schroeter recognized innumerable irregularities—isolated peaks, small craters, and the so-called "wrinkle ridges," which in fact he was the first to describe. He concluded that all of these

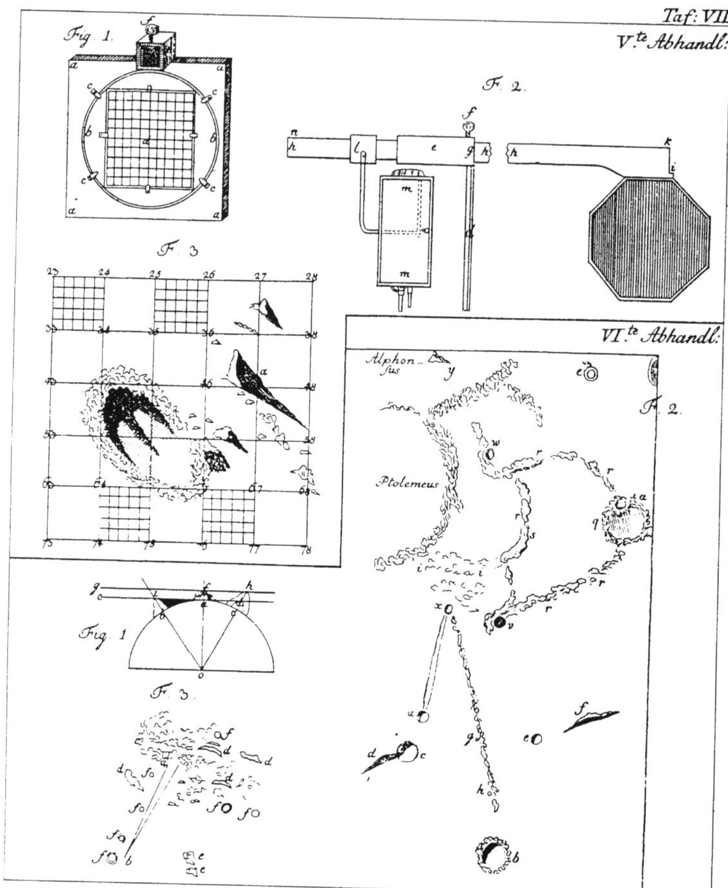

Figure 6.2 The upper part of this figure from Schroeter's *Selenotopographische Fragmente* of 1791 depicts the "projection machine" he used for measuring the dimensions of lunar formations. Although this device was relatively easy to construct, it was far more difficult to use than the conventional eyepiece micrometer. The lower part of the figure illustrates Schroeter's method of determining the heights of lunar mountains. Courtesy Yerkes Observatory library.

regions named "seas" by earlier selenographers were in fact completely waterless. In addition, he called attention to several special features of the Moon—the formation now known as Schroeter's Valley, the Hyginus Rille, and the Straight Wall (Figure 6.3), all forgotten since having been recorded by Huygens a century before. He discovered the Ariadaeus cleft, and also was the first to make out many of the lunar domes, which he referred to as "boils" or "swellings."[7]

Besides drawing parts of the Moon on a much larger scale than ever been attempted by previous observers, Schroeter made a fresh set of measurements of the heights of the lunar mountains. Ever since Galileo, it had been the custom of as-

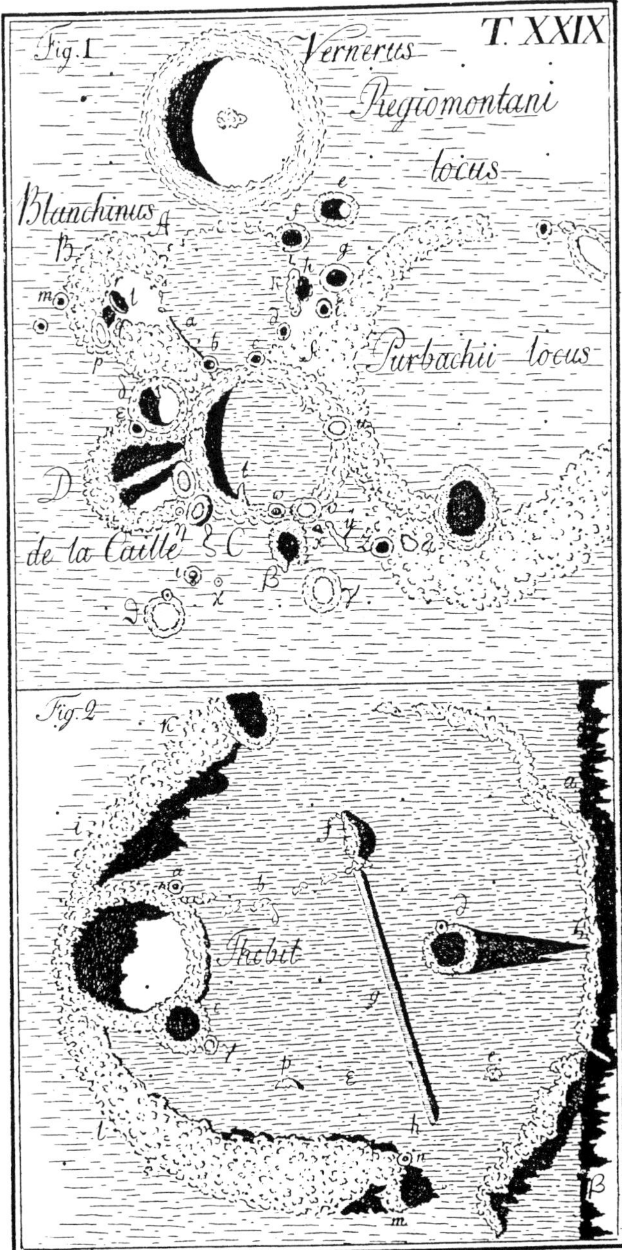

Figure 6.3 Representative drawings by Schroeter, showing (bottom) the Straight Wall, which he rediscovered a century after Christiaan Huygens, and the region surrounding the crater Purbach (top). As these drawings demonstrate, Schroeter was a clumsy draughtsman who represented craters and other lunar features in a highly stylized form. Nevertheless, he was a very capable observer who seldom made a grievous error. From *Selenotopographische Fragmente* (1791). Courtesy Yerkes Observatory library.

tronomers to attempt to measure the heights of the mountains exclusively where they caught light just beyond the shadows of the terminator. But Schroeter also carried out measures where the mountains projected beyond the limb, and—supplied with mathematical formulas worked out for him by Olbers—estimated their heights from the lengths of the shadows they cast across the surface. He made thousands of these measures, most surprisingly accurate—certainly much more so than Herschel's. He also deserves credit for being the first to realize the great heights of the mountains near the south pole, which he named the Leibniz and Doerfel ranges. (We now know that these mountains, which tower to heights of more than six miles, form parts of the outer rings of the huge South Pole-Aitken basin, located chiefly on the lunar far side.)

Armed with an unprecedented mass of data, Schroeter attempted detailed studies of the morphology of lunar craters. He found that when represented in profile the largest of them are very shallow, their walls rising only gently above their interiors, while in general the smallest craters feature the steepest walls. Compared with terrestrial volcanoes, the craters of the Moon are on a much grander scale—tens and even hundreds of miles in diameter. By contrast, the crater of Aetna is only 4,000 feet and that of Vesuvius 1,800 feet. In fact, these terrestrial volcanoes are so tiny that if transferred to the Moon they would have been at the very limit of visibility with Schroeter's telescopes.

Terrestrial volcanoes of the Vesuvian type are built up through a series of eruptions forming a steep mountain with a very small crater at the top—a form markedly unlike the shallow basins of the lunar formations. Schroeter, however, having approvingly read both Sir William Hamilton's *Campi Phlegraei* and Franz Aepinus's essay on the volcanic origin of lunar craters, did not hesitate to pronounce the lunar craters volcanic. He even thought he had definite proof of the fact. From his crater profiles, he fashioned scale models of some out of sand, and compared the amount of material making up the walls with that needed to fill the crater's interior. In general, he found the wall materials would just fill the pit. This observation, which came to be known as "Schroeter's Rule," satisfied him that they could not have been formed through subsidence or collapse. Instead the walls must have been thrown out in an eruptive process, meaning the craters were, at least broadly speaking, volcanic.[8]

Though he rejected Herschel's reports of erupting volcanoes on the Moon, Schroeter never doubted that the Moon was geologically active, and spent most of his time compiling evidence of possible changes. One of his cases involved the crater Hevel, near the Moon's limb. He had made one of his special charts of this region on October 24, 1787. Returning to it not quite a year later, he found that something remarkable had apparently taken place:

> On August 27, 1788, the compulsion to observe awoke me at about 3 A.M. The sky was overcast, but a short time later the clouds dispersed, and for the next few hours the air became extraordinarily pure and serene. Around 6:40 A.M., by bright sunshine, I noted the libration of the Moon with 134X on my 4-foot reflector... On the eastern limb [*western* according to the modern astronautical convention] near Grimaldi some

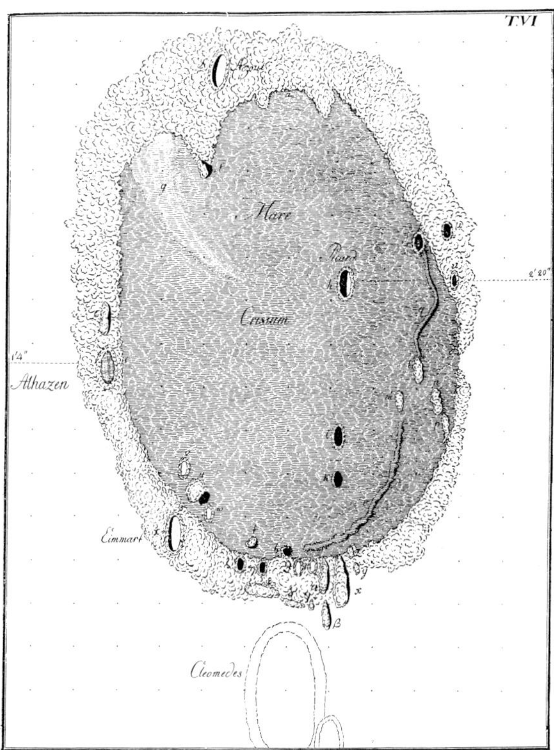

Figure 6.4 Schroeter's drawing of Mare Crisium, indicating the position of the crater Alhazen at left, one of the features in which he suspected changes. From *Selenotographische Fragmente* (1791). Courtesy Yerkes Observatory library.

> very high lunar Cordilleras looked greatly altered under the changed conditions of libration from the appearance they had shown only the day before. My eye also happened to fall upon [the crater] Hevel, and I noticed a dark round spot, unknown from all previous observations, and so peculiar in form that I at once drew up a special chart of the region.[9]

This small crater, he believed, had been newly formed by volcanic action since his previous observation ten months earlier.

Another formation that Schroeter believed was the site of marked changes was the crater Alhazen, on the Mare Crisium (Figure 6.4). It was prominent enough for him to employ it as a reference point for his measures of the lunar librations, but later it seemed to undergo such extreme alterations in form that it was frequently unrecognizable. Elsewhere on the Moon he found other cases of variable visibility, which he attributed to obscurations by veils of cloud and mist, and still other changes affecting the colors of the seas and certain dark-floored craters he ascribed to cycles of vegetative growth and decay.

Rejecting the conclusions of Boscovich and Tobias Mayer, Schroeter refused to believe in the airlessness of the Moon. Admittedly, the lunar atmosphere was thinner and drier than our own, and it was likely to be confined only to very low regions such as the interiors of the deeper craters—but he had no doubt it ex-

SELENOTOPOGRAPHISCHE

FRAGMENTE

ZUR

GENAUERN KENNTNISS DER MONDFLÄCHE,

IHRER

ERLITTENEN VERÄNDERUNGEN UND ATMOSPHÄRE,

SAMMT DEN

DAZU GEHÖRIGEN SPECIALCHARTEN UND ZEICHNUNGEN,

VON

JOHANN HIERONYMUS SCHROETER

KÖN. GROSSBR. UND CHURF. BRAUNSCHW. LÜN. OBERAMTMANNE, DER KÖN. SOC. DER WISSENSCH. ZU GÖTTINGEN CORRESPONDENTEN, DER CHURF. MAYNZ. AKAD. NÜTZL. WISSENSCH. ZU ERFURT, UND DER BERL. GES. NATURF. FREUNDE MITGLIEDE.

Mit 43 Kupfertafeln.

Auf Kosten des Verfassers.

LILIENTHAL bey demselben und in Commission bey CARL GOTTFR. FLECKEISEN, Universitäts-Buchhändler in Helmstädt.

Gedruckt Göttingen bey JOH. GEORG ROSENBUSCH, Univ. Buchdr. 1791.

Figure 6.5 Title page of the first volume of Schroeter's *Selenotopographische Fragmente* of 1791. Courtesy Yerkes Observatory library.

isted, and even thought it was sufficient to support advanced forms of life. The strongest evidence in favor of a lunar atmosphere were his observations of September 16, 1789 and February 16, 1790, when he traced thin dull attenuations of the cusps of the Moon far around the dark side—the faint glowing, he surmised, of a lunar twilight.[10] This was the first (albeit mistaken) confirmation of a prediction by the French astronomer Adrien Auzout (1622–1691), who had surmised that if the Moon possesses an atmosphere, the effects of a lunar twilight should be visible at the tips of the cusps of the crescent.

We now know that Schroeter was actually observing only a narrow strip of the averted hemisphere under oblique solar illumination—the Sun being almost on the far side of the Moon. Such apparent "twilight attenuations" are especially apparent at the southern horn—the region of the Doerfel and Leibniz ranges which Schroeter himself was, ironically, the first to describe—at about two days before New Moon, when optimum conditions of libration occur to bring out such effects.

Schroeter summarized his results concerning the Moon in his *Selenotopographische Fragmente* (Selenotopographical Fragments)[11] (Figure 6.5), a monumental work in two volumes published at his own expense in 1791 and 1802. The text of Volume One runs to 676 pages and includes 43 copper plates. In addition to a wealth of observational material, Schroeter offered his conclusions—presented in "a fantastically unreadable style," says Ashbrook[12]—about the physical conditions of the Moon. Despite its ponderous, rambling prose, replete with undigested thoughts and redundancies, the book immediately established his reputation. Even Goethe purchased a copy, and later obtained a small telescope from

Schroeter himself which he often pointed at the Moon.

In the last few pages of Volume One of the *Fragments*, Schroeter turned to the question of lunar inhabitants—Hevelius's "Selenites." "Should such a world," he asked, "be suited, like the Earth, as a habitation of life, and in particular of rational creatures?"[13] His answer was an emphatic yes:

> I am fully convinced that the Almighty may so order the physical conditions on every celestial body as to conform to the requirements of living creatures filling it and praising the power and goodness of God, and that the unlimited greatness of the Creator may thus be glorified in the analogous diversity of the physical arrangement of the celestial bodies, even as it is revealed in the infinite variety of their living creatures.[14]

Despite Schroeter's traditional pluralist arguments, his wry old friend Lichtenberg remained skeptical. "An astronomer knows whether the Moon is inhabited," he wrote, "with approximately the same probability as he knows who his father was; but not with the same degree of certainty as he knows who his mother was."[15]

Schroeter suggested that the vista from a lunar peak such as Mont Blanc (Figure 6.6), one of the tallest members of the lunar Alps whose height he had measured at 14,000 feet, might very nearly resemble that from the Earth's Mount Aetna or Vesuvius:

> I at least imagine... the region of *Plato* and *Newton* [the latter is not the same feature designated by the name today] together with the adjacent grey surface of Mare Imbrium to be quite as fruitful as the Campanian plain. Here nature has ceased to rage, there is a mild and pleasant tract which is given over to the quiet cultivation of rational creatures, who... give thanks for the fruit of the field, their complete happiness diminished only by their fear that Mont Blanc... may erupt again, with renewed destruction overrunning the many moon cottages. At least the southern region of Mont Blanc has many analogies with the Phlegrean fields, and the lunar Alps terminate frequently with [a] new small crater... as the Apennines of Italy do with Vesuvius.[16]

Schroeter's later career was both colorful and eventful. The year after the publication of his first lunar volume, he became acquainted with Johann Gottlieb Friedrich Schrader, a professor of mathematics and physics at the University of Kiel. The nearly deaf Schrader was a skillful optician, and from him Schroeter acquired a 9.5-inch (241mm) reflector of 13 feet (4 meters) focal length with which he made many of his best observations. The French astronomer Lalande called this instrument "the finest telescope in existence."

Still, Schroeter was unsatisfied. With the help of his gardener, Harm Gefken, he built an even larger telescope, with a 19.25-inch (489-mm) diameter mirror of 6.9 meters focal length (Figure 6.7). It was erected in Schroeter's garden in 1793 and soon attracted the curious from throughout Germany. A contemporary guidebook for tourists mentions it under the heading "Lilienthal":

> An hour further over the river Wümme, one can take a lovely walk in the woods, among beech trees measuring two feet around, and marvel at Herr Oberamtmann Schroeter's newly built observatory, with its 27-foot telescope built by himself, unique in Gemany.

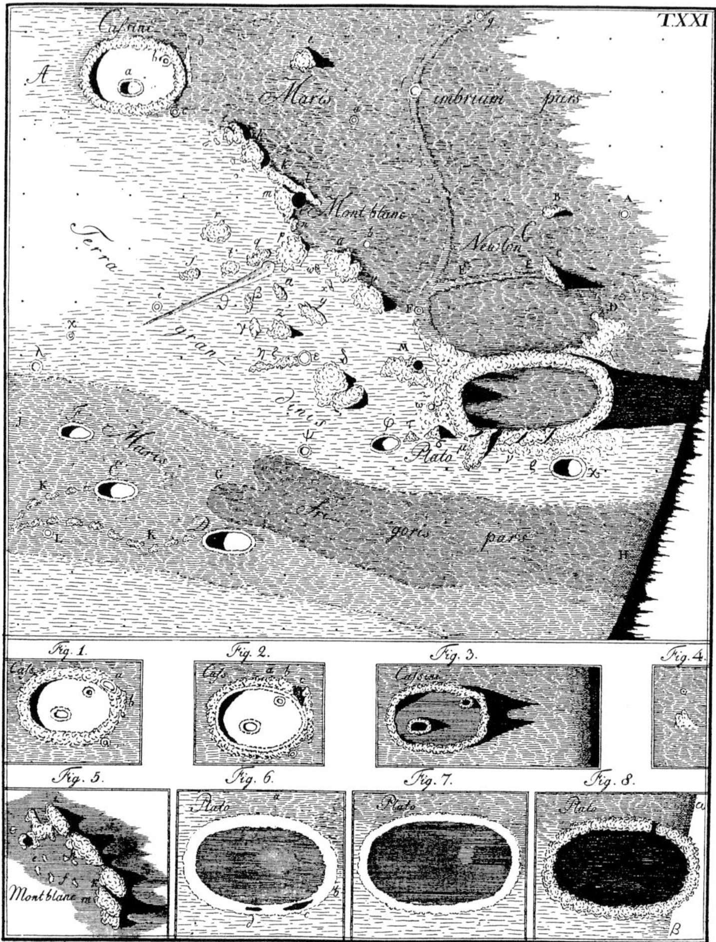

Figure 6.6 Schroeter's depiction of the lunar Alps, with detailed renderings of Mont Blanc and the craters Cassini and Plato. The grey surface in Mare Imbrium, a portion of which is depicted here, he fancied to be "quite as fruitful as the Campanian plain" of Italy. From *Selenotopographische Fragmente* (1791). Courtesy Yerkes Observatory library.

Following an idea of Schrader's, Gefken gave the speculum a coating of vaporized arsenic to enhance its reflectivity, and later made many mirrors using this technique in his optical workshop. (He died in 1811, at the age of 55, quite possibly of arsenic poisoning.)

Until 1797 the Moon remained Schroeter's chief interest, and in 1802 he published a second volume of lunar "Fragments."[17] Meanwhile, he had begun a series of systematic observations of the planets, which established him as a pioneer in the comparative study of worlds—*planetology*, as the discipline later came to be called by Percival Lowell.[18] By 1799, Schroeter's compulsion to acquire new telescopes had nearly exhausted his means. He was rescued from insolvency by

Figure 6.7 The 27-foot reflector constructed for Schroeter by his gardener and self-taught instrument-maker Harm Gefken. Although it was Schroeter's largest telescope (and the largest telescope anywhere outside of England), it was more impressive to look at than to look through. Its definition proved inferior to that of his Schraeder telescope of only half the aperture. From A. von Schweiger-Lerchenfeld's *Atlas der Himmelskunde* (1898).

Herschel's patron, George III, who gave him a large sum for his instruments and allowed him to use them for the remainder of his life, with the stipulation that after his death they were to be turned over to the Georg-August University in Göttingen. The King also provided Schroeter with the funds to hire an assistant, Karl Ludwig Harding (1765–1834).

In 1800 the observatory at Lilienthal served as a meeting place for a group of astronomers that included Harding, Olbers, and the Hungarian Franz Xaver, Baron von Zach (1754–1846), the director of the Seeberg Observatory near Gotha. Together they laid plans for a systematic search for a "missing" planet that von Zach believed was orbiting between Mars and Jupiter (Figure 6.8).

In the event, they were forestalled by Giuseppe Piazzi (1746–1826), a Sicilian monk who was director of the Palermo Observatory. On January 1, 1801, he discovered a small planet answering to this description, which he named Ceres. Nevertheless, the Lilienthal group decided not to disband, and went ahead with their search. Olbers found another small planet "Pallas" in 1802; this was followed by Harding's discovery of Juno in 1804, and of Vesta by Olbers in 1807.

After discovering Juno, Harding left Lilienthal to become director at the observatory in Göttingen. He was replaced by Friedrich Wilhelm Bessel (1784–1846), a young merchant whose interest in astronomy had been awakened by Olbers and who later became one of the greatest figures in astronomy. He remained

Figure 6.8 The meeting of the "Celestial Police" at Schroeter's Lilienthal Observatory, September 20, 1800. This 1988 painting by Wolfgang Vogt-Vilsec shows the astronomers Harding, von Zach and Olbers in the group on the left behind the bowing Schroeter, greeting the visiting Adolph Friedrich, Prince of England and Hannover. Following the meeting, Schroeter and the other members of the "police" began to search among the stars of the zodiac for a "missing" planet between Mars and Jupiter. Courtesy Dieter Gerdes.

in Lilienthal until 1810, when he left to take over the directorship of the new observatory at Königsberg.

By then, there were troubling developments in Germany. The Napoleonic Wars were underway, and the peaceful "Valley of the Lilies" was occupied by the French in 1806. Henceforth Schroeter was cut off from his support from George III, and the French did not pay him for his work, which included the unenviable duty of collecting taxes for them. Soon his income was so limited that he was barely able to keep up the observatory, and his situation became even more desperate in 1810, when he was dismissed from his position. Lilienthal became, by Napoleon's decree, part of the "Departement de le Bouche de Weser," with Bremen as its capital. Still, Schroeter managed to make extensive observations of the Great Comet of 1811.

The worst was yet to come. In April 1813, as the French were reeling back after their disastrous winter campaign in Russia, a skirmish took place near Lilienthal between a French detachment and a small band of Cossacks. A French officer wounded in the fray reported that his detachment had been fired upon by the local peasantry. Without further notice, the French commanding general, Vandamme, gave orders to set fire to Lilienthal in reprisal. A strong wind aided the conflagration, which destroyed the government buildings where Schroeter kept many of his books and manuscripts. Schroeter was "obliged to fly with my family, in our night dresses, to my farm at Adolphsdorf." His observatory escaped the conflagration, but several days later it was broken into and plundered by the French troops, who

> with a fury the most unprovoked and irrational destroyed or carried off the most valuable clocks, telescopes, and other astronomical instruments. My servants afterwards bought a few of these articles from the robbers; and I got back a 3-foot achromatic telescope, which the *honest* officer of the gang had secured, but most of the other instruments were lost. The loss falls properly upon the University of Göttingen, but I am willing to consider it my own… Many of the poor sufferers in this country, whom the French have robbed of their all, have applied to me, that I might represent their case, and procure some charitable relief for them in England by a collection among the benevolent people there, but I do not know whether the particular case of Lilienthal will excite much attention.[19]

With expulsion of the French, Schroeter was reinstated as chief magistrate and did what he could to try to rebuild Lilienthal. But it was too late for him to rebuild his observatory. Sadly, with the help of his son, he began transferring what little had survived of his instruments—the 3-foot Dollond refractor, the 7-foot Herschel reflector, the mirror and oculars of the 13-foot Schrader reflector, and the mirror of the 27-foot Schrader reflector—to the University of Göttingen. To keep despair at bay he wrote up his observations of the great comet of 1811, and published his "fragments" on Mercury. He turned at last to Mars, but by now his eyesight was failing and he was unable to complete the task—this work was finally published only in 1881. He wrote on January 22, 1815:

> At length after the most touching afflictions of mortality, I once more awake in my temple consecrated to the Eternal Godhead, and am again able after a total derangement of my affairs to edit these collections concerning the great comet of 1811… Under the endurance of these troubles all my scientific patrons and friends will doubtless, as far as possible, excuse me, if through melancholy and on account of the extraordinary high rate of postage I have been compelled to put out of sight so many obligations of courtesy; for to the present time everything is so straitened with me that my observatory from want of time and heavy expenses is for the most part a confusion.[20]

Schroeter did not long survive the shock of these terrible events; he died on the eve of his 70th birthday, August 29, 1816.

As tragic as was the ruin of his observatory and instruments, fate would prove even more unkind to Schroeter's reputation. During his lifetime, he had achieved an enviable fame, and his keen eye and large instruments had made him the "Herschel of Germany."[21] The demolition of Schroeter's reputation was largely accomplished by his countryman Johann Heinrich Mädler, according to whom Schroeter's attempt to establish evidence of changes on the lunar surface

> compromised the value of his painstaking observations and drawings. As the outlines of those particular landscapes which seemed to offer nothing pertinent to his pursuit

> of this theme were returned to only superficially, he himself could not determine, with certainty, whenever a new object appeared, whether it had been visible before. Moreover... it was only too obvious that he *wanted* to notice changes.[22]

The evil that men do lives after them. So it was with Schroeter. His lunar observations would be remembered mainly for their illusory claims of changes. His planetary observations would become notorious for the incorrect rotation periods he derived for Mercury and Venus and the impossibly high mountains he imagined near the south poles of these planets, for his strange delusion that Mars is enshrouded in a dense blanket of clouds, and for his curious belief that Saturn's ring was a solid structure studded with mountains and possessing its own atmosphere. Joseph Ashbrook characterized his work as "an odd blend of honest observing records and extravagantly wrong conclusions from them."[23]

More charitable, perhaps, was the Reverend T. W. Webb's assessment: "Schroeter was a bad draughtsman, used an inferior measuring apparatus, and now and then made considerable mistakes; but I have never closed the simple and candid record of his most zealous labours with any feeling approaching to contempt."[24] Though in the end what he achieved was indeed fragmentary, as the titles of his works acknowledged, he has been greatly underrated as an astronomer. Some of his records of time-dependent features, such as the dark spots on Jupiter, remain of great scientific value.

That he was a giant is attested by how far the shadow he cast extended into the next century. He saw an opportunity which none before him had recognized, and he made the most of it, with the result that his thirty years of watchful attention of the Moon and planets opened a new era in their study. Above all he was the first of the great selenographers, and all who followed—including even Mädler—were more deeply in his debt than they sometimes cared to remember.

References

1. A brief sketch appears in Agnes M. Clerke, *Popular History of Astronomy in the Nineteenth Century*, 3rd ed. (London: Adam & Charles Black, 1893), pp. 299–300. For a more recent summing up, see Michael J. Crowe, *The Extraterrestrial Life Debate 1750–1900* (Cambridge: Cambridge University Press, 1986), pp. 70–73, and W. Sheehan and R. Baum, "Observation and Inference: Johann Hieronymus Schroeter, 1745–1816," *J. Brit. Astronomical Association*, 105 (1995), pp. 171–175. For readers of German, more detailed information can be found in the privately printed works of Dieter Gerdes, *Die Lilienthaler Sternwarte 1781 bis 1818* (Lilienthal: Heimatverein Lilienthal, 1991) and *Die Geschichte der Astronomischen Gesellschaft* (Lilienthal: Heimatverein Lilienthal, 1990).
2. Schroeter's results were summarized in *Beobachtungen verschiedener schwarzdunkler Flecken des Jupiters* (Lilienthal, 1786) and *Beobachtungen über die Sonnenfackeln und Sonnenflecken* (Erfurt, 1789). Both works have been handsomely reprinted recently by the late Dieter Gerdes, who was curator of the Heimatverein Museum, Lilienthal.
3. Quoted in E. E. Both, *A History of Lunar Studies* (Buffalo, New York: Buffalo Museum of Science, 1961), p. 7.
4. Entry for July 30, 1776. *Herschel MSS*, 10–11.
5. Both, *A History of Lunar Studies*, p. 8 suggests that Schroeter's engraver, Georg Tischbein, may bear most of the blame, citing a comment by Olbers that he was "certainly no great artist." However, some of Schroeter's original drawings of Mars have survived. When they are compared to

Tischbein's engravings, one can readily see the fault was not the engraver's.

6. Described in J. H. Schroeter, *Selenotopographische Fragmente* (Helmstedt: Johann Georg Rosenbusch, 1791), p. 63. William Herschel implied, however, that he had priority in the invention, writing in "Observations of the Planet Venus," *Philosophical Transactions of the Royal Society of London* (1793), pp. 201–219 about Schroeter's projection table that "those who do not know its construction may have a very perfect idea, when they read the descriptions of my lamp, disk, and periphery micrometers."
7. James H. Phillips, "Lunar Domes—first observed by Johann Hieronymous Schroter," *Journal of the Association of Lunar and Planetary Observers*, vol. 32, no. 3–4 (1987), 67–73.
8. J. H. Schroeter, *Selenotopographische Fragmente*, Band 1, 614–619.
9. Ibid., 414.
10. As was so often the case, his observations were correct, although he drew the wrong conclusions from them. Thus, Harold Hill writes that "about two days before New Moon, when optimum conditions of libration occur, we are able to view a narrow strip of the averted hemisphere under oblique solar illumination—the Sun being almost on the far side of the Moon—and that under these circumstances it is not difficult to imagine that Schroeter would conceive such thin dull attenuations as being due to atmospheric effects." Harold Hill, personal correspondence to W. Sheehan, July 13, 1990.
11. In full: *Selenotopographische Fragmente zur Genauern Kenntniss der Mondfläche, ihrer Erlittenen Veränderungen und Atmosphäre, sammt den dazu gehörigen Specialcharten und Zeichnungen.*
12. See J. Ashbrook, "Schröter's Observations of Mars," *Sky & Telescope,* 14 (1955), p. 140.
13. Schroeter, *Selenotopographische Fragmente*, Band 1, p. 670.
14. Ibid.
15. Quoted in Rudolf Kippenhahn, *Bound to the Sun* (New York: W. H. Freeman and Company, 1990), pp. 87–88.
16. Schroeter, *Selenotopographische Fragmente*, Band 1, p. 673.
17. Schroeter, *Selenotopographische Fragmente*, Band 2 (Göttingen, 1802).
18. The main planetary works of Schroeter are: *Beobachtungen über die sehr beträchtlichen Gebirge und die Rotation der Venus* (Erfurt, 1793); *Aphroditographische Fragmente zur genauern Kenntniss des Planeten Venus* (Helmstedt, 1796); *Kronographische Fragmente zur genauern Kenntniss des Planeten Saturn* (Göttingen, 1808); *Hermographische Fragmente zur genauren Kenntniss des Planeten Mercur* (Göttingen, 1816); and *Areographische Beiträge zur genauern Kenntniss und Beurtheilung des Planeten Mars*, which was left in manuscript form at the time of his death and finally published by H. G. van de Sande-Bakhuyzen in Leiden in 1881.
19. Quoted in Richard Baum, "The Lilienthal Tragedy," *Journal of the British Astronomical Association,* 101 (1991), 369–370, who gives as his source a letter by W. F. Denning to *The Observatory*, July 4, 1904. Denning in turn cited J. Hemingway, *The Northern Campaigns and History of the War, from the Invasion of Russia, in 1812* (Manchester, 1815).
20. Ibid., 370.
21. Clerke, *Popular History of Astronomy*, p. 299.
22. W. Beer and J. H. Mädler, *Der Mond nach seinen kosmichen und individuellen Verhaeltnissen oder allgemeine vergleichende Selenographie* (Berlin: Simon Schropp, 1837), p. 185.
23. Joseph Ashbrook, "The Disappearance of a Marking on Mars," in *The Astronomical Scrapbook,* pp. 287–290.
24. Webb, *Celestial Objects*, vol 1, p. 106.

Chapter 7:

The Moon's White Cities

One of Schroeter's most ardent admirers was a Bavarian, Franz von Paula Gruithuisen. (Figure 7.1) Possessed of the same romantic attitude, he followed developments at Lilienthal with keen interest, drawing inspiration for his own scrutiny of lunar landscapes. He was well advanced in this apprenticeship when the astronomical idyll of Lilienthal ended in ghastly conflagration. Endowed with exceptionally keen eyesight, he went on to discover a wealth of fine detail that had escaped even the master's attention. His first scoutings confirmed Schroeter's view of the physical condition of the Moon, while his later work seemed for a time even to lend credibility to Schroeter's dreams of Selenites.

Gruithuisen—the name is Dutch—was born on March 19, 1774 in Castle Haltenberg near Kaufering am Lech, Bavaria, where his father was a falconer in the employ of the Elector of Bavaria. Despite the family's very modest means, Franz spent his childhood amidst the picturesque surroundings of a medieval castle and alpine forests.[1]

At the tender age of fourteen Gruithuisen was sent to assist an Austrian field surgeon in the Russo-Turkish War of 1787–1792. Later employment as a servant at the court of the Elector Karl-Theodor in Munich opened doors and brought patronage. He acquired a small telescope and by his eighteenth year had used it to identify all the principal formations on the lunar maps of Hevelius and Riccioli. Despite this budding interest in astronomy, he enrolled as a medical student at Landshut University, graduating in 1808, at age 34, as a Doctor of General Medicine. He was named to a position as instructor at the Landarztlichen Schule in Munich, where he exhibited a remarkable versatility; he lectured not only on medical topics, but on physics, chemistry, zoology, and anthropology.

The awe-inspiring sight of the Great Comet of 1811 (Figure 7.2) rekindled Gruithuisen's boyhood interest in astronomy and seems to have cast a lasting spell on him. He soon issued his first astronomical publication, *Über die Natur der Kometen* (On the Nature of Comets), in which he speculated that these cosmic interlopers might well be inhabited.[2]

Possessed of a boundless enthusiasm, Gruithuisen now set out to acquire a better telescope. He was aware of the work of a brilliant Munich optician, Joseph Fraunhofer (1787–1826), who was producing refracting telescopes vastly superior to anything that had been hitherto available. Early refractors had suffered from chromatic aberration caused by light dispersing into a spectrum of colors after

Figure 7.1 This portrait of Franz von Paula Gruithuisen appeared in his journal *Annalekten fur Erd-und Himmelskunde* in 1829. Courtesy Michael J. Crowe.

passing through their simple lenses, further compounded by spherical aberration. As we have seen, attempts to remedy this defect had reached a dead end with the hopelessly clumsy "aerial" telescopes of the seventeenth century. It was for this reason that Herschel and Schroeter had abandoned the refractor and turned their energies toward developing the reflector.

The refractor had acquired a new lease on life in 1729 when an English lawyer, Chester Moor Hall (1704–1771), discovered the principle of the achromatic lens. By making a compound lens of flint and crown glasses, the chromatic and spherical aberrations of one element could be made to nearly cancel those of the other. A high degree of correction could be achieved in such a doublet lens with a focal length only twelve or fifteen times its diameter, a ten to thirtyfold improvement over the simple singlet lenses of the aerial telescopes. While Hall did little to follow up on his idea, by the middle of the eighteenth century his countryman John Dollond (1706–1761) was producing small achromatic refractors of such high quality that he was awarded the Copley Medal of the Royal Society. This work was carried on by his son, Peter (1730–1820) and John Dollond (1733-1804).

But Fraunhofer was the first to combine expert craftsmanship with a comprehensive scientific approach to lens design. He collaborated with the Swiss artisan Pierre Louis Guinand (1747–1824) to develop techniques of producing large, homogeneous pieces of optical glass free of streaks and veins. While the Dollonds had been hard pressed to procure glass discs of the requisite quality 4 inches (102 mm) in diameter, by 1812 Fraunhofer had completed a virtually flawless instrument of 7.5 inches (190 mm) aperture, a feat far surpassing the capabilities of his contemporaries. Even larger lenses soon followed.

These were convenient, practical instruments that produced crisp, clear images of the Moon and planets and their evocative details. Thanks in no small measure to Fraunhofer's mastery, Germans now took a decisive lead in lunar reconnaisance. Gruithuisen was among the first of this new generation of pioneers. In 1812 he purchased from Fraunhofer refractors of 1.6 and 2.4 inches

Figure 7.2 The Comet of 1811, which inspired not only Gruithuisen but Wilhelm Gotthelf Lohrmann, Johann Mädler and many others to careers in astronomy. This was the comet mentioned by Leo Tolstoi in War and Peace: "Above the Prechistenka Boulevard, surrounded on all sides by stars but distinguished from them by its nearness to the earth, its white light, and its long uplifted tail, shown the enormous and brilliant comet... the comet which was said to portend all kinds of woes and the end of the world. In Pierre, however, the comet with its long luminous tail aroused no feeling of fear..." Contemporary engraving from Baron von Zach's *Monatliche Correspondenz*, vol 24 (1811). Courtesy John Koester.

(40 and 60 mm) aperture, with focal lengths of 18 and 30 inches (46 and 76 cm). Impressed with their performance, he soon acquired a third, this one of 4 inches (102 mm) aperture with a focal length of 5 feet (1.5 meters). He set up these telescopes in a small observatory at his house on the Sonnenstrasse in Munich, and went to work.

Success was both immediate and surprising. Schroeter, armed with his large reflectors, had called attention to a class of features that he referred to as *Rille* (from the German word for "groove," and not, like the English "rill," implying the presence of water). He had mapped only half a dozen of these curious features, but Gruithuisen added dozens more (Figure 7.3).[3] Similarly, although Schroeter had taken the greatest pains in examining the floor of the imposing crater Plato, it had always appeared to him perfectly smooth and uniform. Gruithuisen, however, managed to detect five tiny pits. Puzzled by Schroeter's failure to make out these details with larger telescopes, Gruithuisen wondered: "Shall these natural basins not contain water, over which at times fog hovers; shall Plato since 1789 not have lost just so much water as is necessary for the visibility of these craterlets?"[4] We shall see later that the apparent comings and goings of minute features on the floor of Plato would continue to confound lunar observers for many decades to come.

Most of Gruithuisen's observations were of the Moon, but he also paid some attention to the planets. In December 1813 he became the first to call attention to the bright "polar spots" of Venus, which he thought might be snow caps. In this

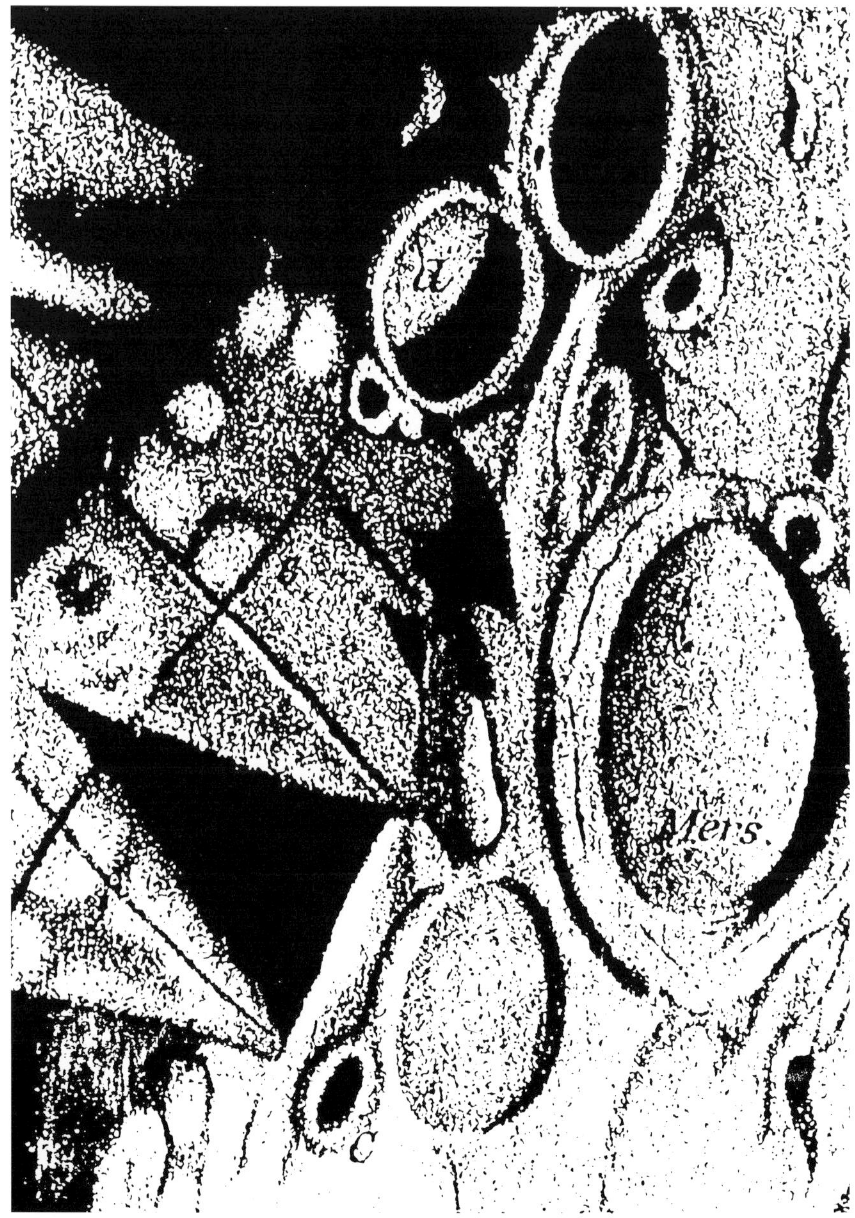

Figure 7.3 Gruithuisen's sketch of the crater Mersenius and its surroundings, showing several of the rilles which he was the first to record. While Gruithuisen's unrestrained speculations about lunar inhabitants and their constructions brought scorn and derision, his skills an observer and draftsman remain largely beyond reproach. Courtesy Michael J. Crowe.

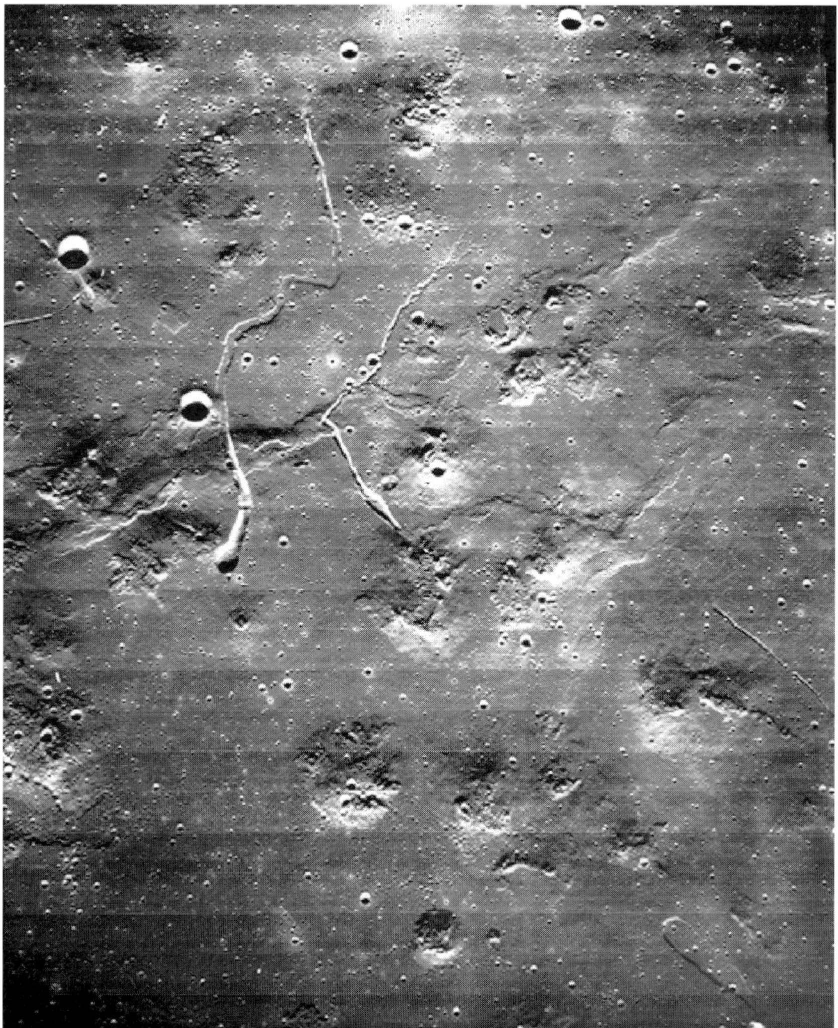

Figure 7.4 The Marius Hills, where Schroeter imagined a kind of lunar city, photographed by the Lunar Orbiter 5 spaceprobe in 1967. Note the rilles winding through this field of volcanic domes and cinder cones. Courtesy NASA.

he was mistaken—they are now known to be bright cloud swirls—but at least he was right in assuming they marked the positions of the planet's poles.[5]

Like Schroeter before him, Gruithuisen organized his work around the theme of the habitability of the Moon and planets. He once confessed: "We are still in love with the beautiful Moon, and dry reports of observations are better able to hold our attention if we can somehow keep alive the possibility of Selenites."[6] Everywhere the influence of Schroeter is unmistakable—hence Gruithuisen gave his first lunar treatise a distinctly Schroeterian title, *Selenognostische Fragmente*.

This work, published in serial form between 1821 and 1823, enthusiastically endorsed Schroeter's views of lunar life (Figure 7.4):

> How often Schröter speaks of the culture and industry of the lunar inhabitants. A summary of such material from his works would itself fill a volume. In the small region north of Marius [50° W and 13° N] he supposes a lunar city, while the canals near Hyginus... appear to him suitable for the Selenites to use for commerce. Yes, and he further says of part of Mare Imbrium that it appears "as fruitful as the Campanian fields."[7]

Gruithuisen confirmed Schroeter's observations of the "twilight" extensions of the Moon's horns and reached the same conclusion—the Moon has a thin but appreciable atmosphere. He believed that the large dark areas of the Moon were the basins of primordial seas which had long since dried up; instead of *maria* or "seas," he referred to them more matter-of-factly as "surfaces." Despite the fact that these areas never showed the slightest trace of green, Gruithuisen regarded them as districts covered by forests and cultivated fields. He was quick to point out that even on Earth, "the color of fir trees becomes much less vivid at two miles, and at four miles is indistinguishable from black. At ten miles, even the brightest leaves show up as little more than a grey, as the painter is careful to render them. How much more must this be the case on the far-off Moon!"[8]

Nevertheless, in addition to the usual neutral-grey tones, Gruithuisen did occasionally detect hints of brown and yellowish-grey, which he believed indicated areas "where leafy plants turn yellow or brown as they emerge from the chill of lunar night into the warm sunshine."[9] Other lunar spots appeared very pale before New Moon, emerging from the shadows of lunar night still blanched and greyish-white, then growing darker as the sun rose higher above them. This behavior, he wrote, was "in contradiction of the laws of photometry," and he argued that it constituted further evidence of the presence of vegetation.[10]

Parenthetically, what Gruithuisen observed was accurate enough. Many dusky markings on the Moon do appear to darken under a high Sun, but the effect is illusory. With increasing solar elevation these features actually brighten, as expected, only their intrinsically more reflective surroundings brighten even more. The intriguing phenomenon turns out to be merely a result of relative contrast.

Such were Gruithuisen's "proofs" of the existence of lunar vegetation. What, then, of animal life?

Here his attention focused on the clefts, rilles, and more delicate structures, which he referred to, collectively, as *Einfurchungen* ("furrows"). Some were true clefts; a few followed winding courses, like desiccated rivers. Unable to distinguish embanked edges to many of the smaller examples, he surmised that they might be broad, tree-lined avenues cleared in forests. "Whatever they are," he wrote, "they must have been produced by a higher order of animal life, traveling frequently between these regions."[11]

Though the broader of these features appeared (at least according to Gruithuisen) to have been formed by herds of animals following meandering paths, many of the smaller ones were either perfectly straight or followed regular curves. Taken as a whole they formed a network connecting the large dark areas with one another, and Gruithuisen did not hesitate to pronounce that in all probability they were roads. (Had he lived a few decades later, he might well have decided they represented a lunar canal system.) "Such an extensive transportation system," he added, "could be completed only with shrewd planning and concerted effort, and would be inconceivable without a civilization of Selenites."[12] The interiors of dark-floored craters were also rich in these road-like structures. This was hardly surprising, since the Selenites would undoubtedly select just such sheltered hollows in which to build dwellings protected from the chill lunar winds.

Gruithuisen never expected to make out the individual Selenites in their journeys—though perhaps, he suggested, their brightly colored caravans might be recognized where they met and separated again![13] This gives some idea of the resolving power that Gruithuisen imagined he was capable of achieving with his small telescopes and conveys a sense of his excitable state of mind on the eve of his most remarkable "discovery."

For some time Gruithuisen had wanted to find a way to honor "the highly deserving Schröter" by naming a suitable lunar formation after him. He believed that the Moon, as a whole, was a fearfully cold place, and decided to limit his attention to features as close to the warmer lunar equator as possible. At first he considered giving Schroeter's name to Hevelius's crater Mysius, now known as Triesnecker. Here, in 1815, Gruithuisen had described an intricate system of clefts (the Triesnecker rille system) that is one of the classic formations on the Moon. However, deciding that this crater was too small a feature to do justice to his hero, he turned his attention to the largest continuous dark area in this part of the Moon, the small mare now known as Sinus Aestuum. Being, to his mind, more appropriate for his purpose, he designated it the "Principality of Schroeter," and made a first sketch of this landscape when it was close to the terminator on March 2, 1822. It shows nothing unusual, and further observations made on March 15 and 16, 1822, were equally unremarkable.[14]

Yet amazing things were soon to come. In the early morning hours of July 12, 1822, Gruithuisen turned his 2.4-inch refractor toward the Moon, then at Last Quarter, with the terminator crossing over the western rim of Clavius in the one direction and Newton in the other. (Schroeter's "Newton," not the formation later designated by the name; it consisted of the part of Mare Imbrium enclosed between the Alps and the Montes Teneriffe.) With a magnification of 90X, he carefully studied the area near his "Principality of Schroeter." In the gloomy region to the southeast, where the mountains cast long dusky spires of shadow across the plains, he was astonished to find a collection of "dark ramparts" whose arrangement impressed him as more orderly than haphazard. The significance of what he was viewing seemed immediately apparent. Trembling with excitement, he exclaimed: "O Schroeter, here is that for which you always searched in vain!"[15]

"At first sight of this object," he recalled, "I fancied I was looking down from the height of a steep mountain, through all the seething ocean of the air, and had

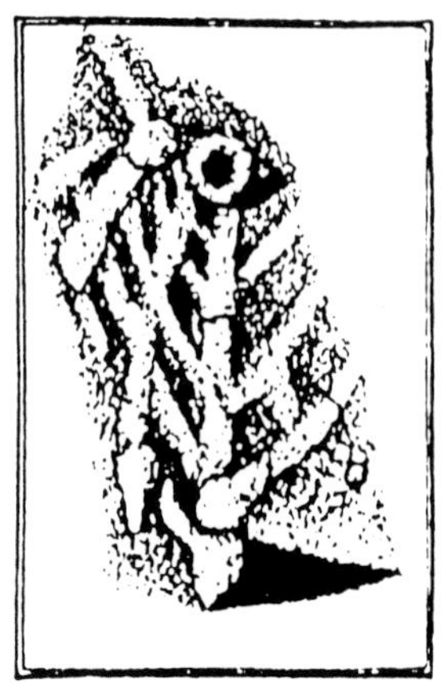

Figure 7.5 Gruithuisen's depictions of the celebrated *Wallwerk* and, in the lower drawing, the "star temple," which he regarded as colossal monuments erected by the Selenites. Courtesy Michael J. Crowe.

the birds-eye perspective of a city before me."[16] Overpowered with emotion, his hands trembling with excitement, he did not even attempt a drawing. The air, moreover, was as unsteady as he was, so he did not immediately resort to his larger achromat for a more detailed look.

He next examined the region on October 23, 1822, at which time he made a sketch (Figure 7.5) and gave a fuller description. The *Wallwerk*, as he called it, was arranged around a main wall 5 miles long that followed almost exactly along a north-south lunar meridian. From this wall, branching off on either side at an angle of about 45°, were several short parallel walls, "like the veins of an alder- or a rose-leaf." Perhaps even stranger than the *Wallwerk* itself was a nearby structure which he had failed to notice in his earlier view. Adjoining the northwesternmost of the ramparts, it resembled an irregularly-shaped star.

Gruithuisen did not write up a full account of his discoveries until 1824, when he published his best-known work, entitled *Entdeckung vieler deutlichen Spuren der Mondbewohner, besonders eines collossalen Kunstgebäudes derselben* ("The Discovery of many distinct Evidences of Lunar Inhabitants, in particular a colossal Artificial Structure by the same"). Under three headings—(1) Evidence of Vegetation on the surface of the Moon, (2) Evidence of Animal Life on the Moon, and (3) Artificial Works on the Surface of the Moon's Visible Hemisphere—he presented his case for an inhabited Moon, building steadily toward his climax, the discovery of a "colossal structure, not dissimilar to one of our cities." In his mind there was not a shred of doubt that the structure was artificial:

> No one, no matter how fanciful, would consider it possible that Nature alone could bring forth such a structure. Could crystal druses 5 geographical miles in breadth arise on the Moon, when on Earth they excite the greatest wonder when they reach 5 feet?... Could the structure have been built by termites? Certainly, for all we know to the contrary, giant wingless insects, with their instincts for such construction, might dwell in the Moon. But would they have the understanding of a man? The answer is surely no.[17]

The strength of gravity on the Moon being only one-sixth of that on Earth, the Selenites would find it comparatively easy to construct buildings even on such an enormous scale. Gruithuisen was confident that if the lunar inhabitants had telescopes as powerful as our own, they would be able to make out such man-made structures as the Great Wall of China—perhaps, he even mused, they built their *Wallwerk* "in emulation of our pyramids." Its location near the lunar equator, in one of the deepest and darkest landscapes on the Moon (under high Sun the area coincides with the largest of three dusky patches that appear superimposed on the bright rays of Copernicus),[18] indicated perhaps a situation of unusual fecundity, and the fact that the structure's surface was nearly as dark as its surroundings suggested that its roof was probably overgrown with vegetation. In any event, there could be no doubt that it served as a habitation of some sort. Other habitations on the Moon might exist in sublunarian vaults, for only the "troglodytic way of life" would allow the Selenites to endure their bitterly cold lunar nights and oppressively hot days. Gruithuisen wondered if the structure he had found was actually the extension of some enormous underground city—an unmistakable echo of Kepler's *Somnium*.

The regularity of the *Wallwerk*, with the main components tilted at angles of 45° and 90°, attested to the work of a supreme artificer. Moreover, the whole structure occupied a singular position on the Moon: the main wall ran exactly along a south-north meridian, and the structure as a whole stood exactly in the meridian of the nutation of the Moon's axis. This could hardly be a coincidence.

Finally, Gruithuisen discussed the significance of the nearby star-building. Constructed along classic lines, it was a structure worthy of an essay by Winckelmann. What could it be but a Selenite temple? Gruithuisen surmised from its shape that the Selenites were star-worshippers. This surmise, he added, was supported by the extreme clarity of the thin lunar air. After all, from the Moon the stars would be distinctly visible even in broad daylight.

Inevitably, Gruithuisen's announcement of a city in the Moon created an immediate sensation. Olbers wrote to Karl Friedrich Gauss, the great mathematician at the University of Göttingen, on June 22, 1824:

> Have you seen the representation of Gruithuisen's alleged lunar city and his avenues of trees and his roads in the Moon? What an imagination he has! And yet what he describes as a city, even though it has no real similarity to one, is certainly noteworthy, provided his drawing is even reasonably accurate—and I have no reason to doubt that it is.[19]

Now astronomers and would-be astronomers scurried to see for themselves the "fortress-like ramparts." The effect on the popular imagination was electrifying—and not only in Germany. In England, young Alfred Tennyson wrote:

> I saw
> The Moon's white cities, and the opal width
> Of her small, glowing lakes, her silver heights
> Unvisited with dew of vagrant cloud,
> And the unsounded, undescended depth of

her black hollows. Nay—the hum of men
Or other things talking in unknown tongue,
And notes of busy Life in distant worlds.[20]

As early as December 1822, Gruithuisen had revealed the strange formations on the Moon to two visitors to his observatory, Professors Oersted and Grouner. However, the first independent confirmation came from an unexpected quarter. The Austrian Prime Minister, Prince Clemens Metternich, paused from his efforts to maintain the balance of powers in Europe long enough to report its detection from the Vienna Observatory. Many others, however, sought for it in vain.

Astronomers had hardly absorbed Gruithuisen's first announcement when he dropped another bombshell. There were apparent changes in the form of the *Wallwerk*. On September 16, 1824, the main wall running through the middle of the structure "appeared cloudy," and some of the walls and the star-temple were likewise obscured.[21] Indeed, the regularity which had impressed him at first was no longer evident; only disconnected traces of the ramparts could be seen. Moreover, since his observations of 1822, an entirely new system of walls seemed to have gone up. Could it be that the Selenites were still actively building?

In June 1825 Gruithuisen left Munich for an extended tour of northern Germany.[22] In Tübingen he pointed out the *Wallwerk* to a Professor Bohnenberger, who could not contain his excitement and exclaimed aloud when he saw it. In Bonn he showed it to Professors Esenbeck and Münchow. In Göttingen he derived the greatest pleasure in demonstrating his lunar discoveries to Schroeter's former assistant, Karl Ludwig Harding, through the seven-foot Herschel reflector once used by Schroeter himself. There he discussed with Gauss the possibility of communicating with the lunar inhabitants. Gruithuisen's own idea was to lay out on the Earth's surface a large figure demonstrating the Pythagorean Theorem. Gauss, who once said that it "would be a discovery even greater than that of America, if we could get in touch with our neighbors on the Moon,"[23] not only endorsed the idea, but offered a plan of his own. He suggested using heliotropes—mirrors designed to track on the Sun, and able to reflect its light over vast distances.[24]

That autumn Gruithuisen traveled to Weimar, where he visited the aged poet Goethe. After wintering in Nürnberg, he continued on to Vienna, where he met Metternich and Karl Ludwig von Littrow, the director of the Vienna Observatory. He finally returned to Munich in September 1826.

By the time he returned, the university at Landshut had been relocated to Munich and merged with the Munich Academy. By combining the two institutions into the University of Munich, King Ludwig I of Bavaria hoped to create a scientific center for predominantly Catholic southern Germany capable of rivalling Berlin in the Protestant north. In Ludwig, Gruithuisen could not have hoped for a more congenial patron. "It was in the world of ideas and of dreams," wrote the historian Heinrich von Treitschke, "that the new king felt most at home," he who said of himself:

Let me yearn, dream, be an enthusiast;
Imagination alone gives satisfaction and delight."[25]

At first the discovery of the city in the Moon contributed to the advancement of Gruithuisen's career. He was offered but turned down professorships at Freiburg and Breslau. In 1826, the physician was appointed Professor of Astronomy at the University of Munich. Relieved of all administrative duties, he could concentrate exclusively on his research.

But in time an inevitable reaction set in. Gauss complained of Gruithuisen's "mad chatter," while Olbers referred to him as "peculiar." The authoritative German journal of astronomy, *Astronomische Nachrichten*, eventually refused to accept any more of his submissions. Unwilling to submit to a "scientific censor," he resorted to publishing his own journals—*Analekten für Erd- und Himmelskunde* (1829–1831), *Neue Analekten für Erd- und Himmelskunde* (1832–1836), and *Naturwissenschaftlich-astronomischen Jahrbuche* (1838–1847).[26] A casualty of his own brief celebrity, he lived to see much of his later work buried in obscurity. Even his announcement in 1828 that he had discovered yet another *Wallwerk* aroused scarcely a stir.[27]

It cannot be denied that some of Gruithuisen's ideas were, as Camille Flammarion put it, "most fantastic." Concerning the so-called ashen light of Venus—the occasionally reported glowing of its night side, an appearance similar to the phenomenon of earthshine on the crescent moon (a very real mystery, by the way, since Venus has no satellite of its own and sunlight reflected by the Earth cannot illuminate the planet perceptibly)—Gruithuisen wrote in 1833: "I believe the best explanation is that we are seeing a festival illumination put on by the inhabitants of Venus in honor of the ascension of a new emperor to the throne of the planet."[28] Later he had second thoughts, and wondered instead if the glow might be caused by the burning of large tracts of forest to make more land available for agriculture.

Yet not everything he wrote was the stuff of fantasy, and some wheat was almost invariably mixed in with the chaff. For example, he changed his mind about the "cold" Moon and surmised that lunar surface temperatures must exceed the hottest ever experienced on Earth and he was also more than a century and a half ahead of his time in speculating that "watery comets" might have deposited snow in the Moon's polar regions, an idea that has been revived in the light of recent data returned from the Clementine and Lunar Prospector satellites.[29]

His most important idea was the impact theory of the origin of the lunar craters. About this we ought to say a bit more, since it is one of the master ideas in the study of the Solar System. In a real sense, it was a natural conclusion following from the realization that stones and pieces of iron from time to time fall to Earth from the heavens.

There had been many anecdotal reports of such falls. They are mentioned in the Bible, and in more recent times falls of large stones had been reported in Alsace in 1492 and in Croatia in 1751. The cognoscenti were incredulous, since many of the reports came from ignorant peasants. After a hail of stones was reported in the Gascogne region of France in 1790, the Académie des Sciences in Paris refused even to accept a formal testimonial from the local citizens in order to avoid encouraging a "superstition unworthy of these enlightened times."[30] Apparently no one thought to connect these falls with meteors, which astronomers, in a last stubborn vestige of Aristotelianism, continued to regard as "fiery exhalations" in the Earth's atmosphere.

Fortunately, Lichtenberg, the professor of physics at the University of Göttingen, was more open-minded. Soon after the Gascogne fall, he suggested to a colleague, Ernst Florens Friedrich Chladni (1756–1827), that a thorough investigation ought to be carried out. Chladni combed the scientific literature of the previous two centuries and came to a startling conclusion: not only stones but chunks of iron fell from the sky. He went on to argue that the planets had formed from the aggregation of particles scattered throughout space, or from fragments of larger planets "broken to pieces, either perhaps by some external shock, or by an internal explosion."[31] In either case, some of this debris must have remained at large, and on making a chance approach to a planet and coming within the sphere of its gravitational influence it might be pulled toward it at high velocity. Encountering the planet's atmosphere, the hurtling fragments would be heated by friction. Though smaller particles would burn up altogether, the larger ones would melt, shatter, and fall in shards to the surface.

Chladni's book appeared in 1794. On April 26, 1803, his claims were substantiated when a brilliant fireball was seen by thousands of witnesses in the early afternoon sky at Caën and elsewhere in Normandy; an apparition followed by the sound of thunderous explosions and by a hail of stones falling on L'Aigle in the Département de l'Orne, France. At last the Académie des Sciences decided to investigate. One of its young luminaries, Jean Baptiste Biot (1774–1862), was sent to L'Aigle to gather information and to write up a report. Biot decided in the affirmative—stones had actually fallen from the sky.

Meanwhile, there had been other developments. In 1801, Piazzi had discovered the asteroid Ceres. When Olbers added Pallas a year later, he suggested that the two small planets might once have made up a larger body which had broken up.

Once it was realized that interplanetary space contained bits of matter whirling around the Sun, it was only a short step to postulate an impact origin for the craters of the Moon. Hooke, as we have seen, had toyed with the notion as far back as the seventeenth century after noticing the similarity between lunar craters and the structures formed when bullets were dropped into a viscous mixture of pipeclay and water. But unable to imagine what the source of such projectiles might be in the case of the Moon, he abandoned the idea. Now, however, a source of impacting bodies was clear—meteorites.

Chladni's theory that the Moon and planets were formed by the accretion of meteoritic particles was embraced by the brothers Karl Wilhelm and Friedrich August Marchall von Bieberstein in 1802,[32] and it was further elaborated by Karl Ehrenbert von Moll in 1815.[33] These writings captured Gruithuisen's imagination and he soon became an enthusiastic supporter of the "aggregation theory." As early as 1821 he had concluded that the Moon's craters, as well as certain terrestrial formations, were formed by impacts during the last stages of the aggregation process.[34] Later he attempted to work out additional details about the mechanics of impact. Somewhat implausibly, he envisioned concentrically layered spheres, formed in space by the accretion of meteoric particles, sinking into the still-plastic surface of the Moon. On arrival the sides of a sphere would be sheared off, leaving the visible crater walls, while the rest of the mass sank beneath the surface, pushing crushed and fused rock upwards to further bolster the upraised rim.[35] Alas, few paid any attention—it was just more "mad chatter." Though later revived,[36] the impact theory in the form advanced by Gruithuisen commanded scant attention.

Gruithuisen's last years must have been lonely ones. The astronomical world had turned against him, Julius Schmidt being perhaps typical in citing Gruithuisen's "fantasy" as the main reason that "today, one directs to astronomers questions about the Moon only in jest."[37] In the last five years of his life, Gruithuisen suffered from poor health and published little. He died in Munich on June 21, 1852, a figure who, if remembered at all, almost invariably evoked ridicule.

And yet, slowly and grudgingly, the keenness of Gruithuisen's observations, if not always his inferences from them, came to be acknowledged. It had been no idle boast when Gruithuisen wrote: "In the matter of visual acuity, I have known only one, now no longer living, who was my equal. As a rule, most observers need a telescope twice as powerful in order to see features as fine or as numerous as I can see."[38]

The experience of Heinrich Samuel Schwabe (1789–1875) is worth recalling. A Dessau pharmacist and skilled amateur astronomer who is best remembered for his discovery of the eleven-year sunspot cycle, Schwabe attempted to find the *Wallwerk* as soon as he read Gruithuisen's account in 1824. For a long time he was unsuccessful, and was on the verge of concluding that Gruithuisen had been the victim of "an optical deception or an overactive fantasy." Then, on February 15, 1826, as the terminator of the Moon lay just to the east of Eratosthenes, he was astonished to make out the whole structure perfectly with his 3.5-inch (90-mm) Fraunhofer refractor. With a magnification of 126X, "the walls were extraordinarily clear and sharply bounded, indeed just as Gruithuisen has shown them."[39] Schwabe also confirmed Gruithuisen's claims about the infrequent visibility of the ramparts, but he refuted his explanation that they were sometimes hidden by local obscurations; instead, their visibility seemed to depend critically upon the elevation of the Sun. As long as the terminator lay far enough to the east of Eratosthenes, the ramparts were distinct, but let the terminator advance just west of Eratosthenes and they vanished, the Sun being already too high over the region.

Others had similar experiences. Beer and Mädler, for example, repeatedly failed to find any trace of the *Wallwerk*. Yet Gruithuisen's "comparative accuracy as an observer," as Rev. T. W. Webb would write, "was to be established, even on their own not very willing testimony"; at any rate, on Mädler's testimony. On May 2 and 3, 1838, using the 9.6-inch (244-mm) Fraunhofer refractor in Berlin, he finally succeeded in making out the structure described by Gruithuisen (Figure 7.6). "The uniform height of the side and cross walls," he noted, "and the similarity in form and size of the hollows between them, give an aspect of regularity to this fig-

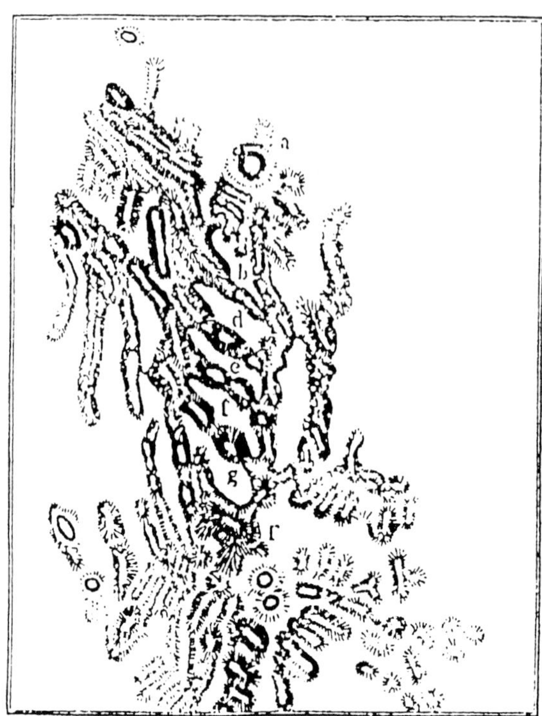

Figure 7.6 The Wallwerk rendered in hachures on the 4-section 1834 map of Beer and Mädler. Sheehan's collection.

ure, which is increased when they [the hollows] are filled with shadow, and the dividing ridges appear as bright, straight, narrow lines. It is, however, evident that we have only a product of nature in view."

After citing Mädler's view, Webb described his own observation on February 7, 1862, using a 5.5-inch (140mm) refractor and a magnifying power of 170X. At the time the structure was not well placed for viewing because the terminator lay far to the west, running through Timocharis and the western edge of Eratosthenes. Nevertheless, Webb found that "the whole agreed more with Gruithuisen's description than [Mädler's] figure, in the progressive shortening of the oblique walls, towards the N., so as to give fairly one-half the design of a tapering leaf. But my remark at the time was, 'the whole object looked coarse, and though curiously arranged, would never have given me the idea of an artificial production.'"[40]

Of course the walled city is only a chance arrangment of low ridges that gives the impression of artificiality when seen with a small telescope under fleeting conditions of glancing illumination. The parallel ridges and hillocks on this patch of jumbled terrain point radially away from the Mare Imbrium, as do countless other landforms in the region (for reasons that did not become apparent until G. K. Gilbert described the Imbrium "sculpture" in the 1890s. Today we realize that they represent the scars of the massive ejecta thrown from the colossal impact that formed the Mare Imbrium.)

Figure 7.7 Gruithuisen's defender Thomas Gwyn Elger, the pre-eminent British Selenographer of the late nineteenth century. A Liverpool engineer, Elger published a series of almost 200 "Selenographical Notes" in *The Observatory*, consisting of perceptive descriptions of lunar formations based on observations with his 8.5-inch reflector. He became the first director of the Britsih Astronomical Association's Lunar Section in 1890. Courtesy John D. Koester.

Caught up in a web of preconception and his own overactive imagination, Gruithuisen had misinterpreted features he strained to glimpse at the very limit of resolution of his small telescopes. "It is to be regretted," wrote Webb, "that the extravagance of his fancy should have brought discredit upon the unquestionable precision of his sight."[41]

Yet not all of Gruithuisen's observations were so tainted. The accuracy with which he recorded delicate rilles, for instance, was vouchsafed by the eminent observer Thomas Gwyn Empey Elger (1838–1897), (Figure 7.7) who wrote in 1887:

> A slur has often been cast upon Gruithuisen's lunar observations on the ground that he was inclined to set down more than he actually saw, and... it has been considered that a lively imagination frequently led him astray. The more careful and systematic study of these delicate features during the last twenty years has, however, sufficed to show that his records are by no means to be depreciated.[42]

Gruithuisen's interpretations read so much like fiction that it was, however, only a matter of time before a work of fiction would be produced that would pass for fact. This brings us to the great "Moon Hoax," which was written very much in the spirit of Gruithuisen and published in serial form in August 1835 in a penny paper, the *New York Sun*. It gave the *Sun*, briefly, "the largest circulation of any paper on this planet."[43]

The series of articles purported to be reprinted from "Great Astronomical Discoveries Lately Made by Sir John Herschel, LL.D. F.R.S. & c. at the Cape of Good Hope" (Figure 7.8), a special Supplement of the *Edinburgh Journal of Science* (in fact, a non-existent journal) by Herschel's amanuensis Dr. Andrew Grant, describing the discoveries made by Sir John with a super-telescope of his own devising.[44] Most of the alleged observations were of the Moon, whose surface details

Figure 7.8 Sir John Herschel, the only son of William, was the most prominent British scientist of his day. His 1864 catalog of over 5,000 nebulae and clusters was the basis of the New General Catalogue or "NGC" that astronomers still use today. Courtesy Peter Hingley, librarian of the Royal Astronomical Society.

Sir John was supposed to have examined with the same definition as the naked eye commanded of terrestrial objects viewed from a distance of a hundred yards—observations which had "affirmatively settled the question whether this satellite be inhabited, and by what orders of being."

At that moment, Herschel actually was at the Cape making astronomical observations. However, that was the only part of this remarkable document that was true. The "Great Astronomical Discoveries" was actually the work of Richard Adams Locke, a non-lineal descendent of the philosopher John Locke. He had been born in the United States, claimed to have studied at Cambridge, and worked as a freelance writer in England and New York before joining the *Sun*. His appearance, according to Herschel historian David S. Evans, was that needed by "a successful con-man: 'Five feet 7 inches tall, with a noble air of genius, a pocked face, a fine forehead and obliquity in the eyes.'"[45]

That Locke intended to write a "hoax" has been contested by historian Michael J. Crowe, who argues that he only meant to satirize the views of Rev. Thomas Dick, a well-known advocate of the existence of life elsewhere in the universe.[46] However that may be, we do know that in the summer of 1835 Locke had been reading back issues of the *Edinburgh New Philosophical Journal*, which contained reports of Schroeter's observations, of Gruithuisen's supposed discovery of structures built by Selenites, and of the proposals by Gauss and Gruithuisen to erect immense geometrical structures on the Earth in order to signal the Moon. If Locke's original intention had been satiric, even Crowe admits he "underestimated the gullibility of a generation,"[47] since his stories were widely accepted as eyewitness accounts of actual discoveries. Edgar Allan Poe, himself the author of

Figure 7.9 The lunar unicorns and bat-winged humanoids of Richard Locke's celebrated Moon Hoax, as depicted in a lithograph which first appeared in the New York Sun. Courtesy Ruth S. Freitag, Library of Congress.

a fictional Moon voyage, wrote, "not one person in ten discredited it," and added: "A grave professor of mathematics in a Virginia college told me seriously that he had no doubt of the truth of the whole affair!"[48] Among the *Sun*'s competitors, the *New York Times* pronounced the discoveries "probable and possible," while the *New Yorker* announced that they created "a new era in astronomy."[49]

According to Locke's account, Herschel had observed on the Moon:

> A lofty chain of obelisk-shaped, or very slender pyramids… They were of a faint lilac hue, and very resplendent. I now thought we had assuredly fallen on productions of art; but Dr. Herschel shrewdly remarked that if the Lunarians could build thirty or forty miles of such monuments as these, we should ere now have discovered others of a less equivocal character.

There were reports of tidal water marks on the Moon, and of living creatures including lunar quadrupeds and bearded goats with a unicorn-like horn. The climax of Locke's fabrication came with the discovery of intelligent creatures, much larger than birds, which hovered in the thin lunar air on bat-like wings and had faces like orang-outangs but "more open and intelligent" because of their expansive foreheads (Figure 7.9). Their intelligence was demonstrated by the fact that they were seen to be

> … engaged in conversation; their gesticulation, more particularly the varied action of their hands and arms, appeared impassioned and emphatic. We hence inferred that

> they were rational beings, and, although not perhaps of so high an order as others which we discovered the next month on the shores of the Bay of Rainbows, that they were capable of producing works of art and contrivance.[50]

Even these "Vespertilio-homo, or man-bat" creatures were surpassed by yet a higher order of beings, also winged but "of infinitely greater personal beauty"; they appeared, indeed, "scarcely less lovely than the general representation of angels by the more imaginative school of painters."[51] Their handiwork on the Moon included a "magnificent... temple—a fane of devotion, or of science," intended, no doubt, to recall the star-temple Gruithuisen had reported close by his lunar city.

Before the hoax was openly admitted, it had been reprinted as a pamphlet, which sold 60,000 copies. Not everyone was taken in. François Arago, Director of the Paris Observatory, declared it "utterly incredible." When Sir John Herschel learned about it from a visiting American, he was "in general amused." Lady Herschel later wrote to Sir John's aunt Caroline:

> Have you seen a very clever piece of imagination in an American newspaper, giving an account of Herschel's Voyage to the Cape... & of his wonderful lunar discoveries Birds, beasts & fishes of strange shape, landscapes of every colouring, extraordinary scenes of lunar vegetation, & groups of the reasonable inhabitants of the Moon... It is only a great pity that it *is not true* but if grandsons stride on as grandfathers *have* done, as wonderful things may yet be accomplished."[52]

Sir John's initial amusement ebbed rapidly before an avalanche of correspondence about his supposed discoveries, and he complained bitterly in early 1837: "I have been pestered from all quarters with that ridiculous hoax about the Moon—in English, French, Italian and German!!"[53]

Locke's story had more signficant fall-out, for there is no doubt that along with Gruithuisen's "mad chatter" it damaged the reception given to serious studies of the Moon. Originating in America, it also, according to the German rocket pioneer and astronomical historian Willy Ley, damaged the cause of American astronomy; "for decades to come astronomical news from America was received with great caution in Europe. The skepticism did not die out because American astronomers made more and more genuine and valuable contributions; it ended because the Moon hoax itself was gradually forgotten."[54]

References

1. For source material on the life of Franz von Paula Gruithuisen, the authors have relied mainly on Siegmund Günther, "Kosmo- und geophysikalische Anschauungen eines vergessenen bayerischen Gelehrten", *Festrede gehalten in der öffentlichen Sitzung der K. Akademie der Wissenschafen* (München, 1914). His most important works concerning the Moon are: "Selenognostische Fragmente", *Nova Acta Physico-Medica Academiae Caesareae Leopoldino Carolinae Naturae Curiosorum*, 10 (Bonn, 1821), 635–691; "Selenognostische Fragmente," *Verhandlungen der Kaiserlichen Leopoldinisch-Carolinischen Akademie der Naturforscher* (Bonn, 1823); "Entdeckung vieler deutlichen Spuren der Mondbewohner, besonders eines collossalen Kunstgebäudes derselben," *Archiv für die gesammte Naturlehre*, ed. K. W. G. Kastner, 1 (1824), 129–171, 257–322; *Der Mond und seine Natur* (München, 1844). His observing books were published by H. J. Klein, *Wochenschrift für Astronomie*, vols. 22–24, 1879–1881.

2. Gruithuisen, *Über die Natur der Kometen mit Reflexionen auf ihre Bewohnbarkeit und Schicksale, bei Gelegenheit des Kometen von 1811* (München, 1811).
3. The most complete catalog of these discoveries is found in Gruithuisen, "Naturwissenschaftlich-astronomische Aufsätze und Beobachtungen," *Naturwissenschaftlich-astronomisches Jahrbuch für physiche und naturhistorische Himmelsforscher und Geologen*, 10 (1847) 1–76.
4. Gruithuisen, "Entdeckung," p. 133n.
5. Gruithuisen, "Physikalisch-Astronomische Beobachtungen," *Nova Acta Physico-Medica Academiae Caesareae Leopoldino Carolinae Naturae Curiosorum* (Bonn, 1821), 10, 239–256. Also Richard Baum, "Franz von Paula Gruithuisen and the Discovery of the Polar Spots of Venus", *Journal of the British Astronomical Association*, 105 (1995) 144–147.
6. Gruithuisen, "Selenognostische Fragmente", p. 653.
7. Ibid., p. 650–651.
8. Gruithuisen, "Entdeckung," pp. 133–134.
9. Ibid., p. 134.
10. Ibid.,, p. 138.
11. Gruithisen, "Entdeckung," p.151.
12. Ibid., p. 159.
13. Ibid., p.160.
14. Gruithuisen, "Aufsätze und Beobachtungen," p. 39.
15. Gruithuisen, "Entdeckung," p. 163.
16. Ibid.
17. Ibid., p. 164.
18. Alan MacRobert, "The Fabled City on the Moon," *Sky & Telescope*, 84, 4 (1992), 424–425.
19. C. Schilling, *Wilhelm Olbers: Sein Leben und seine Werke, Band II* (Berlin, 1909), p. 321.
20. "Armageddon", a dream-vision, was first published in 1931 by Sir Charles Tennyson, who ascribed it to "the poet's fifteenth or sixteenth year." This would date it to 1824 or 1825, soon after Gruithuisen's discovery. "Armageddon," including the lines quoted, was later adapted by Tennyson for his Cambridge Prize Poem "Timbuctoo" (1829). See Alfred Lord Tennyson, *Poems*, ed. Hallam Lord Tennyson (London, 1908), vol. I, 317–325.
21. Ibid., p. 42.
22. Gruithuisen, "Gruithuisen's naturwissenschaftlicher Reisebericht," *Archiv für die gesammte Naturlehre*, ed. K. W. G. Kästner, 8 (1826), pp. 1–88.
23. Quoted in Crowe, *The Extraterrestrial Life Debate*, p. 207.
24. For Gruithuisen's conversation with Gauss, see Gruithuisen, *Naturwissenschaftlich-Astronomisches Jahrbuch,* 10 (1847), pp. 71–72.
25. Heinrich von Treitschke, *History of Germany in the Nineteenth Century*, ed. Gordon A. Craig (Chicago: University of Chicago Press, 1975), p. 202.
26. Dieter B. Hermann, "Franz von Paula Gruithuisen und seine Analekten fur Erd- und Himmelskunde," *Die Sterne*, 44 (1968), pp. 120–125.
27. Gruithuisen, "Über einige neu entdeckte reguläre Bildungen auf der Mondoberfläche und andere verändlerliche Gegenstände daselbst," *Berliner Astronomisches Jahrbuch 1828*, pp. 101–109.
28. Gruithuisen, *Neue Analekten für Erd und Himmelskunde*, 1 (1833), pp. 40–55.
29. J. Kelly Beatty, "Clementine's Lunar Gold," *Sky & Telescope*, 93 (1997), 24–25.
30. A. Pannekoek, *A History of Astronomy* (New York: Dover, 1989 reprint of 1961 ed.), p. 357.
31. E. F. F. Chladni, *Über den Ursprung der von Pallas gefundenen Eissenmassen* (Riga, 1794); abstracted in the *Philosophical Magazine*, Tillock's Series, 2 (1798), 225.
32. Marschall von Bieberstein, *Untersuchungen über den Ursprung und die Ausbildung der gegenwärtigen Anordnung des Weltgebäudes* (Darmstadt, 1802).
33. Karl E. von Moll, *Über den Zussamenhang der Gebirgsbildung mit dem Erscheinen der Feuerkugeln* (München, 1815).
34. Gruithuisen, "Selenognostische Fragmente," pp. 639–640.

35. Some of Gruithuisen's publications on these subjects are the following: *Gedanken und Ansichten über die Ursachen der Erdbeben nach der Aggregations-Theorie der Erde* (Nürnberg, 1825); "Erster Versuch einer vollstaendigen Darstellung der Grunde fur die Aggregationstheorie," *Neue Analekten*, 1 (1834), 1–17; "Was war der Mond, ehe er sich zur Erde gestellte," *Neue Analekten*, 1 (1834), pp. 18–25.
36. Among the first to revive the idea was the mining expert and ordnance engineer K. L. Althans. Struck by the similarity between the scars left in armor plate by bullets and the craters of the Moon, Althans carried out a series of experiments reminiscent of Hooke's, tossing musket balls and grapeshot into pans of fresh mortar and semi-liquid plaster of Paris. He convinced himself that lunar craters were the product of cosmic impacts and further suggested that such forces had played a role in shaping the Earth's surface. See K. L. Althans, *Grundzuge zur gänzlichen Umgestaltung der bisherigen Geologie* (Koburg, 1839).
37. Quoted in William Graves Hoyt, *Coon Mountain Controversies: Meteor Crater and the Development of Impact Theory* (Tucson: University of Arizona Press, 1987), p. 21.
38. Gruithuisen, "Entdeckung," p. 129.
39. Heinrich Samuel Schwabe, "Gruithuisen's sogenannte Kunstgebäude, nebst anderen merkwürdigen Gebilden der Kunst oder Natur im Monde," *Archiv für die gesammte Naturlehre*, ed. K. W. G. Kästner, 8 (1826), pp. 141–144.
40. Rev. T. W. Webb, "Gruithuisen's City in the Moon," *Intellectual Observer*, 12 (1860), pp. 214–222.
41. Webb, *Celestial Objects for Common Telescopes*, p. 63.
42. T. G. Elger, "Selenographical Notes," *The Observatory*, 10 (1887), pp. 419–421.
43. Crowe, *Extraterrestrial Life Debate*, p. 210.
44. The Moon Hoax literature is extensive and cited in full in Crowe, *Extraterrestrial Life Debate*; but see especially William N. Griggs, *The Celebrated "Moon Story," Its Origin and Incidents with a Memoir of its Author* (New York, 1852).
45. David S. Evans, "The Great Moon Hoax," *Sky & Telescope*, 62 (1981), part 1, pp. 196–198; part 2, pp. 308–311.
46. M. J. Crowe, "New Light on the 'Moon Hoax,'" *Sky and Telescope*, 62 (1981), 428–429.
47. Crowe, *Extraterrestrial Life Debate*, p. 215.
48. Ibid., p. 213.
49. Ibid., p.214.
50. Griggs, "The Celebrated "Moon Story," p. 96.
51. Ibid., pp. 115–116.
52. Sir John and Lady Herschel to Caroline Lucretia Herschel, received in London September 26, in Hanover October 1, 1836. In David S. Evans et. al., eds., *Herschel at the Cape: Diaries and Correspondence of Sir John Herschel, 1834–1838* (Austin: University of Texas Press, 1969), p. 236.
53. J. W. Herschel to Caroline Lucretia Herschel, January 10, 1837; in Evans, *Herschel at the Cape*, p. 282.
54. Willy Ley, *Watchers of the Skies* (New York: Viking, 1963), p. 275. Locke's own later career was checkered. In 1836, he resigned from the *Sun* in order to start his own daily, attempted unsuccessfully to repeat the success of the Moon Hoax with the *Last Manuscript of Mungo Park,* and after the paper failed, went to work as a reporter for the Brooklyn *Daily Eagle*. He died in 1871—two months before Sir John Herschel.

Chapter 8:

The Moon of Beer and Mädler

At the beginning of the nineteenth century, existing maps of the Moon were woefully inadequate and views about lunar conditions were romantic, to say the least. Instead of following the lead of Boscovich and Mayer, Schroeter and Gruithuisen had revived ideas of a lunar atmosphere which, however thin by terrestrial standards, was still adequate to support life. They then went on to fill the lunar landscapes with active volcanoes, cultivated fields, forests, even cities.

The man who led the reaction against these flights of fancy was Wilhelm Gotthelf Lohrmann (Figure 8.1), whom Joseph Ashbrook called "the first true mapper of the Moon."[1] Lohrmann proved to be almost completely insusceptible to the romantic excesses of the age, and as a professional cartographer was ideally suited to map the Moon on a scale and with an accuracy that completely cast into the shade the work of his predecessors. He made a magnificent beginning—though fate, alas, did not allow him to see it through to the end.

Lohrmann was born in 1796 at Dresden in the Kingdom of Saxony. The son of a bricklayer, he early heard stories of Johann Georg Palitzsch, a local hero in the astronomical world. This farmer and amateur astronomer from nearby Prohlis had achieved immortality on Christmas Day, 1758, when he had beaten the rest of the world (including France's preeminent "ferret of comets" Charles Messier) by first spotting Halley's Comet on the return predicted by Edmond Halley long before. Though Lohrmann never met Palitzsch, who died in 1788, his own interest in astronomy was fired by another comet—the Great Comet of 1811, which he wondered at as a youth of fifteen.[2] Discovered by Honoré Flaugergues at Viviers, France, in March of that year, this splendid comet became circumpolar in the northern sky in October and remained above the horizon throughout the night. It was also then at its closest to the Earth and sported two tails, the longer spanning some 24 degrees of sky, corresponding to an actual extension in space of some 100 million miles.

A few years later Lohrmann embarked upon his career as professional surveyor and cartographer, which he would follow for the rest of his life. He was employed in a geodetic survey of the entire Kingdom of Saxony, an occupation which involved extensive travel. His labors were long, and by this time he had a wife and children to support. Nevertheless, he still found time for avocational interests. During a visit to Munich in 1820, he acquired a good telescope—a 3.3-inch (84-mm) refractor by Fraunhofer. His first look at the Moon through the instru-

Figure 8.1 Wilhelm Gotthelf Lohrmann, the meticulous Dresden surveyor and lunar cartographer, was a worthy successor to Tobias Mayer. The publication of the first four sections of his lunar map in 1824 set a new standard. Unfortunately, although the rest of the map was completed by the time Lohrmann died in 1840, it remained unpublished until 1878, when it finally appeared in an edition by Julius Schmidt. Courtesy José Olivarez.

ment was a revelation.

As far as Lohrmann knew, Tobias Mayer's map was still the only one based on actual measurements rather than on estimated positions of lunar features. (A better map than Mayer's, also based on trigonometric measures, had been drawn up by the English portrait artist John Russell in 1805–06; unfortunately, it seems to have been unknown outside of England.) The time for a better map seemed long overdue, and his experience as a cartographer gave him confidence that the Moon could be mapped using the methods developed for terrestrial cartography. He resolved upon a trial of his ideas during the winter of 1821–22, when the weather interrupted his travels for the geodetic survey. The tilt of the ecliptic brought the Moon high into skies of central Germany, favorably placed for observation during the long nights. Laboring diligently, he completed a drawing based on micrometric measurements of a portion of the lunar Apennines and the crater Eratosthenes.

Lohrmann was sufficiently encouraged by this first result to commit himself to a far more ambitious plan. He resolved upon nothing less than "to represent the Moon's surface features, and their gradations of light and shade, with all possible accuracy, and to carry out both measurements and drawings by scientifically sound methods."

The 3.3-inch telescope, excellent as it was, did not seem quite adequate for such an undertaking. Therefore in the summer of 1822—even as Gruithuisen at Munich was first catching sight of his lunar "city"— Lohrmann invested in a 4.8-inch (122-mm) Fraunhofer refractor with a focal length of six feet (1.8 meters). It was more than powerful enough for lunar work, comparing favorably with many professional instruments in use at the time—the Berlin refractor, then the largest in the world, had an aperture of only 9.6 inches (244 mm). Lohrmann mounted his new telescope equatorially on a stone pier in the top floor of his house, located in the Dresden suburb of Pirna. His observatory had no dome; instead the roof over the telescope consisted of planks which could be removed whenever it was in use.

Lohrmann discussed his lunar mapping project with some of the leading professional astronomers, notably Joseph Johann von Littrow (1781–1840), Director of the Vienna Observatory, and Johann Franz Encke (1791–1865), who from 1813 to 1825 directed the Seeberg Observatory at Gotha. While both men provided moral support, Encke also worked out mathematical formulae which Lohrmann could use to convert his micrometric measures of lunar features into lunar coordinates. Thus prepared, Lohrmann set out in earnest mapping the Moon during the winter of 1822–23.

Lohrmann laid the groundwork for his map by making micrometric measures of a set of seventy-nine reference points on the lunar surface. At first he planned on a map of 50 inches to the diameter of the Moon. However, he soon had second thoughts, and settled on a more modest scale of 38.4 inches—still much larger than had ever been attempted (Mayer's map, remember, had measured only 7½!). It would cover the entire visible surface of the Moon. For convenience he divided this into twenty-five sections, thereby establishing a tradition that would endure for more than a century. He first sketched the Moon in pencil directly at the telescope, aided by the dim light of a partially shielded oil-lamp, and adopted the method introduced by the German cartographer J .G. Lehmann of using hachures—short lines denoting relief, drawn in the direction of slopes—to represent topography. The brightness of various features was estimated according to a scale devised by Schroeter, ranging from 0 (black shadow) to 10 (the most brilliant features). Once the pencil drawings were finished, Lohrmann combined them and copied them in ink before handing them over for copperplate engraving.

At first progress was rapid. Already in 1824 Lohrmann published the first installment of a book, *Topographie der sichtbaren Mondoberfläche* ("Topography of the Visible Features of the Moon"), containing discussions of his methods, the first four sections of his map, and detailed descriptions of various features. Section 1 of the map (Figure 8.2) covered the central region of the Moon, including the ridges south of the "bay" Sinus Aestuum—the name adopted by Lohrmann for Gruithuisen's "principality of Schroeter." Lohrmann noted the difficulty of representing the many low hills which appeared when this region was near the terminator, since he found that as the Sun rose higher, still other hills emerged from the shadows. He concluded: "One must with the greatest circumspection, under every condition of lighting, attend to this region if one hopes to depict it accurately." It was among these gentle slopes and shadows that Gruithuisen had glimpsed the outlines of his lunar metropolis. But Lohrmann's naturally cautious temperament made him wary of such extravagant claims. In a footnote to his description of Sinus Aestuum, he wrote:

> In this region… Mr. Gruithuisen believes he has seen a city, a fortress, and other artificial works. He hopes soon to recognize the lunar inhabitants themselves, if they

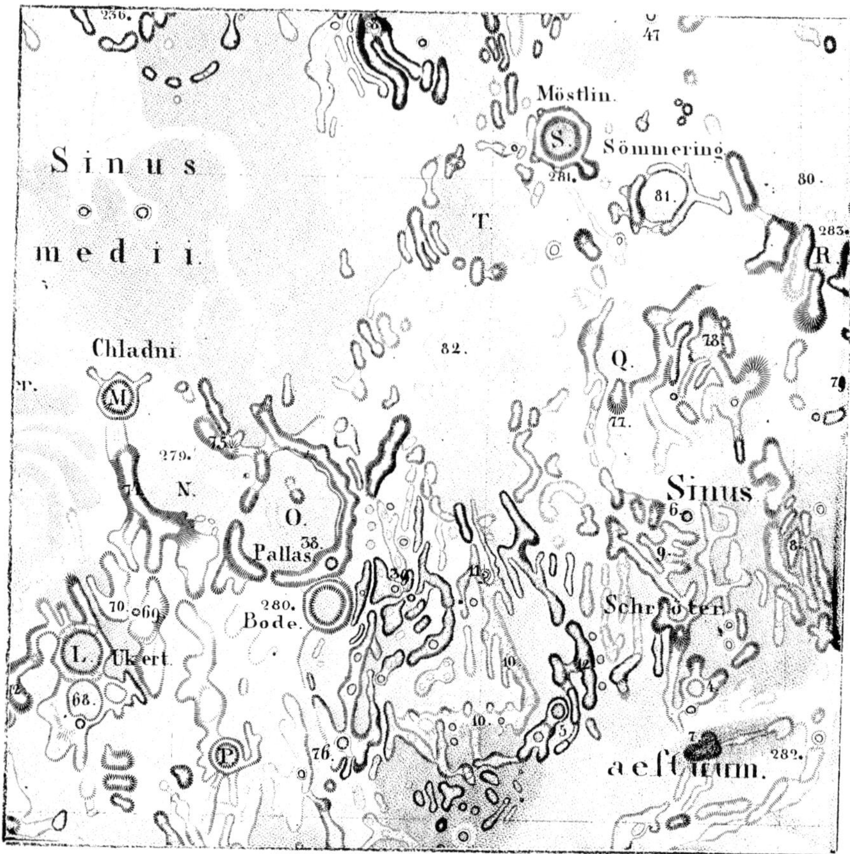

Figure 8.2 Detail of Section 1 of Lohrmann's map, reproduced at full scale. This is from the 1878 edition by J. F. J. Schmidt, who added the labels and names, and shows the central part of the Moon, including the small "bay" of Sinus Aestuum where Gruithuisen had recorded his celebrated *Wallwerk* two years earlier. The location of the *Wallwerk* is the area on the map labeled "Schröter." Courtesy José Olivarez.

> parade en masse through their forest glades, and he tells much in his selenographical writings of hot springs, minerals, animals, and plants. But these famed discoveries and the elaborate hypotheses based on them have no place in a straightforward book on lunar topography.[3]

In Section 4 of his map (Figure 8.3), Lohrmann depicted a small crater in Mare Serenitatis. It was sufficiently well defined to serve as one of his reference points, but apart from this, there was no reason to call special attention to it—indeed, he did not even bother to give it a name, instead designating it only with the letter *A*. His description deserves to be quoted in full: "*A* is the second most conspicuous crater on this plain [Mare Serenitatis]… It has a diameter of rather more than one [Hanoverian] mile [about 4.9 statute miles], is very deep, and can be seen under every illumination."[4] This is the feature to which Lohrmann's successors

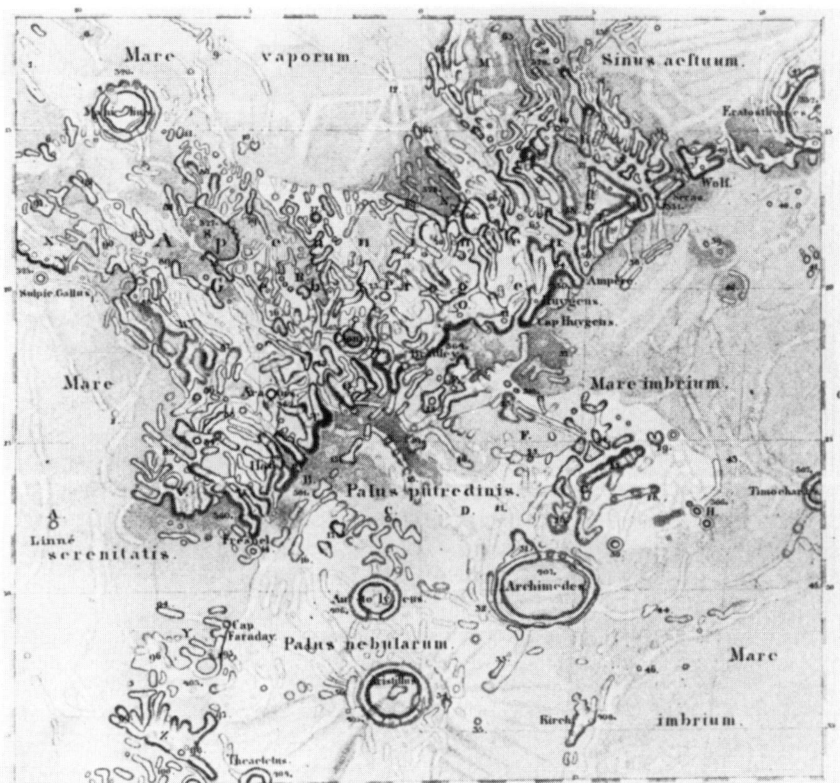

Figure 8.3 Section 4 of Lohrmann's map (J. F. J. Schmidt edition, 1878), demonstrating its unprecedented level of detail and the hachure method he adopted for depicting differences in elevation, which proved so aesthetically pleasing in his hands. Few features within the grasp of his 4.8-inch telescope escaped Lohrmann's attention. The innocuous looking crater *A* or Linné, later the subject of the most celebrated case of a purported lunar change, is shown at lower left. Courtesy José Olivarez.

Beer and Mädler would later attach the name Linné, after the Swedish botanist Carl von Linné or Linnaeus (1707–1778). A generation hence this relatively obscure feature would become, as we shall see, one of the most celebrated formations on the entire surface of the Moon.

The first installment of Lohrmann's *Topographie* received an enthusiastic reception. On July 18, 1825, Heinrich Olbers wrote to Johann Elert Bode at the Berlin Observatory:

> Lohrmann's *Topography of the Moon* I have read with particular interest, and it is deserving of the highest praise. His lunar map is vastly superior to any which has previously appeared… but is only such as the present state of optics and astronomical techniques would lead one to hope for. It may be compared with the worthy achievement of my estimable friend J. H. Schroeter, who contributed only fragments, and satisfied himself with representing each region of the Moon as it appeared to him at

Figure 8.4 Johann Heinrich Mädler, the greatest lunar observer of the first half of the nineteenth century. He succeeded only too well; his map and the accompanying book were regarded as so definitive that they had the unintended consequence of turning many observers away from the Moon. Courtesy Wolfgang Beyer.

> a given moment, under a given angle of illumination and libration. Lohrmann, however, by a judicious method of projection, has eliminated the confounding factors of lighting and libration to show the Moon as it really is. Moreover, it appears that Fraunhofer's capital telescope has given him many opportunities to learn the truth of what Schroeter's telescope showed imperfectly or incorrectly. Indeed, I have compared Lohrmann's chart directly with the Moon itself, and with great pleasure have verified its correctness, precision, and completeness. Now, with the help of this chart, may posterity determine with certainty whether any changes in the lunar surface occur from time to time.[5]

Lohrmann, who expected to finish the complete lunar map in six more years, published installments at regular intervals. At first he was able to keep the pace; he finished the observations for sections V to IX of his map during the winter of 1824–25. Bode wrote to him in March 1826, asking when the next sections could be expected. Lohrmann did not reply at once, as he hoped to send the copperplate engravings of one or two more sections before he did so. But the work had stalled. In August Lohrmann confided to Bode that the engraving was still unfinished. The regions covered by these sections were, he explained, mountainous. This made progress slow, but there were other reasons for the delay. There were the demands of his professional work. More ominously, his eyesight was beginning to fail. Sketching at the telescope under conditions of dim lighting was proving increasingly difficult for him, and he could work for only short intervals before having to rest his eyes.

Despite all this he persisted, and finally finished the observations in 1836. Though he published a small general map of the Moon on a scale of 15 inches to its diameter in 1839,[6] he had not even begun work on the copper-plate engravings for the large map when, in February 1840, he died of typhoid fever.

Historian E. E. Both found it "somewhat difficult to understand why Lohrmann could not have published [his map] unless his increasing professional duties did not allow him to devote the necessary time."[7] Indeed, there were increasing professional duties. Lohrmann was studying English railroads while serving as a consultant during the construction of the rail system in Saxony in 1836, and he had been appointed to the directorship of the geodetic survey of Saxony. All this no doubt weighed heavily upon him. But perhaps the most important reason for the delay was, quite simply, that he had been forestalled. The appearance of the first installment of his *Topographie* in 1824 had created great expectations. However, when later installments failed to appear on schedule, two other observers, Wilhelm Beer and Johann Heinrich Mädler, seized the opportunity to carry on where he had left off. They worked at a break-neck pace, and the publication of their own map between 1834 and 1836 made Lohrmann's map redundant.[8]

Mädler (Figure 8.4), one of the most important selenographers who ever lived, was born in Berlin in 1794, the son of a master tailor.[9] He was two months premature, and appeared to be stillborn, "without the slightest trace of life." As his mother had already had three previous stillbirths, there was little hope for his survival, but survive he did, living to the ripe age of 80.

The precocious Mädler taught himself to read by the age of four. A maternal uncle, Paul Strobach, recognized his ability and pushed for him to get a sound education. At the age of twelve Mädler was enrolled in the Friedrich-Werdersche Gymnasium in Berlin, where training in natural sciences was, by the standards of the day, excellent. Meanwhile his interest in astronomy had surfaced, inspired, as had been the cases with Gruithuisen, Lohrmann and so many others, by the Great Comet of 1811.

At the Gymnasium Mädler proved to be an excellent scholar and seemed destined for an academic career. But then tragic circumstances intervened. When he was nineteen, an outbreak of typhus claimed the lives of both of his parents within a span of only five days, and Paul Strobach died in the epidemic soon afterwards. This left Mädler with the responsibility of providing for four younger siblings. Though sent to live in the home of a distant relative, they still depended upon him financially. Abandoning his dreams of an academic career, Mädler's life now took a different turn. He attended the Küsterschen Seminary, where tuition was free, in order to become an elementary school teacher. At the same time he began to give lessons as a private tutor and in 1819 founded a school in Berlin for children whose parents had limited means—hardly a lucrative way to make a living.

Meanwhile, he began to attend lectures at the University of Berlin, notably P. G. L. Dirichlet's on higher mathematics and Bode's and Encke's on astronomy. In 1824, the year that saw the publication of the first installment of Lohrmann's *Topographie*, he was introduced to Wilhelm Beer by the celebrated explorer and naturalist Alexander von Humboldt (1769–1859) (Figure 8.5).

Humboldt himself was an important inspiration for Beer and Mädler's great study of the Moon. He was then at the height of his fame, as towering a figure in German science as Goethe was in letters. In 1799, at the age of thirty, Humboldt and the botanist Aimé Bonpland, intending to enter Africa through Cadiz, changed direction and instead set out for the Spanish possessions in America, in order, Humboldt afterward recalled, "to gratify the longings for foreign adventure and the scenery of the tropics, which had haunted me since boyhood."[10] They spent the next five years in privately funded explorations in equatorial South America. In

Figure 8.5 The great German explorer Alexander von Humboldt, who introduced Mädler to his patron and collaborator Wilhelm Beer. Their admiration for him was attested to by the name they gave to the newly discovered sea they found on the eastern (modern astronautical convention) limb of the Moon—Mare Humboldtianum. Sheehan's collection.

Ecuador, Humboldt set a world altitude record by planting his boot atop the "colossus of the Andes," Mount Chimborazo; in Venezuela he reached the channel shared by the Orinoco and the Amazon. His was "a passionate quest, at times almost demonic."[11]

After returning to Europe, Humboldt settled in Paris, where he devoted twenty years to writing up his South American travels. Though he never completed this work, his *Interior Provinces of New Spain* made the rest of the world conversant for the first time with the plains, jungles, mountains, and deserts of the Spanish possessions in South and Central America. Meanwhile, Humboldt had already begun to turn increasingly to another work encompassing "the entire material universe, all that we know of the phenomena of heavens and earth, from the nebulae of stars to the geography of mosses and granite rocks—and in a vivid style that will stimulate and elicit feelings."[12] He aimed to produce a magnificent embrace of all the sciences:

> Outer space—all of astronomical physics—the solid body of our earth, interior, exterior, electro-magnetism of the interior. Vulcanism, i.e. the reaction of the interior of a planet at its surface. Arrangement of [rock] masses. A brief geognosy—sea—atmosphere—climate—organic phenomena—plant geography. Geography of animals—human races and language—of which the subsequent organization (articulation of sounds) is governed by intelligence (whose product, manifestation is language).[13]

This farflung work became Humboldt's *Kosmos.* Although he admitted that its "undefined image… floated before my mind for almost half a century," it was only after 1827, his personal fortune exhausted by the drain of financing his extensive travels and voluminous publications, that Humboldt began to work on it in earnest. With the support of a generous pension from Frederick Wilhelm IV of Prussia, he returned from Paris to Berlin, and there he toiled away amidst the myriad distractions that were bound to press on such a man—requests for information on the most diverse matters, schemes for scientific institutes, countless letters of introduction—all the while expanding, revising, and, in the end, failing to complete, his *Kosmos*.

"The little old man sat among a tower of books, cards, letters, articles of every kind sent to him from all quarters of the earth. Facing him on the green wall was a great map of the world. Here he would write far into the night."[14] This was the image Humboldt presented to his contemporaries, including Beer and Mädler: a Promethean figure, attempting to subdue and collate all the phenomena of the whole of the known physical universe. More important than any specific result he achieved was the sheer grandeur of his towering aspirations. "Humboldt," writes historian Stephen Pyne, "attempted to integrate the thousands of newly discovered species and strata through the physical laws of geography, organizing them into ensembles of plants, suites of rocks, communities of human settlement. In doing so, he mathematicized his novel information by hurling it onto a sort of Cartesian coordinate system, the map."[15]

It was Humboldt himself who first put Beer in touch with Mädler. The result was one of the most successful collaborations in the history of astronomy. Beer was then 27 years old. His family was one of the most prominent Jewish families in Berlin. He was the brother of the noted poet Michael Beer and the half-brother of the musician Jakob Beer, who under the name Giacomo Meyerbeer would become famous as the most popular composer of operas of his day. Wilhelm had just then taken over from his father the banking firm of Jakob Herz Beer.

Through Mädler, Beer became fired with enthusiasm to establish his own observatory. The main instrument was a 3.75-inch (95-mm) Fraunhofer refractor with a focal length of 4.5 feet (1.4 meters), purchased from another amateur astronomer, Johann W. Pastorff (1767–1838) of Buchholz. It was mounted equatorially and equipped with a clockwork drive that allowed it to follow the diurnal motion of the stars. Beer set it up in the Tiergarten, and housed it under a dome twelve feet in diameter whose shutters opened on a swath of sky 20° wide.

The telescope became operational in 1828, too late to take advantage of the favorable apparition of Mars that year. Beer and Mädler did make an excellent series of observations of the planet at its next and even more favorable opposition two years later, work which led to the first map of its surface.[16] By then they were also well along with their lunar mapping.

The first four sections of Lohrmann's map of the Moon had made a deep impression on them. Even Beer's decision to purchase Pastorff's telescope may have been partly influenced by Lohrmann's map, for Pastorff had commented in Bode's *Jahrbuch* for 1825 that the instrument's superb optical quality was attested by the fact "it shows on the lunar surface a vast quantity of objects which are not shown in Lohrmann's chart." This proved, if any proof were needed, that much useful work still remained to be done on the Moon.

When Beer and Mädler decided in 1830 to take up the lunar mapping project, six years had passed without anything more being heard from Lohrmann. They explained their motives, and there is no reason to doubt their account. "Lohrmann," they explained, "made the first representation of another world which could rival the best terrestrial charts. We took great pleasure in observing the Moon with the aid of his charts, but when his work was interrupted, we were deprived of the same pleasure for the rest—and greater part—of the lunar surface. After much reflection, we decided to map the Moon ourselves."[17]

So they did. Indeed, it would scarcely be an exaggeration to say that they co-opted Lohrmann's whole program of research. They adopted his methods for calculating selenographical coordinates and the heights of lunar mountains, for estimating the relative brightness of features, and for drawing topographical relief. They even decided upon the same scale—38.4 inches to the diameter of the Moon. The only real difference was that instead of representing the Moon in twenty-five sections, as Lohrmann had proposed to do, they decided to divide their work into four quadrants.

Though the names Beer and Mädler are forever linked in the annals of astronomy, it was Mädler alone who was responsible for most of the actual observing and mapping of the Moon. Beer was the patron who built the observatory and made it available to Mädler. (Mädler himself testified that "though the very busy owner of the Tiergarten observatory joined me whenever he could get away from the office, by far the greater part of the observations were supplied by myself.") Nor did Beer himself ever claim otherwise. Mädler's wife Minna later recalled that Beer "always spoke with the greatest modesty of the role which he personally took in their combined studies."

Mädler's first observation for the lunar map was made on April 29, 1830. Over the next six years, he spent six hundred nights observing and sketching the Moon with the Fraunhofer refractor, customarily employing a magnification of 140X, though with frequent recourse to 300X. As a preliminary to the work of drawing, Mädler did as Lohrmann had done and measured a network of reference points on the lunar surface with a filar micrometer—106 positions in all. Many of these "points of first order" were the central peaks of craters, which he chose because of the precision with which they could be measured.

Only when the measures were completed did he begin, on June 1, 1832, the actual drawing of the map. In preparation for work at the telescope, he took 8 x 11-inch sheets of paper—there would be 104 of these sheets in all—and on them marked the positions of the points of first order. He then connected these points with blue-pencil lines to form a grid of 176 small triangles, and sketched in the most important features to create a skeletal map, which he took with him to the telescope. There he entered smaller features with a red pencil and recorded detailed notes about the appearance of each object. Formations were not always depicted in the same style. For example, one sheet shows the imposing crater Theophilus carefully represented in hachure, but nearby Cyrillus is indicated as a lightly penciled outline, which proves that in addition to the 104 sheets Mädler must have used other drawings of individual features. He worked at great speed and completed his last drawing on March 19, 1836.

Apart from making drawings, Mädler also worked out secondary positions of numerous features from the network of fundamental triangles, and carefully measured the heights of 800 peaks and the diameters of 150 craters. To the rilles discovered by Schroeter and Gruithuisen, he added many more—most exceedingly delicate—bringing the number of those known to 90. He even discovered a new "sea" on the northeast limb of the Moon (according to the now standard astronautical convention) and named it, fittingly enough, "Mare Humboltianum" after their friend Alexander von Humboldt, a departure from the old-fashioned approach of naming such features "after the dreamed-up lunar influences of which Riccioli was so fond." Mädler added that it seemed fitting to give this sea the name of the explorer because its location seemed to link the visible and invisible hemispheres of the Moon, as Humboldt's travels had linked the eastern and western hemispheres of the Earth.[18]

The scale of Mädler's original drawings, 76.8 inches to the diameter of the Moon, was reduced by half for the engraving on copperplates. Already in 1834 the first of the four sections of the *Mappa selenographica totam Lunae hemisphaerum visibilem complectens* had been rushed through the press (Figure 8.6). The great astronomer Friedrich Wilhelm Bessel, once Schroeter's assistant at Lilienthal and now Director of the Königsberg Observatory in Prussia, greeted its appearance enthusiastically. "Selenography has here done what was left to do," he wrote:

> It has undertaken a projection of the Moon's surface which leaves nothing to be desired. The first quadrant of the map has appeared; the second I have had the pleasure of seeing on the stone. I am told that the drawing for the third is finished, or almost finished, and I have assurances that the fourth will be carried forward with uninterrupted zeal on every night the Moon is shining. The enthuasiasm and persistence needed in such an undertaking is remarkable, having in only a few years hastened to its end... and raised selenographical studies to a level which hitherto would have been regarded as unattainable. Beer and Mädler... have erected an imperishable monument.[19]

The Viennese astronomer Littrow, who had earlier encouraged Lohrmann, announced that Beer and Mädler's efforts had rendered almost completely worthless all previous charts. "The earlier maps had often represented the positions and forms of features so crudely as to make identification difficult, but Beer and Mädler have achieved an accuracy such that neighboring objects can be recognized immediately, and with certainty."[20]

In September 1836 the last quadrant of Beer and Mädler's map passed through the press. Needless to say, many of the features depicted on their map had never been named, so that in addition to lending their authority to many of the names introduced by Hevelius, Schroeter, and Lohrmann, they contributed over a hundred new names to selenography and adopted the method of designating small secondary craters with the capital letters of the Latin alphabet (A, B, C, etc.) and indicating individual mountain peaks by lower-case Greek letters (α, β, γ, etc.).

After completing the map, Mädler tied up a few loose ends. He carried out the important task of representing the limb regions of the Moon under various con-

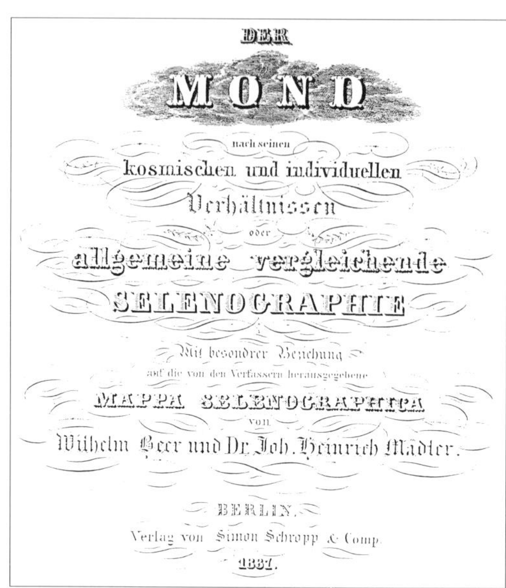
DER
MOND
nach seinen
kosmischen und individuellen
Verhältnissen
oder
allgemeine vergleichende
SELENOGRAPHIE.
Mit besondrer Beziehung
auf die von den Verfassern herausgegebene
MAPPA SELENOGRAPHICA
von
Wilhelm Beer und Dr. Joh. Heinrich Mädler.
BERLIN,
Verlag von Simon Schropp & Comp.
1837.

Figure 8.6 The title page of Beer and Mädler's great book *Der Mond,* which endured as the definitive reference work on selenography for half a century. Sheehan's collection.

ditions of libration—and turned to writing a massive treatise on selenography, "the necessary sequel to the several years of arduous labor which had led to the Moon map." This was the famous book *Der Mond, nach seinen kosmischen und individuellen Verhaltnissen oder allgemeine vergleichende Selenographie* (The Moon, concerning its cosmic and individual Conditions, or general comparative Selenography) (Figure 8.6). Though Beer's name appears on the title page as that of the principal author, the book, like the map, is in reality almost entirely Mädler's work.

The title was modeled on that of the great German geographer Karl Ritter's famous treatise *Die Erdkunde oder allgemeine vergleichende Geographie*; the contents, however, owed much more to Lohrmann's *Topographie*. "Few modern astronomers have mentioned how closely [Lohrmann's] plan was followed, even in minor details," noted Joseph Ashbrook. "The merits of Beer and Mädler's *The Moon* lie much more in its completeness than in its originality."[21] In addition, Humboldt's influence is everywhere implicitly evident. His example must have been very much in Mädler's mind as he set out, from the confines of Beer's small observatory in the Tiergarten, on his own bold exploration of the Chimborazos and Orinocos of the Moon. The task before him was an almost Humboldtean attempt to organize selenography by hurling untold thousands of individual facts about the Moon onto the Cartesian coordinate system of his new lunar map.

Mädler made it clear at the outset that the book was intended for specialists and relied heavily on mathematics. He distanced himself at once from from Schroeter's "fragmentary approach," maintaining that "only by comprehending the whole Moon can the significance of individual features be recognized." He added a further jibe at Schroeter, who had frequently interrupted his rather rambling analysis of lunar surface features with theological meditations: "May no one charge us with an indifference to the unending wisdom and power of the Creator, if we do not see fit to begin and end each section with a pious and edifying reflection."[22]

In just over 400 quarto pages, Mädler discussed every aspect of the lunar studies of his day: the Moon's motions; the methods used to determine the positions of first order (and from them numerous other positions of second order); tables of measured crater diameters and mountain heights; an overview of the physical conditions on the lunar surface; lunar and solar eclipses; the cause of earthshine on the Moon; the long-debated question of a lunar atmosphere. There was even a chapter on the Moon's supposed influence on terrestrial weather, a question which especially intrigued Mädler. He concluded that the Moon does indeed have perceptible influences on the weather, since he found a tendency for the barometric pressure to vary with the phases and also to stand higher when the Moon is farthest from the Earth and to fall when it is nearest. The balance of the book consists of a gazetteer of detailed descriptions of the features on the visible surface of the Moon, which established a precedent followed by almost every subsequent book on the Moon right up to the dawn of the Space Age.

The main findings can be briefly summarized. The *maria* or seas that cover two-fifths of the visible surface of the Moon are only relatively smooth areas which, on closer inspection, are found to be streaked and diversified with numerous irregularities. Thus the "seas" cannot possibly be covered with water as the old observers imagined. The predominant color of the *maria* is flat grey, though there are also more subtle colors, completely washed out near full moon but perceptible when lighting conditions are favorable. The Mare Serenitatis, except for its dark borders, is faintly greenish; Mare Crisium is green mixed with dark grey; Mare Frigoris is a dirty yellowish-green.

The craters can be classified according to size. The largest, measuring up to 150 miles across, are the "walled-plains." Typically these enormous structures have complex wreath-like walls. Sometimes their floors are flat and uniform, as in the cases of Plato and Archimedes; more often they are rough and broken. In the southern hemisphere these formations are so crowded together that their characteristically circular outlines pass over to the polygonal.

Next to the walled-plains in size are the "ring-mountains," formations ranging between 15 and 100 miles in diameter. Their walls are often terraced and their interiors contain one or more central peaks, which are usually lower than the surrounding walls, as shown by the fact that the shadow of the walls usually envelops them long before the Sun has set beneath the local horizon.[23] The ring-mountains in turn grade down into what Mädler called, simply, "craters"—bowl-shaped depressions—and "tiny pits." The latter are extremely widespread, and appear in great numbers even in regions which, at first glance, look perfectly flat; thus the region between Copernicus and Erathosthenes, when studied near the terminator, shows "darkish streaks, which when examined closely under favorable conditions resolve into a host of tiny craters."[24]

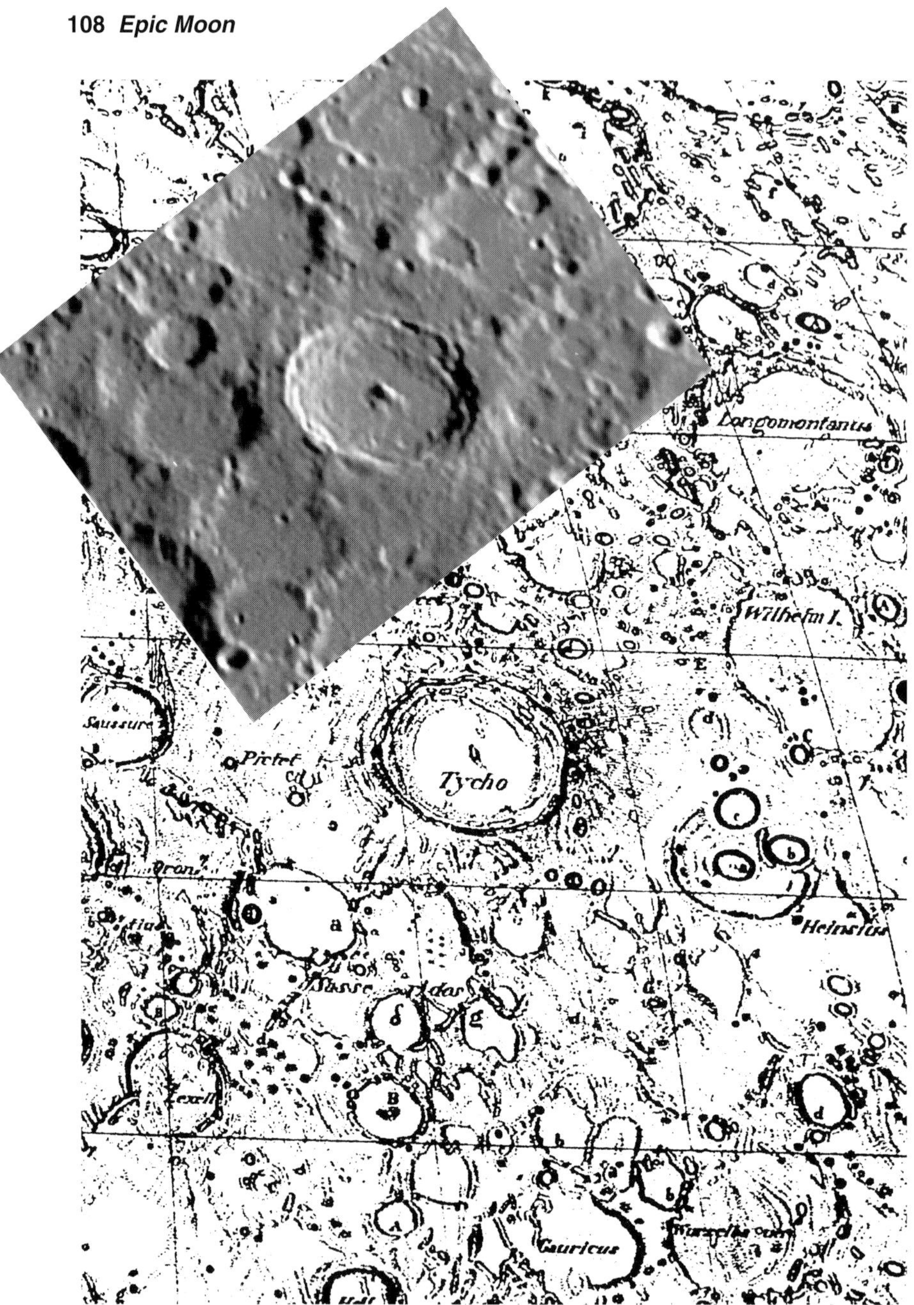

Figure 8.7 Detail of Beer and Mädler's *Mappa Selenographica*, centered on the prominent crater Tycho in the Moon's rugged southern highlands. Tycho is the center of a bright network of rays that give the Full Moon the appearance of a peeled orange through binoculars and small telescopes. Mädler made many attempts to resolve the rays into chains of mountains or other relief features, but without success, leading him to conclude that they could hardly be elevated at all above the surrounding terrain. Courtesy Patrick Moore. For purposes of comparison, the superimposed CCD image shows the Tycho region under afternoon lighting. Image by T. A. Dobbins.

On the other hand, the bright ray systems emanating from certain craters such as Tycho defied even the most resolute attempts at resolution (Figure 8.7). Early observers like Hevelius had represented them as chains of mountains. Mädler demonstrated that, whatever these features might be, they were certainly not elevated. On one occasion he observed one of the rays from Tycho as it reached the terminator and found it vanished immediately, while actual ridges and mountains in the same area continued to stand forth in bold relief.

Naturally Mädler was curious about the rilles, which had figured so prominently in the fanciful speculations of Schroeter and Gruithuisen. They were not really analogous to rivers, since they did not descend from elevations as rivers do. In fact, they seemed to be confined almost entirely to low-lying areas. Some, like the Hyginus Rille, occasionally showed bright reflections, but this was not due to the presence of water but rather to reflection of sunlight from the steep sides of the inner walls. Nor were the rilles likely to be roads. For one thing, they were far too wide. Even the smallest examples, Mädler estimated, were 1,300 to 1,900 feet in breadth. Besides, they seldom met in any particularly distinctive points as would be expected of a road network; mostly they just petered out in the open plains.

Often looking for but never finding Schroeter's alleged twilight extensions of the horns, Mädler concluded that the Lilienthal astronomer had been a victim of illusion. Nor did he find any evidence of obscurations. Instead, regions of the Moon near the limb always appeared with the same sharpness and clarity as those near the center; there was never the slightest suggestion of an imperfectly transparent medium hanging anywhere over the lunar surface. The shadows also appeared perfectly chiseled and jet-black, betokening the same lack of an atmosphere. All these findings betokened alike the lack of an atmosphere. He could find no real evidence of either air or water on the Moon.

Clearly, without these agents of erosion, the Moon would have undergone little or no weathering. Its mountains ought to be sharp and rugged—and so, evidently, they were. The shadows cast by the peaks of the Caucasus range, for instance, were long and narrow, drawn out into fine points which Mädler referred to as "aiguilles"—needles.

Mädler offered no firm opinion as to the origin of the lunar craters, but he did emphasize the almost complete lack of analogy with all terrestrial crater-forms. "The vastest of our volcanic cones," he remarked, "would hardly compare with the smallest pits visible on the lunar surface." Whereas the floors of many of the lunar craters were readily visible, those on the Earth were "black abysses":

> The Lacher See near Andernach, the Alban Hills in the neighborhood of Rome, and a few cauldrons and steppe lakes appear as weak imitations… but on the Earth, these are accidental forms, determined largely by local peculiarities of the terrain; on the Moon, the vast craters are everywhere predominant. A Selenite equipped with telescopes as powerful as our own would strive in vain to discover anything even remotely comparable to Hyginus, Dollond or Silberschlag—much less find anything as grand as Copernicus, Tycho, or Bullialdus.[25]

In every respect, the Moon offered few analogies to the Earth, and Mädler summed up:

> We pause to consider what the Selenite, thus equipped, would make of the difference between land and sea on our own world: the vast oceanic expanses, with their dazzling play of sunlight, appearing one moment somber-grey, the next glimmering white. What of those hair-sharp boundaries between the oceans and the lighter regions of the lands?... What would he make of those long, dark curving streaks, which contract and become ever narrower, then divide and ramify into innumerable branches? Would he know better how to interpret these things than we to interpret the forms of the Moon's surface? It is impossible.[26]

Der Mond completely swept aside previous views of lunar surface conditions. The year of its publication, 1837, has always been regarded as marking an epoch in selenographical studies. It was no longer possible to take for granted, as had Herschel, Schroeter, and Gruithuisen, that our satellite was in any significant respect like our the Earth. It was, in Mädler's famous phrase, "no copy of the Earth."[27] Whereas Beer and Mädler's predecessors had emphasized analogies, they found only differences.

The Moon was airless, moistureless, and—apparently—changeless. Above all, Mädler took exception to Schroeter's views. His work was attacked with such severity that the Lilienthal astronomer's reputation has only recently begun to recover. Mädler scored Schroeter's hopelessly clumsy and defective methods, and dismissed his flimsy and insubstantial results. Mädler admonished:

> How little the investigations of physical selenography have achieved hitherto is well known; the false direction of method being even more deplorable than the defective results which must be rejected in their entirety...
>
> When the positions of individual lunar spots were so poorly determined that one could not accurately predict their ingress and egress [relative to the Earth's shadow] during eclipses—when even the largest and most noteworthy features were apt to be confused with one another—it was in vain to attempt to distinguish changes on the lunar surface brought about by nature or art. Such changes might indeed sometimes catch the eye of Earth inhabitants; but only from the most constant and vigilant study after the whole surface had become known and the tiniest detail duly noted and measured. With the deepest regret we must reflect upon the lost opportunity of Schroeter, whose unproved results have misled a whole generation and provided the basis of countless conjectures each bolder than the last, but here thoroughly refuted. Indeed, his whole method provides a cautionary tale of how not to proceed. We must now be content to leave to posterity the question of whether true changes take place on the Moon.[28]

Mädler failed to concede to Schroeter the enthusiasms of the pioneer. Yet even he had to admit that Schroeter's method for measuring the heights of the lunar mountains was a great improvement over that employed by Hevelius. All in all, Schroeter often succeeded admirably well in making these measures, given the clumsiness of his apparatus. Apart from those sections of Schroeter's work comprising his allegations of changes, Mädler seems to have been only superficially acquainted with his results. In discussing the mountains that Schroeter discovered near the south pole, he completely confused his Doerfel and Leibnitz ranges. The heights of these mountains, as Schroeter had first realized, are the loftiest on the Moon, reaching 25,000 or 26,000, and in a few cases even 30,000 feet. Because of their height and proximity to the lunar south pole, Mädler calculated that the peaks must be bathed forever in sunlight, and even he could not avoid a touch of Schroeter's romance in calling them "Mountains of Eternal Light."

Though Mädler regarded the Moon as unlike the Earth, he stopped short of denying the existence of lunar inhabitants, maintaining only that any inhabitants of the Moon would be very different from ourselves.

> If we rely on the detailed descriptions of the lunar surface based on direct observations, and forego for the most part speculations about their significance, we are well aware that we may be denying the reader what he has most hoped to find there. Our descriptions of individual features may well seem to him dry and uninteresting. But it is our firm conviction that only from sound principles can progress be assured; all else is illusion...[29]

> When the natural philosopher considers the question of the existence of inhabitants not only on the Moon, but on other heavenly bodies, his conviction that they must exist is based entirely on the strong feeling that thinking spirits must be the highest purpose of Creation. This it is which causes them to affirm their existence without hesitation... But if we turn to the observations, we find conditions of life so entirely different, qualitatively and quantitatively, that any inhabitants can have but little in common with us. And this must certainly be so in the case of a world so unlike the Earth as the Moon.[30]

Karl Ludwig von Littrow, who regarded the observations of lunar cities by Schroeter and Gruithuisen as "entirely unproved," strongly endorsed Beer and Mädler's agnosticism about lunar inhabitants. So did the great Bessel, who congratulated them for not spoiling the healthful food of their science by spicing it with baseless speculations about lunar inhabitants. Henceforth the Selenites would be all but relegated, in the phrase of Agnes M. Clerke, "to the shadowy land of the Ivory Gate."[31]

Der Mond was immediately hailed as a scientific classic by none other than the great Humboldt himself. It had been written in a magisterial style, and impressed not least because of its sheer bulk and exhaustive detail. In recognition of his outstanding achievement, Mädler received the honorary title of Royal Professor of Astronomy from the King of Prussia. Since 1831, in addition to pursuing his selenographical research, he had been employed as a teacher of calligraphy at F. A. W. Diesterweg's progressive Royal Teacher's Training College of Prussia. Now at long last he was able to earn a living as an astronomer. At first he assisted Encke at the Berlin Observatory. Then, in 1840, he was invited to the Observatory at Dorpat (now Tartu in Estonia, the northernmost of the Baltic republics, then part of the Russian Empire) (Figure 8.8) to assume the directorship vacated by Friedrich Georg Wilhelm Struve (1793–1864), who had left the previous year to found the Czar's great Imperial Observatory at Pulkowa on the outskirts of St. Pe-

Figure 8.8 The Czar's observatory at Dorpat, depicted in an 1860 lithograph by L. P. Hoflinger, where Mädler worked following the publication of the great Moon map. In general, his days there were not particularly happy. Courtesy Wolfgang Beyer.

tersburg.

By this time Mädler had formed a romantic attachment with a promising poet named Minna von Witte. Her mother had a strong interest in science and had even constructed a relief globe of the Moon. It seems to have been on this basis that Mädler was first introduced to the family, but he soon became more interested in the daughter than the mother. The position at Dorpat meant that he and Minna would be able to marry, but it would take them to a foreign land far from friends and relatives. They found it difficult to make up their minds, but finally decided to accept, and so in 1840 Mädler took charge of the 9.5-inch (240-mm) Dorpat refractor that had been Fraunhofer's masterpiece (Figure 8.9). After Mädler's departure from Berlin, Beer did no further astronomical work of any significance. He died in 1850.

At Dorpat, Mädler carried out his duties in a conscientious manner. At first he spent much of his time furthering his studies of the Moon, and with the great refractor he raised the number of known clefts and rilles from the 90 that he had described in *Der Mond* to 150 (Figure 8.10). He also made the interesting discovery that the Hyginus cleft actually consisted of "a chain of confluent openings" (Figure 8.11). In addition, he made many observations of the planets and double stars, and by plotting stellar proper motions, made the "discovery" that the Galaxy seemed to be rotating around Alcyone, one of the prominent stars in the Pleiades cluster. He has often been ridiculed for this idea of a *Centralsonne*, though in fact he was not far from deducing the correct line leading to the galactic center—he simply chose diametrically the wrong direction, as the center actually lies among the star clouds of Sagittarius 180° from Alcyone.

In addition to astronomical observations, Mädler kept faithful meteorologi-

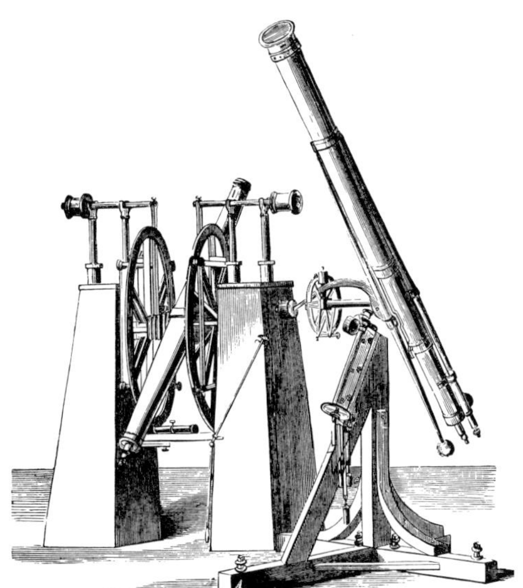

Figure 8.9 A contemporary engraving depicting a meridian transit telescope (left) and the great Dorpat refractor (right), both fabricated by the talented Bavarian artisan Joseph von Fraunhofer. From Hermann Klein, *Katechismus der Astronomie* (1900).

cal records. For nearly six years this driven man recorded the barometric pressure, temperature, precipitation, cloud, fog and atmospheric conditions three times a day (at 9 A.M., 3 P.M., and 9 P.M.) in the hope of demonstrating a correlation between the weather and sunspots.

One senses, nevertheless, that Mädler's years at Dorpat were not particularly happy. He seems to have been a singularly dour individual; cold and humorless, as persons with an obsessive personality often are. He cannot have been easy to get along with, and almost as soon as he arrived in Russia, he began feuding with the Struves, F. G. W. and his son Otto Wilhelm (1819–1905). This was a serious mistake. Since the Struves virtually controlled astronomy in Russia, Mädler found himself unable to gain further advancement and was passed over for membership in the St. Petersburg Academy of Sciences. Minna complained that her husband was appreciated everywhere—except Russia.

Although by 1845 Mädler had examined only a small part of the lunar surface with the great refractor, he had been warned by the elder Struve that "he who would delineate the Moon with this telescope must relinquish every other work, for the details would be too many." The Struves had made double stars their specialty; Mädler, eager to win the approval of other astronomers, now abandoned the Moon and decided that observations of double stars must become his own priority. One suspects a lurking insecurity; at early meetings of the *Astronomische Gesellschaft* he had heard remarks critical of astronomers who lacked a proper education. For all his accomplishments, he had never quite been sure where he stood in the profession.

Apart from this, he had another motive in abandoning the Moon. Selenography in his hands had clearly become the victim of its own success. As Edmund Neison was to put it a few years later: "It was generally regarded as demonstrated

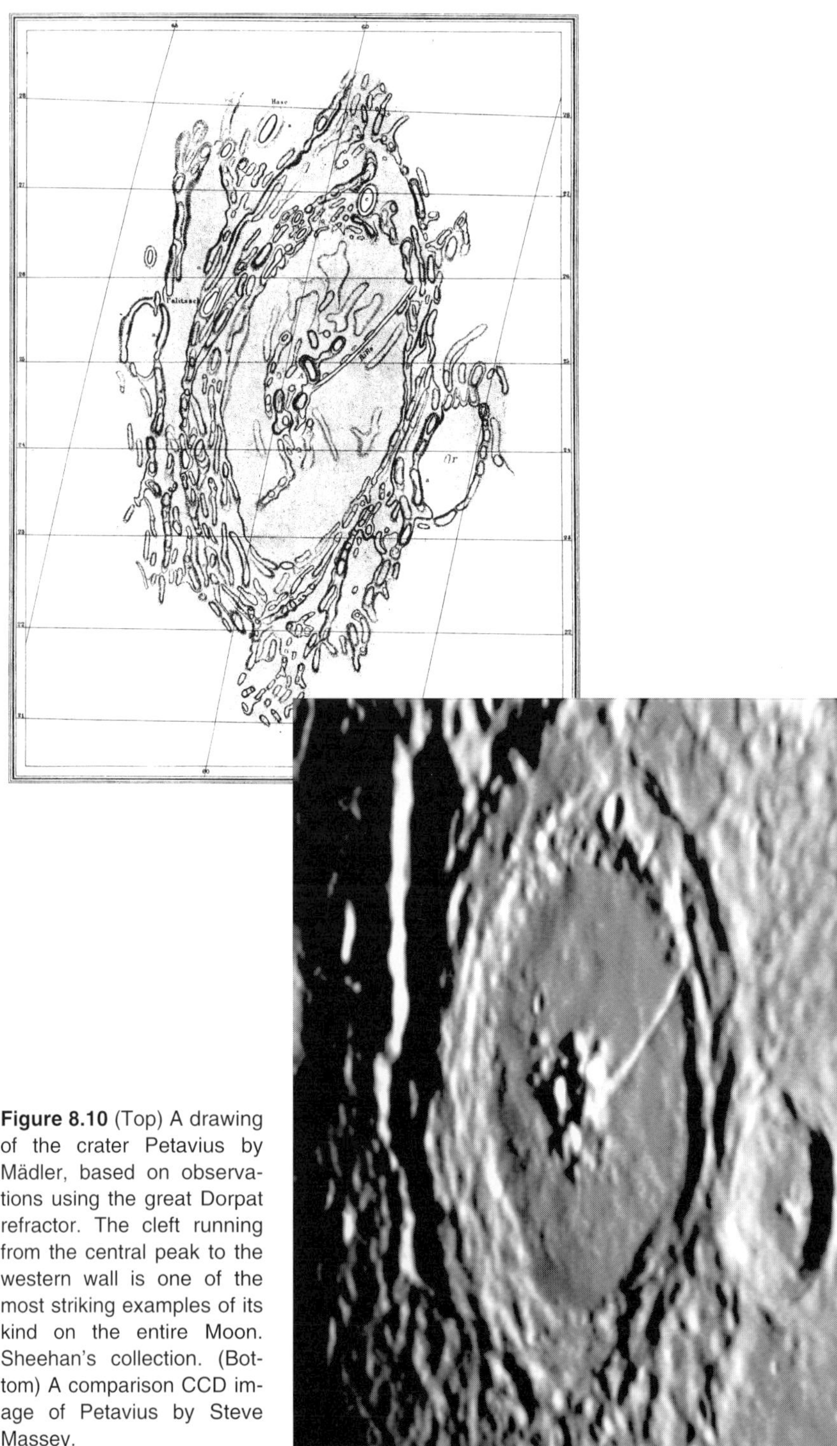

Figure 8.10 (Top) A drawing of the crater Petavius by Mädler, based on observations using the great Dorpat refractor. The cleft running from the central peak to the western wall is one of the most striking examples of its kind on the entire Moon. Sheehan's collection. (Bottom) A comparison CCD image of Petavius by Steve Massey.

Figure 8.11 With the respectable aperture of the Dorpat refractor, Mädler was able to make out the intricate structure of the Hyginus cleft, which consists of a series of confluent openings, as shown in this CCD image by T. A. Dobbins. For decades the Hyginus Rille was pointed to by proponents of the volcanic theory of crater formation as a prime example of a feature that could not possibly have been produced by random impacts. However, it is now widely believed that the confluent openings formed by collapse unaccompanied by any eruptions.

that the Moon was to all intents an airless, waterless, lifeless, unchangeable desert."[32] That being the case, there seemed little point in observing it.

Der Mond itself was largely responsible for this attitude. Though in fact Mädler had left many questions unsettled, the scale of his achievement left the impression that his labors had been exhaustive and that the Moon no longer represented a fertile field of research. What might be called the Mädlerian paralysis now overtook lunar studies. In general, the field became depressed and dormant; it was not Mädler's expectation—nor was it his fault.

Fortunately, the retrenchment was not complete. A few staunch figures remained active, and they prepared the way for the next great awakening.

References

1. Ashbrook, *The Astronomical Scrapbook*, p. 247.
2. The most complete information on Lohrmann's astronomical activities is found in H. Weichold, "Lohrmann als Astronom," in *Wilhelm Gotthelf Lohrmann, 1796–1840: Lebensbild eines Hervorragenden Geodäten, Topographen, Astronomen, Meteorologen und Forderer der Technik* (Leipzig: Johann Ambrosius Barth, 1978), pp. 369–394.
3. Wilhelm Gotthelf Lohrmann, *Topographie der sichtbaren Mondoberfläche* (Dresden and Leipzig: J. F. Hartnoch, 1824), p. 42. In his introduction he also alluded to Gruithuisen's observations: "I believe that it is necessary to pass over everything recent concerning… plants and animals in the Moon, and also what may be known concerning the fortress, buildings, streets and inhabitants,

such pictures can best be left to the reader's fantasy." Gruithuisen responded by criticizing the accuracy of Lohrmann's representation. The structure was shown too far south, and was distorted in form and dimensions. See Gruithuisen, "Naturwissenschaftlich-astronomisches Aufsätze und Beobachtungen," *Naturwissenschaftlich-astronomisches Jahrbuch für physiche und naturhistorische Himmelsförscher und Geologen*, X (1847), 1–76:40.

4. Lohrmann, *Topographie*, p. 92. The diameter was given in Hanoverian miles; each of which is equivalent to 4.87 English miles. For a useful table of units, see Dieter Gerdes, *Die Lilienthaler Sternwarte 1781 bis 1818* (Lilienthal, 1991), pp. 292–294.
5. Quoted in Weichold, "Lohrmann also Astronom," p. 372.
6. W. G. Lohrmann, *Karte des Mondes, Mittlere Libration* (Leipzig: Johann Ambrosius Barth, 1839).
7. Both, *A History of Lunar Studies*, p. 14.
8. After his death, Lohrmann's brother gave the materials for the unfinished Moon map to Wilhelm Barth of the Leipzig publishing firm of Johann Ambrosius Barth. Nothing was done with them until 1851, when Barth's son, Adolph Ambrosius Barth, approached the young selenographer Julius Schmidt with a request to help prepare it for publication. Schmidt agreed to do so, but was then busy making observations for his own map of the Moon. In 1858 Schmidt departed for Athens to take charge of the observatory there and did not see Lohrmann's map through the press until 1878, by which time no less than nine different engravers had worked on it. W. G. Lohrmann, *Mondkarte in 25 Sektionen und 2 Erläuterungstafeln*, ed. J. F. Julius Schmidt (Leipzig: Johann Ambrosius Barth, 1878).
9. For information about Mädler, the authors have have relied largely on Heino Eelsalu and Dieter B. Hermann, *Johann Heinrich Mädler, 1794–1874,* (Berlin: Akademie Verlag, 1985).
10. Alexander von Humboldt, *Cosmos: A Sketch of a Physical Description of the Universe,* trans. E. C. Otte (Baltimore and London: The Johns Hopkins University Press, 1997 reprint of 1858 edition), translator's preface, p. 4.
11. Stephen J. Pyne, *Grove Karl Gilbert: A Great Engine of Research* (Austin: University of Texas Press, 1980), p. 13.
12. Ibid., p. 257.
13. *Briefe von Alexander von Humboldt an Varnhagen von Ense aus den Jahren 1827 bis 1858*, 4th edition (Leipzig: Brockhaus, 1860), p. 20.
14. Trietschke, *History of Germany*, p. 367.
15. Pyne, *Grove Karl Gilbert*, p. 16.
16. W. Beer and J. H. Mädler, *Physikalisches Beobachtungen des Mars in der Erdnähe* (Berlin, 1830).
17. Beer and Mädler, *Der Mond*, pp. 186–187.
18. Ibid., 208.
19. Quoted in Weichold, "Lohrmann als Astronom," p. 374.
20. Ibid.
21. Ashbrook, *The Astronomical Scrapbook*, p. 249.
22. Beer and Mädler, *Der Mond*, p. vi.
23. Ibid., p. 130. Conversely, they note: "It is very interesting to see them emerge out of the shadows as starlike points of light, forming a confluence with the progression of lunar sunrise... When the splendor of such a spectacle, even at the tremendous distance from which we must enjoy it, baffles description, one can only guess how much more so this must be true on the Moon itself. What must the scene be like when sunrise bathes each peak with light as seen from Copernicus A!"
24. Ibid. The remarkable aspect of this region, missed by Schroeter, had been discovered by Gruithuisen, who estimated the diameter of these tiny pits at only 400 or 500 feet.
25. Ibid., p.230.
26. Ibid., p. 233.
27. Ibid., p. 234.

28. Ibid., p. viii.
29. Ibid., p. 429.
30. Ibid., p. 404.
31. Clerke, *Popular History of Astronomy*, p.326.
32. Edmund Neison (pseudonym for Edmund Neville Nevill), *The Moon and the Condition and Configurations of its Surface* (London: Longmans, Green and Co., 1876), p. 104.

Chapter 9:
Depression and Paralysis

By 1837, when Beer and Mädler published their great map, geological contexts were becoming more and more important to the investigation of the Moon. Geology was still a youthful science; the very word had been coined only in 1778, and not until 1810 did the entry "Geology" appear in the *Encyclopaedia Britannica.* But the progress of the new science was rapid, and by the 1830s it had become the most popular science in Britain. "It ranked," wrote Sir John Herschel, "next only to astronomy itself in the magnitude and sublimity of its objects." The public fascination with the vast lapses of time revealed in the strata by the geologists—together with a vague sense of terror—was well captured by the poet Tennyson, who asked:

> And he, shall he,
> Man, her last work, who seem'd so fair,
> Such splendid purpose in his eyes...
> Be blown about the desert dust,
> Or sealed within the iron hills?

Until the end of the eighteenth century, geology had been largely held hostage to Bible-derived chronologies of the creation of the world and the flood, which seemed to indicate that the entire history of the world had occupied no more than a few thousand years. Deluges and other catastrophes—events of staggering violence on a scale quite unknown in the modern world—were conjured up to explain the massive changes that had taken place on the surface of the Earth in so short a time. Neptunists and Vulcanists argued over the preferred agent of catastrophe—water or fire.

Eventually, however, it became increasingly difficult to believe that the Deluge of the Old Testament could be sufficient, of itself, to explain the existence of sediments thousands of feet deep and the deposition of marine fossils a thousand miles from the nearest sea. In place of the literal six days of Genesis, geologists attempted to reconcile science and religion by speaking of "extended days of creation"—indefinitely long periods of time, punctuated with repeated cataclysms, before the deluge of Noah.

A more gradual evolution of the Solar System was also invoked by Kant and Laplace. Inspired by Newtonian ideas, they advanced the nebular hypothesis to account for the fact that the orbits of the planets were nearly co-planar and that all these bodies completed their revolutions in the same direction around the Sun.[1] In

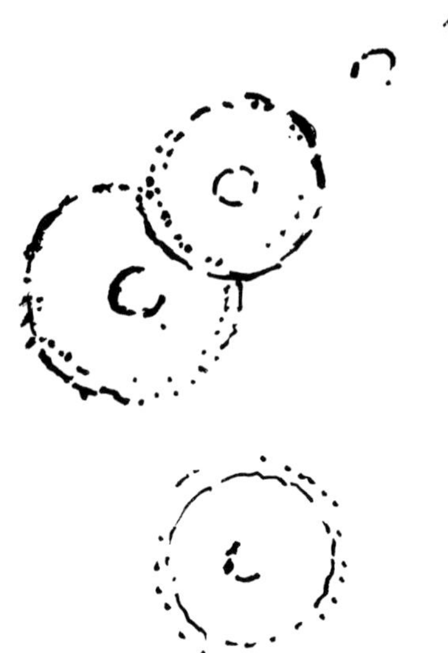

Figure 9.1 A sketch based on the original drawing by James Hutton's associate, John Clerk of Eldin, made at the telescope on May 14, 1785. It shows a younger crater superimposed on an older one (Theophilus on Cyrillus), demonstrating the principle of superposition that Hutton and Clerk had worked out for terrestrial rocks at Glen Tilt in Scotland.

1749 the French naturalist George-Louis Leclerc, Compte de Buffon (1707–1788), worked out a non-scriptural estimate of the Earth's age using evidence of the Earth's internal heat combined with estimates of its rate of cooling. He published a conservative estimate of 75,000 years which, modest as it was, "broke sharply with historical time and liberated the age of the earth from the restrictions of the age of man."[2]

But even vaster—perhaps even infinite—ages were conjured up by the British geological pioneer, James Hutton (1726–1797). A Scottish physician and gentleman-farmer, Hutton wrote in 1788 of the "ruins of a past world" buried in the Earth, and meditated on an "abyss of time" apparently as limitless as the vistas of space that were then being revealed by William Herschel with his giant telescopes. "We find," Hutton wrote, "no vestige of a beginning, no prospect of an end."[3]

Some of Hutton's most critical observations were made in 1785, in the company of his close friend John Clerk of Eldin. At Glen Tilt in Scotland they observed granite veins and dykes breaking and displacing the strata. Later in the Tarf, a tributary of Glen Tilt, they found a boulder in which a vein of granite traversed both the mass of granite and broken schist—an example of the principle of superposition. The granite was younger than the schist it cut; the vein cutting both was younger still. Only four months earlier, in May 1785, Eldin had been observing the craters of the Moon, and made a sketch showing the superposition of craters (probably Theophilus on Cyrillus) (Figure 9.1). "He showed a clear instance of a younger crater superimposed on an older one—just like pits in mud made by rain drops—a later pit partly obscuring an earlier one. Knowing, as they did, that the Glen Tarf boulder records a sequence of events, Clerk and Hutton must surely have seen that they could read the Moon's history as well as the Earth's."[4] They were far ahead of their time; it would not be until the Space Age that the principle of superposition was applied, in any systematic way, to unraveling the complex history of the surface of the Moon.

Apart from Hutton, the other important figure of early British geology was a self-educated land surveyor, William Smith IV (1769–1839), the founder of the science of stratigraphy. He was born at Churchill, a small village in Oxfordshire. His father died when he was eight, and after his mother's re-marriage he was placed in the care of an uncle, who had little appreciation for his early passion for collecting fossils. Fortunately, Smith was not easily daunted. A born observer blessed with what his nephew John Phillips, who would become one of the nineteenth century's leading students of the geology of the Moon, called a "peculiar circumstantiality of memory"—whatever he saw, he remembered forever—Smith bided his time, and at the age of eighteen came to the attention of a local surveyor, who placed him in charge of surveying the parish for enclosure. "This was the crucial moment," recalled Phillips in his biography of Smith; "from this event flowed all the current of his useful life."[5]

Young Smith distinguished himself and was soon entrusted with such important projects as examining soils and the geological circumstances of proposed borings for coal or the digging of canals. In his extensive travels he noticed that "the arrangement of the lias limestone beds in Warwickshire contrasted with the neighboring red marls at Inkborough; he examined the borings for coal in some of the dark lias clays on the road to Warwick, and recorded the absence of [sandy] beds from the limestones of Churchill. These were some of the points traversed in a mind capable of combining them at a future time."[6] He soon discovered that the strata in the different areas were always the same, and inclined to the east, so as to successively terminate on the surface in such a way that their appearance resembled, on a large scale, "superposed slices of bread and butter."[7] He became convinced that each had been in succession a bed of the sea. Finally came the realization that each stratum seemed to have its own characteristic fossils, "the mineralized monuments of races of organic beings" which had flourished at the time the stratum was being laid down.

Smith's carefully colored "Map of the Strata of England and Wales" appeared in 1815. However, it did not long remain the standard. It was superseded by the work of the Reverend William Conybeare (1787–1857), who defined the sequence of the strata as far as the Coal Measures (now referred to as the Carboniferous) and the underlying Old Red Sandstone (Devonian) of south Wales whose characteristic fossils included the placoderms or jawed fishes. Geologists attempted to extend the stratigraphic column to still older layers of rock. Before long they had defined a Silurian system below the Devonian, and a Cambrian below the Silurian, just as chemists of the day pursued the missing elements of the periodic table and astronomers searched for the planet needed to fill the unoccupied void between Mars and Jupiter. In this enterprise, British geologists were most fortunate in having relatively undisturbed regions to work on; there were no great mountain chains to confuse matters, like the Alps on the Continent or the Appalachians of the eastern United States. They were also fortunate, though they could not have realized it then, that the great geological eras in Britain had counterparts elsewhere in the world. The same strata represented by British rocks provided a sequence broadly applicable to sediments and strata elsewhere. Thus it proved possible to speak of a Cambrian era in North America, also of a Silurian, a Devonian, and a Carboniferous, though these names were originally drawn from place-names in the British Isles. (It is sobering to think, by the way, that all these sequences of rocks as far as the Cambrian—the earliest rocks with clearly defined fossils—represent only a tenth of the entire history of the Earth.)

Adam Sedgwick (1785–1873), the Cambridge geologist who identified the Cambrian system of rocks in north Wales, was accompanied in this groundbreaking geological fieldwork by a young man hitherto best known at Cambridge as an inveterate sportsman and beetle-collector, Charles Darwin. Soon after returning from tramping around Wales with Sedgwick, Darwin boarded the HMS *Beagle* as an unpaid naturalist on the ship's round-the-world voyage, armed with the Bible, Milton's *Paradise Lost*, and Alexander von Humboldt's *Personal Narrative* of his travels in Venezuela and the Orinoco basin. He also brought with him a book he had been given by one of his teachers, the botanist J. S. Henslow. It was the first volume of the just published textbook, *Principles of Geology*, by the London barrister turned geologist, Sir Charles Lyell (1797–1875), which was just then creating a sensation. Henslow told Darwin to read it but not to believe a word of it.

Lyell's dream was to put geology on the same footing as the exact sciences of physics and chemistry. He decided that, like these other sciences—and in contrast to the Noachian catastrophism of earlier days—geology ought to invoke no causes other than those still operating at the present time. His basic point was that—given enough time—the daubs of modest everyday forces could accumulate vast results. By way of illustration, he chose for the frontispiece of his first volume a depiction of a Roman temple, long submerged in Naples Bay, which had been thrust above the surface of the water by a recent earthquake. Such earthquakes produced small rises which, he pointed out, if repeated often and accumulated over the lapses of ages, could throw up mighty mountain ranges. The key was the inconceivable vastness of time.

Lyell became the leading exponent of what came to be known as uniformatarianism, the view that "*no causes whatever* have from the earliest time to which we can look back, to the present, ever acted, but those *now acting*; and that they never acted with different degrees of energy from that which they now exert."[8] Admittedly, his views were always rather extreme; as a strict uniformitarian, he was nearly alone, even in Britain where his influence was enormous. On the Continent, he was vigorously opposed by theorists who took their main inspiration from Laplacian cosmology. In their view, the Earth had formed hot and molten and had been cooling ever since by conduction of heat from the molten interior to the surface followed by its dissipation by radiation into space. This theory was summed up in a single arresting phrase by Pierre Louis Cordier (1777–1861), according to whom the Earth was " a cooled star, consolidating from the outside toward the center."[9]

Lyell saw the surface of the Earth as existing in a long-term equilibrium between opposing igneous and aqueous forces, with the rise of a continent here offset by the reclamation of land by the sea there; where even variations in climate were mainly local and produced by the changing distribution of continents and oceans. But France's most eminent geologist, Jean-Baptiste Elie de Beaumont (1798–1874), doubted that Lyell's forces of gradual uplift and the fluctuations of the Earth's surface around a mean could possibly account for such magnificently elevated and folded structures as the Alps. Beaumont, one of the men responsible for the masterly geological survey of France, argued forcefuly for paroxysmal episodes of mountain-building due to crustal movements during periods of sudden convulsive change or, in Cuvier's phrase, *révolutions du globe*.[10] According to Beaumont, the surface of the Earth did not oscillate around a mean; instead there was progressive development dictated by the gradual and progressive cooling (*refroidessement séculaire*) of the Earth and the globe's concomitant contraction. Simply put, as the Earth cooled, its interior shrank and the crust crinkled around it like the skin of a drying apple. These crinkles were the mountains.

William Whewell (1794–1866), who himself coined the term "uniformitarianism" for Lyell's gradualist and steady-state view of changes on the Earth, summed up the issues at stake. "Whether the causes of change do act uniformly; whether they oscillate only within narrow limits;—whether their intensity in former times was nearly the same as it is now;—these are precisely the questions which we wish Science to answer."

Turning from the Earth to the Moon, the igneous and aqueous forces—which, as Lyell argued, could over myriad ages apparently produce scenes of such chaotic disorder, ruin, and havoc on the surface of the Earth—seemed to have left equally legible marks on the shattered surface of the Earth's sister world. There were certainly visible the Huttonian "ruins of former worlds." The plains formerly known as seas suggested geological deposits, the jumbled regions of the lunar southern highlands testified unmistakably to the influence of volcanic forces.

Indeed, the broader debates within geology were clearly relevant to the scientific understanding of the processes and events shaping the Moon. And *vice versa*: an understanding of the conditions at work on the lunar surface could not help but shed a light on those at work on the Earth. Had not the Moon been shaped by a frequent succession of revolutions of its globe? Had these revolutions in its globe ended in a long-ago past?—a result which, if true, would have supported the catastrophist position. Or were there still changes taking place on its surface, as was to be expected if the uniformitarian perspective were true?

The Moon alone, among all objects of astronomical study, came within range of allowing geological scrutiny of its suite of rocks, landforms, and mountains. It was the only object other than the Earth that could be scrutinized in detail as to its physical constitution. Indeed, due to its proximity, finer detail could be seen on the Moon with the optical aid afforded by a pair of field glasses than could be seen on any of the planets with the most powerful telescopes in existence—for that matter, none of the planets appeared any larger than a lunar crater of modest proportions. To put it another way, for a feature on Mars to be detectable by an Earthbound observer, it had to be at least 20 miles in diameter. On Jupiter, a marking would have to be the size of Australia, on Saturn the size of Africa, to have any hope of being seen. Yet even a humble 6-inch (152-mm) telescope was capable of revealing a virtually inexhaustible welter of lunar detail—tens of thousands of craterlets less than a mile in diameter and scores of rilles only hundreds of yards wide.

Though "the Moon was no copy of the Earth," as Beer and Mädler had so aptly put it, there certainly were at least some analogies. As early as 1831, Elie de Beaumont published a paper comparing the craters of the Moon to the terrestrial "calderas" of Ceylon and Osians.[11] And in England—where Lyell was long oddly silent on the geology of the Moon—the search for analogies between the Earth and Moon became the major force driving lunar studies from at least 1832, when Conybeare presented an influential report to the Geological Section of the British Association for the Advancement of Science at Oxford. [12] The Association had been founded at York a year before in order to promote the growth of scientific culture in areas of the British Isles beyond London and the two university towns. Its means of doing so was by peripatetic annual meetings in different cities, which did much to stimulate "the knowledge and enthusiasm of local amateurs or 'cultivators' of science in grand collaborative projects of 'Baconian' science."[13] Conybeare listed, among the still outstanding questions of geology, the nature of the forces which had been the agents in dislocating and elevating the Earth's strata—whether these were earlier geological disturbances of some kind, or still active phenomena such as volcanoes and earthquakes. He referred in passing to the Austrian Leopold von Buch (1774–1853) and his theory of "craters of elevation," according to which the existence of mountains was due to the upwelling of molten matter from within the Earth. According to this still influential view, volcanoes were isolated uplifts in which a rising column of molten rock pushed the crust upward into a circular blister. When the blister broke, its apex was overtopped with flows and explosive gouts of lava. The theory might not seem readily susceptible to verification. But, Conybeare remembered,

> There is one source of analogy which has always appeared to me as likely to throw illustration on this subject... I would premise the observation, that we must surely in no respect consider our planet as an isolated body in nature; it is one only of the general planetary system, and every fair presumption of analogy favours the supposition, that similar general causes have acted in all the members of that system. Now one of these members, our own satellite, is placed sufficiently near us to enable our telescopic observations to distinguish accurately the general outlines of its mountain chains, and other similar features of its physical geography. We have been able to discern even the eruption of volcanos; and any one who has viewed its surface through a telescope, must be struck with the exact identity of the forms which he there contemplates with the maps and descriptions of the volcanic districts of our own globe.[13]

In von Buch's craters of elevation, with their "crateriform amphitheatres of many leagues in diameter, encircling central conical craters," Conybeare could "almost fancy that this description was intended as an exact portrait of what we observe on the lunar surface."[14] He concluded by recommending a program in which drawings of the features of the Moon by the best observers of the past would be systematically compared with the actual forms visible in the telescope, "so that we may be able to detect any changes in them. Is it too much to hope that we may thus effectually extend our knowledge of the general laws of the volcanic forces, which should appear to be among the general planetary phaenomena?"[15]

Conybeare found a kindred spirit in the young Thomas William Webb, later to become one of the leading lunar observers in Britain. Webb submitted a brief paper, "On Lunar Volcanos," to the Association's meeting at Newcastle in 1838, embracing the whole project of searching for evidence of lunar changes. He wrote,

> It is obvious," he wrote, that either the formation of new craters, or the enlargement of those previously existing, would afford convincing proof of the continuance of explosive or eruptive action: and having examined several portions of the Moon with an excellent achromatic of five-feet by Tulley, with the express view of detecting any appearances of the kind, I think I am enabled to assert that both the one and the other of these changes have taken place since the observations of Schroeter at the close of the last century.[16]

Webb noted several lunar formations which did not appear on Schroeter's charts, which he surmised had actually formed only since Schroeter's time. If so, this meant that the Moon's surface was still in a state of flux and that it was still being molded by ongoing volcanic processes.

Webb did not regard his results as conclusive. Nevertheless, he expressed the hope "that a more extended and accurate investigation may, in the course of a few years, not only bring to light the progress of many interesting changes, but may even enable us to form some inferences as to the nature and mode of action of that power which has produced such extensive and multiplied revolutions upon the lunar surface."[17] His call was not immediately heeded. But then he had chosen a singularly inopportune moment to make his plea: the combined effects of the Moon Hoax and the Mädlerian paralysis were just beginning to be felt.

Not until the 1840s was there evidence of a change in attitudes. Even then, it was almost imperceptible. Lyell visited Birr Castle in County Offaly, Ireland, where William Parsons, Lord Rosse (1800–1867) (Figure 9.2), the heir of a wealthy peer, was building a giant reflecting telescope—the 72-inch (1.83-meter) "Leviathan of Parsonstown." There Lyell examined "a model which [Rosse] had made of one of the mountains [craters] of the Moon." Lyell thought its form resembled "not so much… a volcano as one of the largest atolls, its sides being externally so steep and lofty and its crater sixty miles in diameter; but you must suppose the lagoon of enormous depth, and the ocean, of course, to be removed."[18]

In 1846, an American, James Dwight Dana (1813–1895), already well-known for his studies of minerology, published a paper on the "lunar volcanoes." Broadly speaking, Dana accepted a geological history of the Earth characterized by contraction of the globe and steadily declining activity, and on a visit to Hawaii with the Wilkes expedition of 1838–1842, he studied lava pools where the heat de-

Figure 9.2 William Parsons, the third Earl of Rosse, combined the wealth and leisure of a feudal lord with the talents of an engineer and the passion of an astronomer. He was the last individual to construct the world's largest telescope, the 72-inch "Leviathan" of Parsonstown, which was not surpassed until the 100-inch Mt. Wilson reflector went into operation in 1919. Courtesy Peter Hingley, librarian of the Royal Astronomical Society.

creased in circles from the center and mentioned the apparent similarity between such phenomena and the circular formations on the Moon.[19] And three years after Dana published his paper on volcanoes, the ubiquitous Sir John Herschel—no mean geologist himself—addressed the question of the origin of the lunar features in the first of many editions of his *Outlines of Astronomy*:

> The generality of the lunar mountains present a striking uniformity and singularity of aspect... They are wonderfully numerous... and almost universally of an exactly circular or cup-shaped form, foreshortened, however, into ellipses towards the limb; but the larger have for the most part flat bottoms within, from which rises centrally a small, steep, conical hill. They offer, in short, in its highest perfection, the true *volcanic* character, as it may be seen in the crater of Vesuvius, and in a map of the volcanic districts of the Campi Phlegræi or the Puy de Dôme, but with this remarkable peculiarity, viz: that the bottoms of many of the craters are very deeply depressed below the general surface of the Moon, the internal depth being often twice or three times the external height. In some of the principal ones, decisive marks of volcanic stratification, arising from successive deposits of ejected matter, and evident indications of lava currents streaming outwards in all directions, may be clearly traced with powerful telescopes.[20]

In illustrating his views of the lunar volcanoes, Herschel referred to the still active examples near Naples and the extinct ones mapped by the French geologist Nicholas Desmarest in the region of Auvergne. He added that although there were no seas on the Moon, "there are large regions perfectly level, and apparently alluvial in character."[21]

An important watershed in the growing interest in the Moon took place at the British Association's annual meeting at Belfast in September 1852. A "Moon Committee," consisting of Lord Rosse, Rev. Dr. Romney Robinson, and John

Figure 9.3 Lord Rosse's legendary "Leviathan" in a contemporary engraving. Unfortunately the instrument rarely performed well in the damp, cloudy climate of Ireland. From Robert S. Ball, *The Story of the Heavens* (1897).

Phillips, was formed, charged with drawing up a report on the physical character of the Moon's surface as compared with that of the Earth.

In 1845, Rosse had completed his giant telescope (Figure 9.3). The monster's tube was slung from chains between two stonework walls (Figure 9.4) 56 feet high and contained a 6-inch thick, 4-ton speculum metal mirror, the largest ever cast. However, Rosse's time for astronomical work was limited, and he seems to have been more interested in making instruments than in actually using them. Also, in the very year that the "Leviathan" became operational, the unprecedented Irish potato crop failure occurred, followed by a devastating famine. Rosse, as one of the leading landowners in the country, was understandably absorbed in this unparalleled crisis, and not until January 1848 did he resume regular observing. Then Rosse devoted his large telescope almost entirely to observations of nebulae—and rightly so, since this was the work for which it was best suited. Indeed, the most important discovery made with it was the spiral form of some of the nebulae. "When it is remembered," wrote J. L. E. Dreyer, one of the last astronomers to use the telescope, "that not one single observation of special importance was made with Herschel's '40-foot' telescope, it would seem that the maker of the great telescope at Birr Castle had not laboured in vain." [22] Nevertheless, Rosse did at least occasionally turn the telescope toward the Moon and planets.

Along with Lord Rosse himself, John Phillips (Figure 9.5) was a leading figure in the BAAS Moon Committee and served as its secretary. Orphaned at eight, he was adopted by his uncle William "Strata" Smith, from whom he received a

Figure 9.4 A winch was used to raise this platform that carried several observers to the eyepiece of Lord Rosse's giant reflector. Courtesy Peter Hingley, librarian of the Royal Astronomical Society.

sound education in land surveying and fossil-based geology. Later, as Keeper of the Museum of the Yorkshire Philosophical Society in York, he established his reputation as an outstanding field geologist with his *Illustrations of the Geology of Yorkshire* (1829). Naturally, his interest in the Moon was long secondary to his geological preoccupations. He always believed that a close comparison of lunar features with those of the Earth would be rewarded with the discovery of striking analogies, and beginning in 1839, after he moved into St. Mary's Lodge in the Yorkshire Philosophical Museum, he occasionally observed the "walled plains" of the Moon with a 2.4-inch (60-mm) refractor, made by the York firm of Thomas Cooke, set up atop a stone pillar in his garden. However, his serious interest in the Moon began only in 1852, when he had an opportunity to observe the Moon through Lord Rosse's giant telescope.[23] He later recalled: "I have never seen some parts of the Moon so well as on that occasion." Thereafter, astronomy largely engaged his attention.[24] However, he found it impossible to draw accurately what he had seen because of "the difficulty of transferring from the blaze of the picture to the dimly lighted paper, on a high exposed station, with little power of arranging the drawing-apparatus… The effect was altogether disheartening." The drawing that he did produce—of the noble crater Gassendi—was set down, he noted, "ex memoria" (Figure 9.6).

His brief survey of the Moon with the Rosse reflector was enough to convince Phillips that a convenient mounting was far more important than great optical power—the Rosse reflector required a crew of four just to keep it pointed at its target—and that "a plan of continuous work by means of one instrument devoted

Figure 9.5 John Phillips, the nephew of pioneer geologist William "Strata" Smith, was one of the first men to attempt to do field geology of the Moon and planets from the eyepiece of a telescope. From *Geological Magazine*, July 1870.

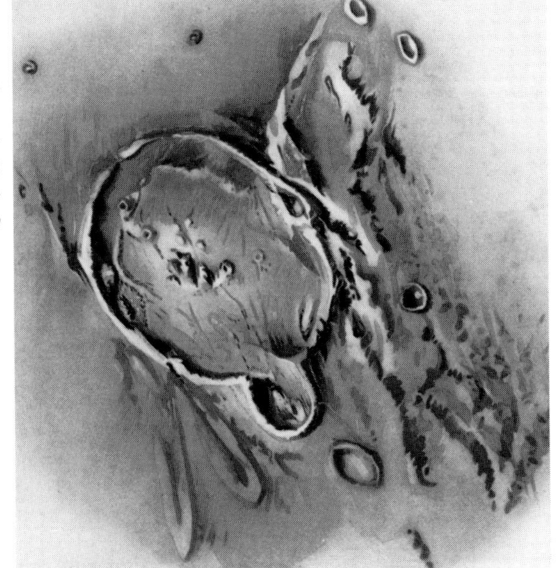

Figure 9.6 John Phillips' drawing of the crater Gassendi, based on memories of his impressions through the great Rosse reflector. From *Philosophical Transactions of the Royal Society of London*, 1868.

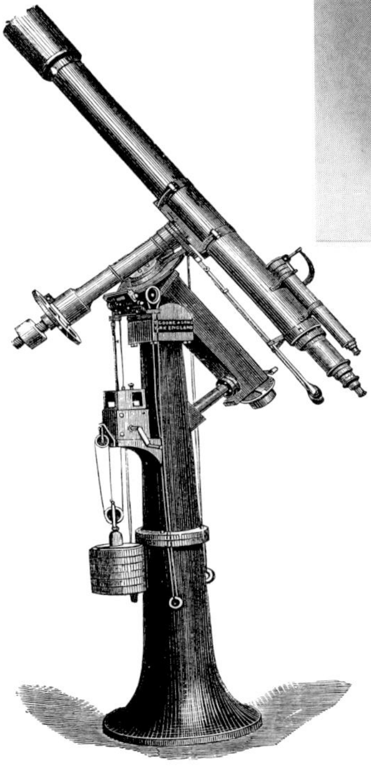

Figure 9.7 A 6-inch equatorial refractor fashioned by the firm of Thomas Cooke and Sons, complete with a clock drive powered by falling weights. It is similar to the instrument used by John Phillips to carry out most of his lunar work during the 1850s and 60s. Courtesy John Koester.

to a survey of selected parts of the Moon"[25] would be far more productive than occasional views with large but unwieldy and inefficient instruments. He therefore ordered, expressly for the purpose, a 6.2-inch (158mm) refractor from Thomas Cooke (Figure 9.7), which he set up in his garden and at once began using to make observations of selected regions of the Moon. Though he was convinced that Beer and Mädler's map was by now hopelessly inadequate, he never contemplated the daunting task of mapping the entire Moon in greater detail. Instead he preferred to do as Schroeter had done: concentrate on specific regions (Figure 9.8), always with the hope that this would eventually allow the decipherment the origin of the lunar features by a comparison with geological forms on the Earth. He suggested:

> The results likely to be attained by such a series of careful drawings of special parts of the Moon's surface... are [already] recognized by Mr. Conybeare in his report on Geology to the British Association in 1832. Indeed, it may be boldly affirmed that a competent theory of volcanic action can hardly be regarded as having been adequately tested, much less completed, without a careful study of the magnificent volcanic surface of the Moon, where for the most part the consolidated products of a long train of igneous eruptions are exhibited as clearly as in the celebrated region of Auvergne.[26]

He always envisaged a collaborative effort: his drawings to be collated and compared with those of other observers engaged on the Moon. Making use of his numerous contacts, Phillips enlisted fourteen observers in Britain and abroad to observe designated areas of the Moon, under as many different conditions of phase and lighting as possible. He also drafted a preliminary "Report on the Physical Character of the Moon's Surface, as compared with that of the Earth," which he presented at the BAAS meeting at Hull in 1853, outlining an impressively comprehensive series of prospective investigations to be undertaken by the Committee: the determination of the steepness of the lunar slopes, the difference of level between the exterior and interior bases of the craters, the nature of the lunar rays, the slopes, heights and breadths of the wrinkle ridges on the maria, and the correctness of "Schroeter's rule"—would the rim materials exactly fill the depressions of the craters?

Phillips was also a visionary when it came to the future promise of photography. The possibility that the Moon might eventually be persuaded to take its own portrait—thereby eliminating the need for those "long and wearisome labours by which men had hitherto sought to chart the Moon"[27]—had been foreseen by François Arago, Director of the Paris Observatory, as early as 1840. In March of that year, the Moon had first registered its image on a silver plate placed at the focus of a 12-inch (305-mm) reflector by Dr. John William Draper (1811–1882) of New York. Only ten years after this hopeful but primitive effort, Harvard astronomer William Cranch Bond (1789–1859), assisted by photographer John Adams Whipple, had captured the 2-inch diameter image of the Moon at the focus of the 15-inch (381-mm) Merz refractor of Harvard College Observatory (Figure 9.9). In this photograph, they succeeded in recording all of the principal lunar features.

Phillips himself obtained images of the Moon barely an inch across with his

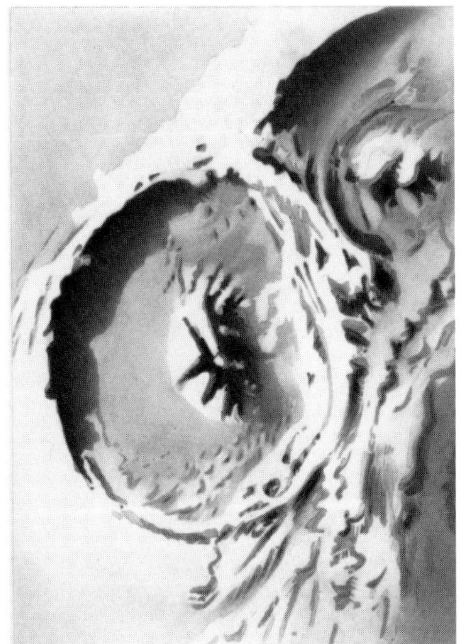

Figure 9.8 The level of detail attained by Phillips and other observers of the mid-nineteenth century is well attested here. (Top left) A 1992 photograph of the First Quarter Moon through a 12-inch Newtonian reflector by the Belgian amateur Leo Aerts shows the usual view of the Moon obtained with a small telescope. The arrow points to the contiguous craters Theophilus and Cyrillus. (Top right) A highly detailed drawing of Theophilus and a portion of Cyrillus by Phillips using his 6.2-inch Cooke refractor; from *Philosophical Transactions of the Royal Society of London*, 1868; (Bottom) An image by T. A. Dobbins of Theophilus and Cyrillus, on roughly the same scale as Phillips' drawing.

Figure 9.9 Co-author William Sheehan at the "business end" of the 15-inch Merz refractor of the Harvard College Observatory, the telescope used in the 1850s by William Cranch Bond and John Adams Whipple to obtain the first good photographs of the Moon. Photograph by Owen Gingerich.

6.2-inch refractor which he exhibited at Hull in 1853. He declared:

> If photography can ever succeed in portraying as much of the Moon as the eye can see and discriminate, we shall be able to leave to future times monuments by which the secular changes of the Moon's physical aspect may be determined. And if this be impracticable—if the utmost success of the photographer should only produce a picture of the larger features of the Moon, [even] this will be a gift of the highest value, since it will be a basis, an accurate and practical foundation of the minuter details, which, with such aid, the artist may confidently sketch.[28]

When Phillips penned these words, photography represented to most astronomers little more than a novelty. But then it must be borne in mind that with an 1840s-vintage daguerrotype plate and a lens of *f*/8, taking the portrait of a subject illuminated by the midday Sun required an exposure lasting from five to six *minutes*. The wet-collodion plate, introduced by F. Scott Archer in 1851, brought only a ten-fold imporvement in sensitivity, and among astronomical subjects only the Sun itself was sufficiently bright that its fine details—the granulation of the photosphere and the structure of sunspots—could be recorded well enough to allow a program of systematic research. (And indeed it would be at the urging of John Herschel and under the auspices of the Royal Society that the solar disk began to be photographed daily from 1855 onward.)

Phillips's photographic experiments and other lunar studies were interrupted in October 1853, when he moved from York to Oxford to assume the position of deputy reader in geology. He arrived to find astronomy at Oxford "seriously in the doldrums."[29] The Radcliffe Observatory (Figure 9.10) had been withdrawn from the university's use in 1839, after a Savilian Professor of Astronomy had been ap-

Figure 9.10 The Radcliffe Observatory at Oxford was housed in these attractive eighteenth-century buildings in John Phillips' day. The observatory no longer exists and the buildings now serve as an infirmary. 1987 photograph by William Sheehan.

pointed without the Radcliffe trustees being consulted. The university consequently had no observatory at all apart from the roof of the Tower of Five Orders in the Bodleian Quadrangle. Meanwhile, Phillips, who had taken up temporary quarters in a house near the Botanic Garden and Magdalen Bridge, was having a new house built near the Ashmolean Museum, of which he had been named Keeper. Until the move was completed, he did not attempt to resume his lunar work. As he later wrote, "my plan for continual work on the Moon was entirely cut through; it was impossible to mount a large instrument near my dwelling till… the ground was arranged around the museum, so as to give me the requisite space and security close to the house which had been appointed to me by the University."[30]

Nevertheless, others were by now joining his general program. Warren de la Rue (1815–1889), a gentleman of independent means in London, had equipped himself with a 13-inch (330-mm) Newtonian reflector and taken an interest in lunar photography. Unfortunately, he did not achieve good results until he equipped the instrument with a clockwork drive in 1857. By 1865 he was able to publish a set of twelve photographs of the Moon at various phases, each enlarged to a diameter of almost four inches and showing "nearly every mountain or object of importance… with a minuteness of perfection that would scarcely have been deemed possible."[31] And yet despite their excellence, de la Rue's images were scarcely half the size of Tobias Mayer's map of 1775. But Lewis Morris Rutherfurd (1816–1892), a New York lawyer, was using 11- and 13-inch (279- and 330-mm) refractors in an observatory located in the garden of his Manhatten residence to obtain photographs of the Moon able to bear enlargement to a diameter of 21 inches. Thus photography was slowly narrowing the gap—here were sharp images at better than half the scale of Beer and Mädler's great chart. Still, the most detailed representations of the Moon would long continue to be those obtained by visual observers equipped with pencils and sketchpads.

Apart from Phillips' "ex memoria" depiction of Gassendi, the best and most detailed representation was an extraordinary drawing of the great crater Copernicus (Figure 9.11) carried out in 1856 by the eminent Jesuit astronomer Angelo Secchi (1818–1878) of the Collegio Romano. Secchi used the 9-inch (227mm) refractor of the Roman Observatory—a fine instrument made by Fraunhofer's successor Merz—with magnifying powers of 760X and 1000X. "As it was impossible to carry through such a work in a single night, on the first night of good opportunity a general outline was taken," wrote Secchi, "and on the other evenings particular drawings were made, and all these parts, taken in different grades of light and shadow, were afterwards harmonized together and compared against the Moon itself."[32] In all, Secchi occupied himself at all favorable positions of the Moon for six months. The finished drawing was on a scale of 10 geographical miles to the inch—"such," Phillips commented, "as only the larger telescopes can command."[33] Both Secchi's Copernicus and his own Gassendi were clearly superior to Mädler's representations of these features. They encouraged "the hope that we are now fairly entered on the long career of discoveries in the Moon," for "in proportion as the power of the telescope rises, the seemingly simple 'ring mountains' of the Moon exhibit as much diversity of outline and structure as the larger terrestrial volcanoes when accurately mapped." Phillips illustrated this by some comments that clearly reveal the acute perceptions of the trained geologist's eye:

> While Gassendi, with peaks 9000 feet high, projects like a huge narrow wall into the Mare Humorum, and hangs over the interior plain in precipices as steep and many times as high as those over the Atrio del Cavallo [near Vesuvius], Copernicus, seated in the midst of broad land... rises in many broken stages, bristling with a thousand silver-bright crests,—a perfect network of rough and complicated ground, crossed by lights and shades, which have a history of their own,—and toward the inside falls off by many irregular terraces, down to an interior plain, as if the whole area had yielded, and the surface had been formed by enormous land-slips.[34]

Gassendi had at least two and, Phillips suspected, several more small craters within its wide plain; Copernicus, as Secchi portrayed it, had none. In some "lunar mountains," he noted, the center is occupied by "a crater-formed hill, as Vesuvius stands within Somma; in others the hill remains a smooth rounded mass, but its crater is lost; and a further stage of decay seems to be seen in Gassendi and Copernicus, where the central mass is broken into fragments and sculptured by ramified hollows." Phillips seemed to discern a definite progression in the stages of erosion and degradation; perhaps evidence of the former action of a lunar atmosphere, subsequently absorbed into the oxidized crust of the Moon, or of features formed by running water similar to the degraded craters of the Eifel, a region of extinct volcanoes in western Germany. But he could not be sure. With his own researches "cut through" by his move to Oxford, he could take the matter no further. Moreover, the efforts of the other members of the "Moon Committee" had been disappointing—"favourable as to good intentions; but in several cases want of ad-

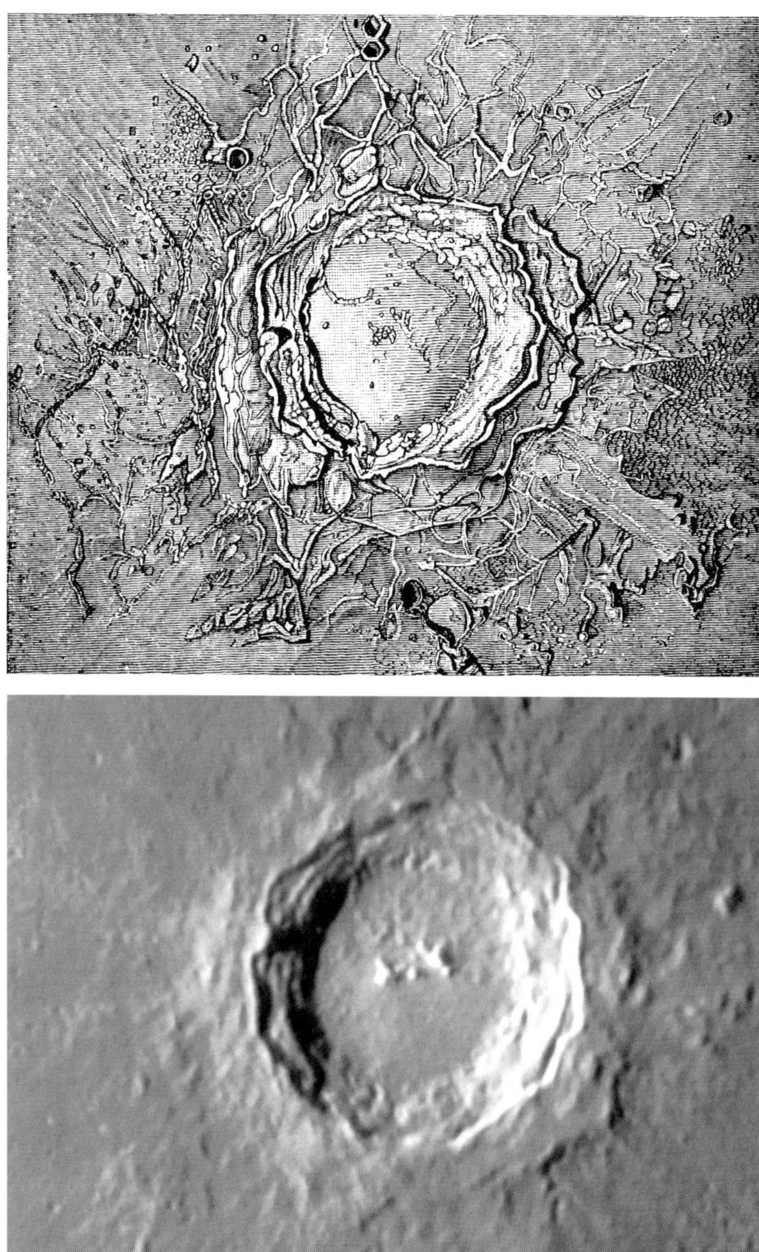

Figure 9.11 The widely reproduced lithograph (top) was based on the Jesuit astronomer Angelo Secchi's celebrated drawing of Copernicus. From W. H. Warren, *Recreations in Astronomy* (1886). It bears careful comparison with the CCD image of Copernicus under late morning lighting by T. A. Dobbins (bottom).

equate lesiure, sometimes want of health, sometimes other causes had prevented the fulfillment of the Committee's wishes, even as preliminary surveys."[35]

Thus the 1850s—though not characterized by the depression and paralysis of the 1840s—were marked more by aspiration than achievement. Not entirely lost for lunar observation, they would nevertheless be, in retrospect, little more than a somewhat disorganized prelude to the 1860s. Then the Moon became the object of widespread attention such as it has seldom commanded before or since.

References

1. See Phillip Lawrence, "Heaven and Earth: the Relation of the Nebular Hypothesis to Geology," in *Cosmology, History and Theology*, eds. Wolfgang Yourgrau and Allen D. Breck (New York: Plenum, 1977), pp. 253–281. Also: Stephen G. Brush, "The Nebular Hypothesis and the Evolutionary Worldview," *History of Science*, 25 (1987), pp. 245–278.
2. Joe D. Burchfield, *Lord Kelvin and the Age of the Earth* (Chicago: University of Chicago Press, 2nd ed., 1990), p. 5.
3. James Hutton, "Theory of the Earth; or an Investigation of the Laws Observable in the Composition, Dissolution and Restoration of Land upon the Globe," *Transactions of the Royal Society of Edinburgh*, 1 (1788), pp. 209–304.
4. Donald B. McIntyre and Alan McKirdy, *James Hutton: The Founder of Modern Geology* (Edinburgh: The Stationary Office, 1997), p. 35.
5. John Phillips, *Memoirs of William Smith* (London: John Murray, 1844), p. 5.
6. Ibid., p. 6.
7. Ibid., p. 8.
8. Charles Lyell, *Principles of Geology*, with a new introduction by Martin J. S. Rudwick (Chicago: University of Chicago Press, 1990 reprinting of 1st edition, 1830), vol. 1, p. 73.
9. P. L. A. Cordier, "Essai sur la température de l'intérieur de la terre," *Mémoires de l'Académie des Sciences de l'Institut de France*, 7 (1827), pp. 473–556.
10. On Elie de Beaumont, see "Élie de Beaumont and the First Global Tectonics," in Mott T. Greene, *Geology in the Nineteenth Century* (Ithaca: Cornell University Press, 1982), pp. 69–92.
11. Élie de Beaumont, "Sur les rapports qui existent entre le relief du sol de l'ile de Ceylan et de celui de certaines masses de montagnes qu'on aperáoit sur la surface de la lune," *Annales Sciences Naturel*, 22 (1831), pp. 88–96. These terrestrial formations have been shown to be erosional basins in gneissic complexes, not volcanic features, as de Beaumont believed.
12. Martin J. S. Rudwick, *The Great Devonian Controversy: The Shaping of Scientific Knowledge Among Gentlemanly Specialists* (Chicago: University of Chicago Press, 1985), p. 30.
13. W. D. Conybeare, "Report on the Progress, Actual State, and Ulterior Prospects of Geological Science," *Transactions of the British Association for the Advancement of Science—2nd Report*, 1832, p. 409.
14. Ibid. pp. 409–410.
15. Ibid., p. 410.
16. T. W. Webb, "On Lunar Volcanos," *Transactions of the British Association for the Advancement of Science—8th Report*, 1838, p. 93.
17. Ibid.
18. Charles Lyell to his sister, September 4, 1843; in K. Lyell, *Letters and Journals of Sir Charles Lyell, Bart.* (London: John Murray, 1881), vol. 2, p. 76. Interestingly, Lyell had already accepted Charles Darwin's theory of the coral atolls, as he informed Sir John Herschel, May 24, 1837: "The annular shape and central lagoon have nothing to do with volcanos, nor even with a crateriform bottom. Perhaps Darwin told you when at the Cape what he considers the true cause? Let any mountain be submerged gradually, and coral grow in the sea in which it is sinking, and there will

be a ring of coral, and finally only a lagoon in the centre. Why? For the same reason that a barrier reef of coral grows along certain coasts, Australia, & c. Coral islands are the last efforts of drowning continents to lift their heads above water."

19. James Dwight Dana, "On the Volcanoes of the Moon," *American Journal of Science,* 2nd series, 2 (1846), pp. 335–353. Dana's views were not to change over the duration of a long career. In his *Manual of Geology*, the standard textbook of American geology for forty years, he declared flatly: "The moon presents to the telescope a surface covered with the craters of volcanoes, having forms that are well illustrated by some of the earth's volcanoes, although of immense size. The principles exemplified on the earth are but repeated in her satellite." Dana, *Manual of Geology* (New York: American Book Company, 4th ed., 1895), p. 11.
20. Sir John F. W. Herschel, *Outlines of Astronomy* (London: John Murray, 10th ed., 1869), p. 283. The statement appears without change in the many editions of Herschel's *Outlines* during the second half of the century. The emphasis on *volcanic* is Herschel's own.
21. Ibid.
22. J. L. E. Dreyer, "Lord Rosse's Six-Foot Reflector," *The Observatory*, 37 (1914), p. 401.
23. John Phillips, "Notices of some Parts of the Surface of the Moon," *Philosophical Transactions of the Royal Society*, 158 (1868), p. 333.
24. [Anon.] "Notice of the Death of Prof. John Phillips," *Annual Report of the Yorkshire Philosophical Society for 1874* (York, 1875), 14–19:16.
25. Phillips, "Suggestions for the Attainment of a Systematic Representation of the Physical Aspect of the Moon," *Proceedings of the Royal Society*, 12 (1862), 31–37:34.
26. Ibid., p. 32.
27. Proctor, *The Moon*, p. 150.
28. *BAAS Report for 1853*, Researches in Science, p. xxxiii.
29. Roger Hutchins, "John Phillips, 'Geologist-Astronomer,' and the Origins of the Oxford University Observatory, 1853–1875," in *History of Universities*, ed. Peter Denley (Oxford: Oxford University Press, 1994), 193–249:194.
30. Phillips, "Notices," p. 334.
31. Note in the *Intellectual Observer*, 3 (1863), p. 228.
32. P. A. Secchi, "Extract of a Letter to George Rennie, Esq., F.R.S., containing explanatory remarks on a drawing of the Lunar Spot 'Copernicus,' presented by him to the Royal Society," March 13, 1856, *Proceedings of the Royal Society of London*, viii (1856–57), 72.
33. Phillips, "Notes on the Drawing of 'Copernicus,' presented to the Royal Society by P. A. Secchi," *Proceedings of the Royal Society of London,* 8 (1856–1857), 73–75:73.
34. Ibid., p. 74. The Atrio del Cavallo, so-called because travelers left their horses and mules there before ascending the cone on foot, is described thus by Lyell in his *Principles of Geology*, vol. 1, p. 344: "The slanting beds of the cone of Vesuvius become horizontal in the Atrio del Cavallo… where the base of the new cone meets the precipitous escarpment of Somma; for when the lava flows down to this point, as happened in 1822, its descending course is arrested, and it then runs in another direction along this small valley, circling round the base of the cone. Sand and scoriae, also, blown by the winds, collect at the base of the cone, and are then swept away by torrents; so there is always here a flattish plain."

Chapter 10:
A Growing Reaction

What Phillips proposed in his program of systematic study of selected regions of the Moon was analogous to the geological fieldwork long pursued by amateurs, whose familiarity with local terrain gave them "a positive advantage over the metropolitan or university geologists who might occasionally visit their home areas. Within those areas, amateur geologists had the *time* that was needed to find fossils by patient hammering in local quarries, cliffs or cuttings, for fossils were rarely to be found without such time-consuming work."[1] For that matter, geology's popularity among amateur scientists of the nineteenth century has been explained, in part, by the sheer accessibility of the subject, since little more was required than a hammer, a magnetic compass, a clinometer for measuring the tilt or dip of strata, and perhaps a hand lens. Lunar observation was similarly accessible. One did not need mammoth instruments like the reflectors of the Herschels or Lord Rosse; indeed, such large and cumbersome instruments were actually a disadvantage, as Phillips had learned from his work with the Rosse "Leviathan." Apart from a modest telescope (Figure 10.1), one needed only diligence, a pencil, and a sketchpad.

There were romantic aspects to both geological fieldwork and astronomical observation. In both disciplines, as Martin J. S. Rudwick has said of geology, "the primacy of fieldwork reflected a recognition that the empirical locus of a science like geology was bound to lie 'in the field.'"[2] This meant that amateurs could contribute more substantially to geology and astronomy than to most other sciences. There would long remain a distinctly Baconian flavor to these investigations, where the collection of facts was generally preferred to far-flung theoretical speculations. And of course it was in the collection of facts that amateurs excelled. They were the beetle-collectors and crab-chasers of science—precisely the kind of men and women who could, with the proper direction, undertake the important role of carrying out the arduous and painstaking fieldwork at local quarries and cliffs or the comparable surveillance of the minor surface features of the Moon.

Inevitably, the Baconian project was far from unproblematical, for the observations themselves were often theory-laden. The situation was no different from that Charles Darwin noted in his ramblings among the Cambrian rocks with Sedgwick in which he proved oblivious to the abundant evidence of past glacial action all around him: one saw as much—and only as much—as one was prepared to see. As he later wrote:

> On this tour I had a striking instance how easy it is to overlook phenomena, however

Figure 10.1 Many nineteenth-century British amateurs employed refractors of modest aperture like this instrument by Thomas Cooke, supported by an altazimuth mounting atop a tripod fitted with steadying rods. Such telescopes were comparable in power to those wielded by Lohrmann and Beer and Mädler, opening up a wealth of research opportunities to a diligent observer. From W. F. Denning, *Telescopic Work for Starlight Evenings* (1891).

> conspicuous, before they have been observed by anyone. We spent many hours at Cwn Idwal, examining all the rocks with extreme care as Sedgwick was anxious to find fossils in them; but neither of us saw a trace of the wonderful glacial phenomena all around us; we did not notice the plainly scored rocks, the perched boulders, the lateral and terminal moraines. Yet these phenomena are so distinct that... a house burned down by fire did not tell its story more plainly than did this valley.[3]

In this respect, it is important to note that the theory or at least assumption that lay in the background of many of the leading amateur observers of the Moon was that of ongoing change. The Moon was not quite dead; it was still the scene of atmospheric obscurations and perhaps even of occasional volcanic eruptions.

The romantic appeal of exploring the remote quarries and cliffs of the Moon—of wandering, visually, over fields that presented objects as impressive as, possibly analogous to, the extinct volcano fields of Auvergne—was not lost on observers of that distant orb. Its features were veritably charged with interest. As the Reverend T. W. Webb pointed out in his classic *Celestial Objects for Common Telescopes*, the Moon presented

> a surface convulsed, upturned, and desolated by forces of the highest activity, the results of whose earliest outbreaks remain, not like those of the Earth, levelled by the fury of tempests, and smoothed by the flow of waters, but comparatively undegraded from their primitive sharpness even to the present hour. The ruggedness of the details, as old Hevelius had anticipated, becomes more evident with each increase of optical power, and we cannot doubt that we look upon the unchanged results of those gigantic operations which have stereotyped their record on nearly every region of the lunar globe.[4]

Webb (Figure 10.2) was himself one of the most energetic amateurs ever to study the Moon. He was born on December 14, 1806, the only son of a clergyman, the Reverend John Webb, Rector of Tretire cum Michaelchurch in the county of

Figure 10.2 One of the great "Moon men" of the nineteenth century, the Reverend Thomas William Webb, shown with his wife Henrietta. Courtesy Peter Hingley, librarian of the Royal Astronomical Society.

Hereford. The elder Webb was a sound classical scholar and an eminent authority on Norman French who was frequently called upon to give evidence in Courts of Law on the interpretation of early documents. But he was more particularly devoted to researches on the history of the west of England during the Civil War, and for the greater part of his long life—he lived to the ripe old age of 93—he was occupied with preparing a history of Herefordshire during the Civil War. But his dread of inaccuracy or precipitate thought slowed the work, which was left to his son to complete many years later.

Thomas was educated entirely by his father. Brought up among books and manuscripts, he became precise, studious, and mature for his years. While he showed some aptitude for experimental science, especially electricity, his father insisted that he become a classical scholar. Thus he was dissuaded from studying Euclid; otherwise it is almost certain that he would have become a mathematician. His efforts at Latin and Greek, though carried out with diligence, did not progress to his father's satisfaction.

In 1826, Thomas entered Oxford as a Gentlemen Commoner at Magdalen Hall. He took his degree in 1829, with second class honors in mathematics. He might have done better had he applied himself more earnestly, for he later admitted that he spent most of his time in the "idleness" of desultory reading in the library.

On leaving Oxford, he was licensed to the curacy of Pencoyd where he remained for two years. He took holy orders in August 1831 and became a minor canon of Gloucester Cathedral. Later he returned to Tretire and worked as curate under his father. Only in 1856 did he attain to the living of Hardwick, a large but thinly populated parish near the Welsh border. While diligent in the pursuit of his clerical duties—he set out early each afternoon with a knapsack on his back to visit the more remote members of his parish—he never lost his early interest in science.

Proficient in German, Webb read Beer and Mädler in the original and became the most reliable source of information about German selenographic studies, then the standard, in the English-speaking world. At Gloucester he began routine observations of the Sun, Moon, and planets with a rather modest instrument—a 3.7-inch (94-mm) achromatic refractor by the English maker Tulley. He immediately realized that there was still much useful work to be done, especially on the Moon. "A little experience," he wrote, "showed that Beer and Mädler have not represented all that may sometimes be seen with a good common telescope. My own opportunities, even when limited to a 3⁷⁄₁₀-in. aperture, satisfied me, not only how much remains to be done, but how much a little willing perseverance might do."[5]

Webb's *Celestial Objects* appeared in 1859, the same year Darwin published *On the Origin of Species* advancing his theory of natural selection (or, as John Herschel quipped, "the law of higgledy-piggledy").[6] The lengthly chapter on the Moon presented a complete, if for the time conventional, view of the lunar surface. The "crater mountains" were "the grand peculiarities of the Moon: commonly, and probably with correctness, ascribed to volcanic agency."[7] Webb mentioned their resemblance to rings left by gaseous bubbles, first ingeniously invoked by Hooke, only to dismiss the analogy on the grounds that no materials could have sufficient cohesiveness to support structures on such an enormous scale. Instead he preferred a modified bubble theory—the idea that the lunar craters were the remains of molten lakes on a once-fiery globe.

What fascinated Webb above all else was the exciting possibility that surface of the Moon was still in active condition. As we have seen, his first essay, "On the Lunar Volcanos," had been presented to the British Association as far back as 1838, but not until 1859 did he follow it up with "Notice of Traces of Eruptive Action in the Moon." Here he pointed out that while astronomers were generally agreed as to the cessation of such action on a large scale, "this would not necessarily infer the impossibility, or even improbability, of minor eruptions, which might still continue to result from a diminished but not wholly extinguished force."[8] He thus re-introduced the Schroeterian project of documenting lunar changes, which had been all but banished since the publication of *Der Mond* in 1837.

Webb recommended two specific regions for further detailed study. In Cichus, a crater in southern Mare Nubium, a small crater seemed to have grown larger as compared to the earlier representations of Schroeter and Mädler. This was an admittedly weak case; but the other site was more convincing, and indeed would become one of the most scrutinized places on the Moon by students pursuing evidence of change.

In Mare Foecunditatis, the name Messier had been given to one of two small craters lying side by side at the end of parallel white streaks "resembling the divided tail of a comet." Schroeter had drawn the region. So had Beer and Mädler, who claimed to have inspected the formation no fewer than 300 times between 1829 and 1837. They always found the two craters "as like as two peas in a pod—perfectly and singularly alike in size, shape, height of ring, depth of cavity, and even in the position of some peaks upon the ring." Nothing, apparently, could be more definite; but Webb, first on November 14, 1855, and on many occasions afterwards, found that whatever had been their previous resemblance, the two craters were no longer peas in a pod. The difference between them was in fact so great "as to indicate a permanent alteration of the surface during the space of twenty years."[9]

Such comments show that seven years before the Linné affair, the notion of lunar change was once more in the air. Indeed, it is hardly surprising that astronomers had rebelled against—or at least found themselves unwilling to embrace—the static desert world of Beer and Mädler. As Richard Anthony Proctor would sum up a few eventful years later:

> The examination of mere peculiarities of physical condition is, after all, but barren labour, if it lead to no discovery of physical condition. The principal charm of astronomy, as indeed of all observational science, lies in the study of change—of progress, development, and decay, and specially of systematic variations taking place in regularly-recurring cycles... In this relation the Moon has been a most disappointing object of astronomical observation. For two centuries and a half, her face has been scanned with the closest possible scrutiny; her features have been portrayed in elaborate maps; many an astronomer has given a large portion of his life to the work of examining craters, plains, mountains and valleys for the signs of change; but hitherto no certain evidence... has been afforded that the Moon is other than "a dead and useless waste of volcanoes."[10]

The charm of geological ruins on the Moon—level wastelands, craggy wildernesses—was never so great as the earlier vision of a still-burning and active planet, with volcanoes, exhalations, erosive forces; perhaps even life of its own. So the romantic vision of the Moon died hard. It is this psychological fact which probably explains the mid-nineteenth century's growing preoccupation with minor eruptions as evidence of the Moon's ongoing geological activity, along with the resurgence of the possibility that the Moon was a living world—"habitable," as Webb expressed the hope, "in some way of its own."[11]

Support for the habitable Moon—the possibility which had so fired the imaginations of William Herschel, Schroeter, and Gruithuisen in a bygone age—came, indeed, from a most unlikely quarter: celestial mechanics, the detailed mathematical study of the intricate motions of heavenly bodies.

The classic three-body problem of the Moon's motion—the attempt to formulate the ways in which the Moon's elliptical orbit around the Earth is perturbed by the Sun—had exercised the genius of Newton and his most brilliant successors, and had by the early nineteenth century grown into a highly specialized area of research. Building on foundations laid by Euler, Clairaut, and d'Alembert, the great Marquis Simon de Laplace had succeeded in whittling the error of the Moon's motion to a mere half a minute of arc. A particularly important advance had been his apparent solution of the problem of the Moon's secular acceleration. Comparing modern observations with old eclipse records, Edmond Halley had noted as long ago as 1693 that the Moon's orbital period seemed to be decreasing by some 10 seconds of arc per century. That is, the Moon appeared to be accelerating. The problem seemed intractable at the time and long remained so. Doubts even began to creep in about the exactness of Newton's inverse-square law itself—Euler went so far as to postulate the existence of an ill-defined resisting medium in space.

Laplace's calculations of 1787 indicated that the secular acceleration could be explained by slow oscillations of the eccentricity of the Earth's orbit. Still, the lunar theory continued to leave something to be desired, and it was later shown that the mechanism proposed by Laplace accounted for only half of the observed acceleration of the Moon. (The excess was later accounted for by tidal friction, of which more anon; for now suffice it to say that tidal friction slows the Earth's rotation, so as to produce an *apparent* acceleration of the Moon.)

In due course, the British Admiralty became involved with the problem. Its interest in lunar tables went back, of course, to Newton's time, when it had been tied to the problem of finding terrestrial longitudes by the method of "lunars." The new initiative was headed by the Astronomer Royal at Greenwich, George Biddell Airy (1801–1892), and involved the reduction of some 8,000 observations of the Moon made at Greenwich Observatory between 1750 and 1830. These tedious calculations kept sixteen human computers busy for eight years. The task of using these data to work out a new theory of the Moon was assigned to Peter Andreas Hansen (1795–1874) (Figure 10.3), a Danish watchmaker, who taught himself mathematics and secured a position at the Seeberg Observatory in Gotha, where in 1825 he succeeded Encke as director and established himself as a prodigy of celestial mechanics.

Hansen's massive reworking of lunar theory, though it could be followed in the details only by specialists, established him as one of the towering figures of science. The Royal Academy of Sciences of Saxony made a standing offer to publish all his work; he was twice awarded the Gold Medal of the Royal Astronomical Society, and his lunar tables were published at the expense of the British government and adopted for use in the *Nautical Almanac*. A younger rival, the brilliant Canadian-born mathematician and astronomer Simon Newcomb (1835–1909), referred to him as "the greatest master of celestial mechanics since Laplace."[12]

The only result of Hansen's that attracted a broader interest was his 1854 calculation of the figure of the Moon; he found that it was not spherical but ellipsoidal, the major axis being directed toward the Earth. Moreover, the Moon's center of gravity was located at a point 33½ miles farther from the Earth than the geometric center of its figure.[13] In itself, this was hardly the sort of thing to set the astronomical world on its ear. What made Hansen's paper of wider interest was not his calculation but his inference that the Moon's water and oceans might have escaped to the far side. The near side of the ellipsoid was undoubtedly "a sterile land, deprived of an atmosphere and of all life," but, wrote Hansen, "one can no longer conclude that the other hemisphere does not have an atmosphere, nor that it has no

Figure 10.3 The Danish astronomer Peter Andreas Hansen, one of the nineteenth century's leading experts on celestial mechanics. For an interlude of several years, Hansen's arcane mathematical studies of the Moon's motion lent credence to notions that its averted hemisphere was habitable. Courtesy Dorothy Schaumberg, Mary Lea Shane Archives of the Lick Observatory.

vegetation or living things."[14]

With one stroke, Hansen seemed to revive notions that had seemingly died with Gruithuisen. For a moment, the Subvolvans and Privolvans of Kepler's *Somnium* lived again—or the Privolvans, at any rate, the inhabitants of the far side. Hansen's speculation was received with enthusiasm by no less an authority than Sir John Herschel, who in the 1858 edition of his *Outlines of Astronomy* examined the possibility of a lunar ocean, atmosphere, and life. Herschel explained that any air, water, or other fluid on the surface of such a globe would

> run towards the lowest place, that is to say, not the nearest to the centre of figure, or to the central point of the mere space occupied by the Moon, but to the centre of mass, or the centre of gravity. There will be formed an ocean of more or less extent, according to the quantity of fluid directly over the nucleus, while the lighter portion of the solid material will stand out as a continent on the opposite side... In what regards its assumption of a definite level, air obeys precisely the same hydrostatical laws as water. The lunar atmosphere would rest upon the lunar ocean, and form in its basin a *lake of air*, whose upper portions, at an altitude such as we are now contemplating, would be of excessive tenuity, especially should the lunar provision of air be less abundant in proportion than our own. It by no means follows, then, from the absence of visible indications of water or air on this side of the Moon, that the other is equally destitute of them, and equally unfitted for maintaining animal or vegetable life. Some slight approach to such state of things actually obtains on the Earth itself. Nearly all the land is collected in one of its hemispheres, and much the larger portion of the sea in the opposite.[15]

A year later the respected Cambridge geologist and philospher of science William Whewell also granted provisional credence to Hansen's "highly entertaining suggestion."[16]

Hansen's theory did not rely entirely on the abstruse demonstrations of theoretical mathematical astronomy, however. Acting on a suggestion by the physicist and inventor Sir Charles Wheatstone (1802–1875), Warren de la Rue obtained stereoscopic photographs of the Moon at epochs of its extreme eastern and western librations in longitude. These images suggested to the photographer that the ray system around Tycho consisted of "ridges and furrows,"[17] and more importantly, as noted in an article in the *Cornhill Magazine* (probably by Sir John Herschel himself), "no one who has seen Mr. De La Rue's beautiful stereoscopes of the full Moon... can fail to have been struck by the marked and undeniable deviation from the spherical form... It is quite obvious, in a certain mode of presenting the images to the eyes, that, were it really a solid object so presented to our view, no one would hesitate to pronounce it rather egg-shaped than spherical."[18]

Measures of de la Rue's images by Matvei Gussew, Director of the Imperial Observatory at Wilna (then part of Poland, now Vilnius, Lithuania), indicated the "apparent anomaly of figure" was indeed real, and in close agreement with Hansen's calculations. However, Gusev suggested that instead of conforming to an ellipsoid, "the portion of the Moon... turned toward the Earth may be considered as a continuous mountain mass, in the form of a meniscus lens, capping the sphere of the Moon, and rising in its middle to a height of about fifty-nine English miles above the general level of its figure of equilibrium."[19] This made the difference between the Moon's geometric center and its center of gravity twice as great as Hansen had derived, and accounted even better "for the total absence both of air and of water on the side of the Moon turned towards us, and would be quite compatible with the abundance of both, and of a habitable hemisphere, on the opposite side."[20]

The Hansen-Gusev theory rushed in to fill a void which had existed when Beer and Mädler had produced their seemingly definitive account of the Moon. Their survey, of course, had applied only to the visible hemisphere; necessarily they had not concerned themselves with the lunar antipodes. Indeed, their map bore the explicit inscription *Mappa selenographica totam Lunae hemisphaeram visibilem complectens*. While the visible hemisphere of the Moon was a barren wasteland, its far-side held out at least slender hopes of an utterly different habitable world.

Just as the antique maps of the Earth had pushed the domain of sea-monsters and other wonders to the bordering regions outside the margins of the known world, so Beer and Mädler's map had banished oceans, air, and possible inhabitants from the Moon's visible side. Long ago, the Greek historian Herodotus had related the belief that the northern parts of Europe, off the limits of the maps of his day, were "much richer in gold than any other... and that the one-eyed Arimaspi purloin it from the griffins." He expressed some incredulity about such stories, but added: "It seems to be true that the extreme regions of the Earth... produce the things which are rarest, and which men reckon the most beautiful."[21] This, of course, has been the ardent hope of all far adventurers. The wish was now father to the wonder-world on the far side of the Moon.

Webb glanced at "the features of the averted hemisphere, on which, as

G[ruithuisen] and Hansen have suggested, other relations may exist."[22] And the theory inspired a memorable passage in Jules Verne's romance *Around the Moon* (1865). Rounding the back side of the Moon cloaked in darkness, his intrepid lunar voyagers see it briefly illuminated by incandescent meteors thrown from an erupting volcano—Verne was nothing if not topical. During the brief illumination of the lunar landscape, they glimpse

> like the fitful flashes of a thunderstorm immense spaces, certainly not arid plains but seas, real oceans, vast and calm, reflecting from their placid depths the dazzling fireworks of the weird and wildly flashing meteors. Farther on, but very darkly as if behind a screen, shadowy continents revealed themselves, their surfaces flecked with black cloudy masses, probably great forests.[23]

Astronomers had long wondered where the Moon's vital atmosphere had fled, and what had befallen those considerable oceans which had scoured the features of the visible face with diluvial action. Phillips, as we have seen, had tentatively identified traces of sculpting by a lunar atmosphere in the central peaks of Gassendi and Copernicus. His view was seconded by the French astronomer Jean Chacornac (1823–1873), who had seen in the fragmentary walls of Fracastorious, extending "like isolated headlands" into the Sea of Nectar, "profound effects of erosion." Chacornac had further argued that the shorelines of such features contained sedimentary deposits—thereby proving that the Moon's oceans once stood at much higher levels than the present arid plains.

Inevitably, like so many things bright and strange, the far-side Moonworld did not for long survive critical examination. A more reflective John Herschel, baited by Richard Anthony Proctor, acknowledged by 1870 that a significant atmosphere on the far side of the lunar globe ought to be distinctly visible along the rim of the Moon at extreme librations.[24] But it had never been detected. It became abundantly clear, moreover, that even if the Moon's presumed lopsided shape had drawn its atmosphere to the far side, the force of lunar gravity was too feeble to hold it there. In the end, a re-examination of the motions of the Moon by Simon Newcomb undermined even the theoretical underpinnings of Hansen's hypothesis by demonstrating conclusively that the Moon was at least as spherical as the Earth—and probably more so.[25] Old Hansen, unable to follow Newcomb's logic, refused to give up, but other astronomers conceded the far-side Moon to realms of fantasy. Thus W. R. Birt would write in 1871: "The question of a lunar atmosphere or the existence of a lunar flora are not worth agitating in 'our' columns, which may be better filled with the publication of *facts*."[26]

And yet despite the downfall of the Hansen-Gussew theory, the idea that the Moon might not be entirely dead, in a geologic sense, at any rate, was clearly gaining ground. This concern was once more foremost in the consciousness of many astronomers concerned with the Moon, who increasingly perceived the need for an improved map of the lunar surface to rigorously test its implications. In March 1862 Phillips presented to the British Association for the Advancement of Science (BAAS) "Suggestions for the Attainment of a Systematic Representation of the Physical Aspect of the Moon." He began by pointing out that even the map of Beer

Figure 10.4 The noted meteorologist and selenographer William Radcliff Birt, shown beside the pier of the Victoria Park Observatory in 1867. The most influential British student of the Moon of the 1860s and 70s, Birt was a proponent of the idea that the surface features of the Moon were still undergoing minor changes. He organized a committee to map the Moon on an unprecedented scale, but the project floundered and was never completed—perhaps largely because it was a project by committee. Courtesy of Peter Hingley, librarian of the Royal Astronomical Society.

and Mädler met the current needs of selenography only about as well as maps of England of the last century satisfied the requirements of physical geography. He called for renewed efforts at mapping the Moon's surface. "In the same proportion as the great one-inch Ordnance Map of 1862 is superior to the old Chart of 1800," he urged, "so should be the new drawings of the features of the Moon to the older delineations."[27]

Phillips had argued the need for such work to be accomplished with one instrument, which might be used in round robin fashion by several observers seriously interested in contributing to selenography. He offered his own 6.2-inch Cooke refractor to the cause in the hope that it might "become the property of some scientific body constituted for long endurance." In addition to his activity within the BAAS, Phillips was also soliciting the support of the Royal Society. He submitted a request for a grant for the telescope to General Edward Sabine (1788–1883), Secretary of the Royal Society and Chairman of the Government Grant Committee. Sabine acknowledged "the importance of giving a new and well considered impulse to Selenography," and recognized "the value of your opinions on... points of detail on which you have obviously thought so much."[28] Nevertheless, the Committee declined Phillips's first request. Later it reversed itself and granted him the sum of £100 to help defray the costs of setting up his telescope in an observatory next to his new house in Oxford, a task completed in July 1862.

Phillips' initiative was taken up by William Radcliff Birt (1804–1881) (Figure 10.4), who soon emerged as the outstanding figure in British selenography. From 1839 to 1843 Birt had served as John Herschel's assistant in reducing the latter's barometric observations made in various parts of the world. After 1843,

Herschel withdrew, and Birt alone continued the investigation, publishing five reports on atmospheric pressure waves in the volumes of the Association for 1844–48. In 1853 he also published a *Handbook on the Law of Storms*. However, from 1859 onward the pioneer meteorologist's interest was almost entirely in the study of the surface of the Moon.

Above all he shared Webb's burning interest in the question: Is the Moon's surface presently in an active or inactive condition? He admitted the possibility that all volcanic activity had long ago ceased—that the largest lunar forms had been the result of the most violent outbreaks while the smaller mountains, especially in the larger craters, indicated "the last expiring efforts of this action." However, he added, this theory would not "satisfy all minds, and accordingly astronomers were not wanting who leaned to the hypothesis that eruptive action still exists, although in a subdued form."[29]

Birt himself was among the most hopeful of the hopeful astronomers who leaned strongly toward the hypothesis that eruptions still took place from time to time. He attempted to draw up a careful list of craters mapped by Schroeter and others which had been missed by Beer and Mädler, "the acknowledged authority in lunar matters," finding no fewer than 368 objects. Many were very small and he readily admitted that while they raised the suspicion of change, "our existing records were inadequate to determine the question."[30] Webb, too, compiled a list.[31] Among Webb's features was a small pit, near the ring of Le Verrier (Beer and Mädler's Helicon A), which he described as "extremely minute, about the *minimum visibile*," in his notebook on May 26, 1836, but recorded as "very conspicuous" on April 20, 1861. (By way of alternative explanation to an actual change, he conceded that he had by then graduated to a larger telescope). Another of his suspicious objects was "a minute, but unquestionable, crater on the summit of the [W.] wall of Helicon"; yet another was a small crater in Mare Imbrium, "more conspicuous than several small craters in the neighbourhood figured by Beer and Mädler."

The vast majority of Birt and Webb's suspects were of this nature—minutiae of the Moon. However, on the south and southeast slope of Copernicus, Webb pointed out a region of greater interest, "thickly studded with very minute craters, not represented by Beer and Mädler" (Figure 10.5). His conclusion shows both the nature and scope of his expectations—widespread eruptions might still take place on the Moon:

> The omission is chiefly remarkable, as [these craters] form a continuation of the extraordinary assemblage of similar foci of eruption lying between *Copernicus* and *Eratosthenes*. This latter wonderful district, it may be observed by the way, has probably assumed its present honeycombed aspect during the present century, as it is hardly conceivable that in such a situation it should have escaped the persevering scrutiny of Schroeter, and been left for the eye of Gruithuisen in 1815.[32]

To settle the matter once and for all required nothing less than a complete remapping of the lunar surface. It had become glaringly obvious that as a basis for discussing the vexed question of change, Beer and Mädler's chart was no less un-

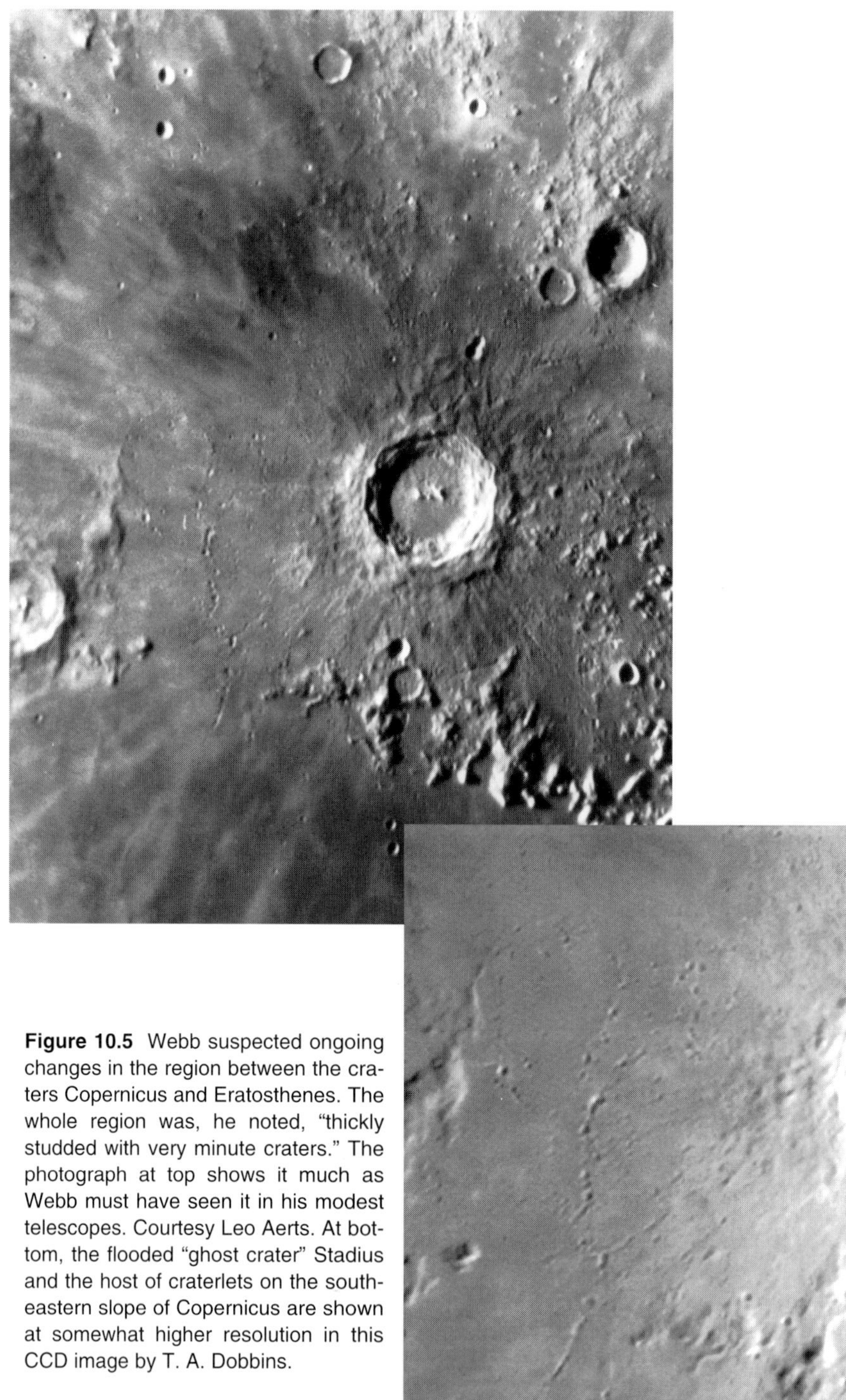

Figure 10.5 Webb suspected ongoing changes in the region between the craters Copernicus and Eratosthenes. The whole region was, he noted, "thickly studded with very minute craters." The photograph at top shows it much as Webb must have seen it in his modest telescopes. Courtesy Leo Aerts. At bottom, the flooded "ghost crater" Stadius and the host of craterlets on the southeastern slope of Copernicus are shown at somewhat higher resolution in this CCD image by T. A. Dobbins.

satisfactory than Schroeter's fragments had been. Recalling his 1859 paper on Cichus and Messier, and regretting that he had not, in the meantime, entered any further into "this curious subject," Webb gave various excuses—not least the perpetual British affliction of cloudy skies. But an equal hindrance had been his growing dissatisfaction with Beer and Mädler's map:

> I have uniformly referred, for the purpose of comparison, to the great map of Beer and Mädler. In doing this, I have been obliged to notice, with concern, occasional indications of a want of that high accuracy which might have been expected from the general character of the work. I have never entertained a doubt as to the correctness of their triangulation, or of the laying down of... primary points... but it has appeared to me that an equal amount of dependence cannot always be placed upon the subsequent filling in... of lesser details. This is much to be regretted... It seems generally admitted that no alteration of any considerable magnitude has taken place upon the lunar surface since the date of anything which can be called accurate observation; it is only in the smallest class of craters... that we can reasonably look for traces of continued activity; and it is precisely *there* that we become sensible of a deficiency in the work of the illustrious German astronomers.[33]

Webb recalled Beer and Mädler's assertion (later retracted by Mädler) that no craters were discernible in Sinus Aestuum. It would never have been made, he declared, if only they had consulted Lohrmann's chart. In this and many other cases, Webb believed that the accuracy of their map suffered "from their unwillingness to be under obligation to [others]."[34]

It seems, from our remote vantage point, that Webb protested too much. And yet the tone of his comments serves to underscore how great had been the aura of infallibility which had so long surrounded the work of Beer and Mädler. For thirty years their map had been the last word. Now active lunar observers such as Webb and Birt were discovering what should have been apparent all along—Beer and Mädler had not mapped, once and for all, the lunar surface, even to the extent allowed by their limited means. Though theirs was the highest standard available, it was still a human standard, attained in a comparatively short time; it demonstrated all the failings and imprecisions of their methods of visual observation and draftsmanship and attested to the small aperture of their telescope. "It is the unfortunate want of confidence" in Beer and Mädler, Webb lamented, "which throws some doubt upon the evidence of change... and renders an appeal to the future still necessary to render it conclusive."[35]

To define the limits of previous work was, however, an essential part of preparing the groundwork for the future, which seemed bullish—and British. The BAAS responded favorably to Birt's proposal by appointing a new "Committee for Mapping the Surface of the Moon," with Birt himself serving as secretary. The other members included Phillips, Sir John Herschel, Warren de la Rue, Lord Rosse, and Webb. Birt's plan of action called for a collaborative and coordinated venture on a very ambitious scale, the goal being nothing less than to produce a detailed lunar map on a scale of 100 inches to the diameter of the Moon—immense compared with the 37.5 inch diameter of Beer and Mädler's map—to be achieved by dividing the Moon's entire visible surface into the four quadrants (1-4) of Beer and Mädler; each quadrant in turn to be subdivided into sixteen "grand divisions," designated by capital letters from A to Q; each "grand division" further subdivided into 25 areas of 5° square (denoted by Greek letters α to ω; the last space being left blank). Any object within a specified region was to be distinguished by a number given together with the quadrant and symbol for the small 5° area. Thus, for instance, the designation IAσ40 would indicate the 40th object in area IAσ.

Any collaborative effort takes time to plan and coordinate. One recalls the "Celestial Police" who attempted to organize at Schroeter's observatory in 1800 for the purpose of prospecting trans-Martian space for small planets. Birt's project was conceived on a grand scale, so it would take much time to complete, especially since it involved the efforts of so many observers. Using photographs provided by de la Rue and Lewis Rutherfurd, Birt prepared outline maps of three 5° zones, located between 0° and 6°W and 0° and 10°S, corresponding to his regions IVAα through IVAγ. He had these outline maps printed and distributed. As a further encouragement, he received on loan a 7.3-inch (185-mm) refractor from Edward Crossley, a wealthy amateur astronomer at Halifax, which was to be used exclusively for the work of the Moon Committee. Birt set it up at "Cynthia Villa," his private observatory at Walthamstow, northeast of London.

Thus began the Moon Committee's great project to scrutinize "lunar objects suspected of change." It was a carefully planned enterprise, promoting the establishment of a basis of observation of the physical aspect of the Moon which, in true Baconian spirit, was to be "laid broad and deep," its "superstructure... characterized by accuracy and precision," so that the results arrived at might be "beyond dispute and fully capable of testing any question that may arise as to the state of the Moon's surface."[36]

The trap was laid, the bow was strung. Under Birt's enthusiastic leadership, the Moon Committee set out on its ambitious quest for the Grail of incontrovertible lunar change.

References

1. Rudwick, *Great Devonian Controversy*, p. 40.
2. Ibid., p. 37.
3. Charles Darwin, *The Autobiography of Charles Darwin and Selected Letters*, ed. Francis Darwin (New York: Dover, 1958 reprint of 1892 edition), p. 26.
4. Webb, *Celestial Objects for Common Telescopes*, Vol. 1, p. 79.
5. Ibid., p. 101.
6. Ibid., p. 81.
7. Ibid.
8. T. W. Webb, "Notice of Traces of Eruptive Action in the Moon,"*Monthly Notices of the Royal Astronomical Society*, 19 (1859), pp. 234–235.
9. Ibid., p. 234.
10. Proctor, *The Moon*, pp. 183–184.
11. Webb, *Celestial Objects*, p. 79.

12. Simon Newcomb, *The Reminiscences of an Astronomer* (Boston: Houghton, Mifflin and Co., 1903), p. 315.
13. P. A. Hansen, "Sur la figure de la lune," *Memoirs of the Royal Astronomical Society*, 24 (1856), pp. 29–90.
14. Ibid., p. 31ff. For a valuable discussion, see Daniel A. Beck, "Life on the Moon? A Short History of the Hansen Hypothesis," *Annals of Science,* 41 (1984), pp. 463–470.
15. Sir John Herschel, *Outlines of Astronomy* (London, 1858), para. 436 a–b.
16. William Whewell, *Of the Plurality of Worlds, An Essay* (London, 5th ed. 1859), pp. 412ff.
17. Warren de la Rue, *BAAS Report for 1861*, Report on the Present State of Celestial Photography in England.
18. [Sir John Herschel], "Figure of the Moon and of the Earth," *Cornhill Magazine*, 6 (1862), pp. 548–551.
19. M. Gussew, "Über die Gestalt des Mondes," *Bulletin de l'Académie des Sciences de St.-Petersbourg* (1860); see ibid., p. 549.
20. Ibid.
21. Herodotus, *History*, 3, p. 116.
22. Webb, *Celestial Objects*, Vol. 1, p. 78.
23. Jules Verne, *Around the Moon* (New York: Airmont Publishing, 1963), p. 163.
24. R. A. Proctor, "Note on Mr. Plummer's Reply," *Monthly Notices of the Royal Astronomical Society of London,* 23 (1873), 419–421: 421.
25. Simon Newcomb, "On Hansen's Theory of the Physical Constitution of the Moon," *American Association for the Advancement of Science Proceedings,* 17 (1868), 167–171. Newcomb was not without his own shortcomings. Shortly before the Wright brothers took to the skies at Kitty Hawk, he "proved" that heavier-than-air flying machines were an impossibility. His arguments consisted of a curious admixture of correct mathematics and incorrect assumptions.
26. W. R. Birt, "Lunar Atmosphere and Vegetation," *English Mechanic*, 14 (1871), 248.
27. John Phillips, "Suggestions for the Attainment of a Systematic Representation of the Physical Aspect of the Moon" (March 20, 1862), *Proceedings of the Royal Society of London*, 12 (1863), 31–37:31
28. Edward Sabine to John Phillips, March 1, 1862; Oxford University Museum, quoted in Hutchins, "John Phillips,"p. 211.
29. Birt, "On methods of detecting changes on the Moon's surface," *British Association for the Advancement of Science, Report* 34 (1864), 4.
30. Ibid.
31. Webb, "On Certain Suspected Changes in the Lunar Surface," *Monthly Notices of the Royal Astronomical Society*, 24 (1864), 201–206.
32. Ibid., p. 204.
33. Ibid., pp. 201–202.
34. Ibid., p. 202.
35. Ibid.
36. Birt, "Report of the Lunar Committee for Mapping the Surface of the Moon", *Report of British Association for the Advancement of Science*, 38 (1868), 1–45:1–2. See also Birt's reports in *Report*, 35 (1865), 286–308; *Report*, 36 (1866), 214–281, and *Report*, 37 (1867), 1–24 and *Report*, 39 (1869), 76–81.

Chapter 11: Linné

In the standard histories of selenography, the period from 1838 to 1866 is generally regarded as a wasteland. Though it is true that a few observers, such as Phillips, Webb, and Birt, continued to take a keen interest in the Moon, on the whole the period has been characterized as one of paralysis and depression.

The views of Beer and Mädler continued to dominate the field. Their remarks about the changelessness of the lunar surface were, as Edmund Neison so aptly put it, sufficient to "crush in a way all lunar work."[1] The Moon was without oceans or rivers, snow or rain; there were neither clouds nor an appreciable atmosphere—if indeed there was any atmosphere at all. Nor was there any life, animal or vegetable, at least of the kind known to us. Neison declared:

> Approaching the study of the Moon with strongly preconceived notions that the surface is a mere arid desert, nearly red hot at times, and almost immeasurably cold at others, without water, air, or life, the real condition of the Moon is not such as to dissipate these views at first sight. Its cold, still, apparently unchangeable surface, so utterly unlike what the Earth might be supposed to appear as seen from the Moon, convinces the casual observer that the world he then sees is utterly unlike the world he knows. He looks for immense cloud-masses floating in a dense atmosphere, and sees none; for wave-tossed seas and winding broad rivers, and there are none; searches for luxuriant forests and green prairies, and they are absent. This is enough, and he retires from further contemplation of the "airless, waterless, lifeless, volcanic desert" of the text-books.[2]

Perhaps there was a shadow of a doubt about these conclusions. The limitations of Beer and Mädler's map were becoming only too readily apparent, and their convictions about the changelessness of lunar conditions were being actively challenged, by mid-century, by revisionist selenographers such as Webb and Birt.

This revisionist school contributed greatly to breathing the life back into selenographical studies. Another factor, however, in the revivification of the Moon as an object of active research was instrumental, and involved a breakthrough in the making of reflecting telescopes that would prove no less revolutionary than the work of Fraunhofer had been for the refractor half a century earlier.

In 1856–57, the Munich optician Carl August von Steinheil (1801–1870) and the French physicist Léon Foucault (1819–1868) independently produced the first examples of silver-on-glass reflecting telescopes by employing a reaction that had been described by the chemist Justus von Liebig (1803–1873) two years ear-

Figure 11.1 Amateur telescopes of the second half of the nineteenth century. (Left) A silver-on-glass Newtonian reflector on an altazimuth mounting by the English maker George Calver. From T. E. R. Phillips and W. H. Steavenson, *Splendour of the Heavens* (1925). (Right) A more elaborate Calver reflector on an equatorial fork mounting fitted with manual slow motions. From W. F. Denning, *Telescopic Work for Starlight Evenings* (1891). Instruments like these quickly supplanted far more expensive refractors in the hands of most serious British observers and quietly revolutionized amateur lunar studies.

lier. The glass substrate was immersed in a warm ammoniacal solution of silver nitrate and a solution of sugar was added. The sugar acted as a reducing agent, slowly depositing an infinitesimally thin, uniform layer of metallic silver onto the surface of the glass. Gentle burnishing of this fragile coating with a piece of rouge-impregnated chamois resulted in a brilliant mirror that reflected half again as much light as one made of speculum metal. More importantly, although silver tarnished even more rapidly than speculum metal, particularly in the sulfurous air of coal-burning towns, a deteriorated silver coating could simply be dissolved with a mineral acid to which the glass substrate was impervious. No longer did the possessor of a reflecting telescope have to laboriously repolish and refigure his speculum at frequent intervals, the fact which had caused the prominent Dublin telescope maker Sir Howard Grubb to lament that "reflectors very seldom do good work except in the hands of their makers." Only periodic re-silvering in a kitchen basin was required, which hardly demanded the skills of an artisan. Moreover, since the glass substrate served only as a support for the layer of silver, it did not have to be free of bubbles and striae as did the far more expensive glass required to make the lenses of a refractor.

Figure 11.2 Two more reflectors from the second half of the nineteenth century. (Left) This 10-inch Newtonian reflector atop a sturdy altazimuth mounting was fabricated by the craftsman John Browning and sold to the eminent British planetary observer William Frederick Denning. From W. F. Denning, *Telescopic Work for Starlight Evenings* (1891). (Right) A huge silver-on-glass reflector developed for the Paris Observatory by Léon Foucault three years before his death in 1868. Despite their widespread success in the hands of amateurs, reflectors failed to catch on among professionals for another half century. From W. H. Warren, *Recreations in Astronomy* (1886).

Unlike Steinheil, Foucault was not a professional optician interested in protecting trade secrets, and he published detailed descriptions of his simple but remarkably accurate methods of figuring and testing mirrors. With the benefit of Foucault's techniques, in a few years English craftsmen like George With, George Calver, and John Browning were offering good Newtonian reflectors at a small fraction of the price of comparable refractors (Figures 11.1 and 11.2). Soon the pages of popular journals like *The English Mechanic* and *World of Science* were awash in reports by the rapidly growing ranks of amateur astronomers equipped with telescopes that only a few years earlier would have been unheard of in the hands of anyone except professionals or wealthy dilettantes. The widespread use of such telescopes led to markedly renewed interest in lunar studies, especially in Britain. Henceforth most of the leading British observers of the Moon used reflectors—Webb, for instance, supplanted his 3.7-inch refractor with a 9.25-inch silver-on-glass reflector made by George With.

The imaginations of many of these observers were captivated by the exciting quest for evidence of lunar changes. But it was always difficult to separate what

Figure 11.3 An 1882 portrait of J. F. Julius Schmidt, arguably the most obsessive visual observer who has ever lived. Although the Moon was his chief specialty, he also produced painstaking records of sunspots, planetary markings, comets, meteors, and even the zodiacal light. During the last two years of his life, Schmidt made no fewer than 85,000 observations of variable stars. Courtesy Wolfgang Bayer.

was new from what was merely a deficiency in the existing database. As a case in point, consider Birt's observations of the ruined crater Hippalus, on the shores of Mare Nubium. With the 4.3-inch (108-mm) refractor of his Cynthia Villa observatory, he noticed, on September 19, 1866, a rille on the floor of the crater. Consulting Beer and Mädler, he found their delineation deficient—"a result by no means rare... in this as in numerous other instances their delineations are very faulty."[3] Indeed, far from being an exceptional case, Birt admitted that "instances of our general ignorance of minute lunar details [must be] numbered by thousands, if not by tens of thousands." Such features are, moreover, highly subject to vicissitudes of visibility, produced by the ever-changing conditions of libration (especially important for regions near the limb), of solar illumination, and of the conditions of our own atmosphere, so that, as Mädler once remarked: "Even without either clouds or fog being distinctly observed, the transparency of our atmosphere is very different according to time and place, and little faint projecting forms may easily become alternately visible or invisible without adducing any other cause than the changeable diaphanism of our atmosphere."[4]

Among all the countless minutiae of the Moon—a stupefying welter of detail which surpasses the capabilities of human skill and endurance to depict exactly—how would it ever be possible to sort out the real changes from the merely apparent? This was the challenge faced by selenographers of the 1860s. It was an appalling task, like looking for a needle in a haystack. Webb and Birt must have been possessed of uncommonly sanguine personalities—eternal optimists. Like the searchers for new planets in trans-Neptunian space who would begin their work a few years hence, they would hardly have had courage to begin had they but grasped the real magnitude of the challenge.

And yet, within only two months of Birt's paper on Hippalus came the dramatic announcement of Johann Friedrich Julius Schmidt (1825–1884) (Figure 11.3) that a change—and a change apparently quite definite and convincing—had taken place in a small crater in Mare Serenitatis first recorded by Riccioli; the crater designated by Lohrmann as *A* and by Mädler as *Linné*. The announcement could hardly have rested on better authority. Schmidt was an observer of unquestionable skill and almost superhuman diligence. Indeed, as suggested by W. H. Pickering, Schmidt "perhaps devoted more of his life than any other man to the study of the Moon."[5]

Schmidt was a compulsive observer, a devotee of everything in the heavens—meteors, the Zodiacal Light, variable stars, nebulae, and comets. The Moon, however, was always his *chef d'oeuvre*. In an era when the study of the Moon had become increasingly specialized, the knowledge of its myriad features so vast, that it had led to the formation of a whole Committee of British observers to further its advance, Schmidt—like Dr. Johnson on the great English Dictionary—worked unaided and alone.[6]

Schmidt had been born at Eutin in northern Germany.[7] At the age of fourteen he chanced upon a copy of Schroeter's *Selenotopographische Fragmente* and was so fascinated by its pictures of mountains and craters that he was then and there determined in the future direction of his life. At once he began to study the Moon himself, using a small telescope with lenses ground by his father and availing himself of a ready-to-hand lamp-post to steady the instrument—hardly more than a spyglass—and took the first halting steps of his lifelong preoccupation with sketching the Moon.

His first glimpse of the Moon through a good telescope came in July 1841, when A. C. Petersen, director of the Altona Observatory near Hamburg, showed him the imposing craters Bullialdus and Gassendi. He also saw for the first time a copy of Mädler's chart. Soon afterward Schmidt moved to Hamburg and for several years made frequent observations with the various telescopes of the Altona Observatory. There was a strange interlude in 1845, when he accepted a position at the private observatory of J. F. Benzenberg at Bilk, near Düsseldorf. Benzenberg was preoccupied with the search for a possible intra-Mercurial planet and did not allow Schmidt to use his large refractor, apparently for no better reason than that "its outward good looks and polish might not suffer by handling." Instead, Bilk gave Schmidt access to a "wretched instrument." After a few months, Schmidt left Bilk in disgust and took a position under Friedrich Wilhelm Argelander (1799–1875) at the Bonn Observatory. Although most of his time was taken up with entering meridian-circle observations of stars for Argelander's great star catalog, the *Bonner Durchmusterung*, he made as many lunar observations as he could. He also made occasional visits to Berlin, where Galle and Bruhns of the Royal Observatory generously put at his disposal the 9.6-inch (244-mm) Fraunhofer refractor with which Galle had made the celebrated discovery of Neptune in 1846. This was the same instrument Mädler had used to observe the Moon prior to his departure for Dorpat.

In 1853 Schmidt left Bonn for E. von Unkrechtsberg's observatory at Olmütz in Moravia, where he made a series of 3,000 measurements of the heights of lunar mountains with a filar micrometer (Figure 11.4). This work was published

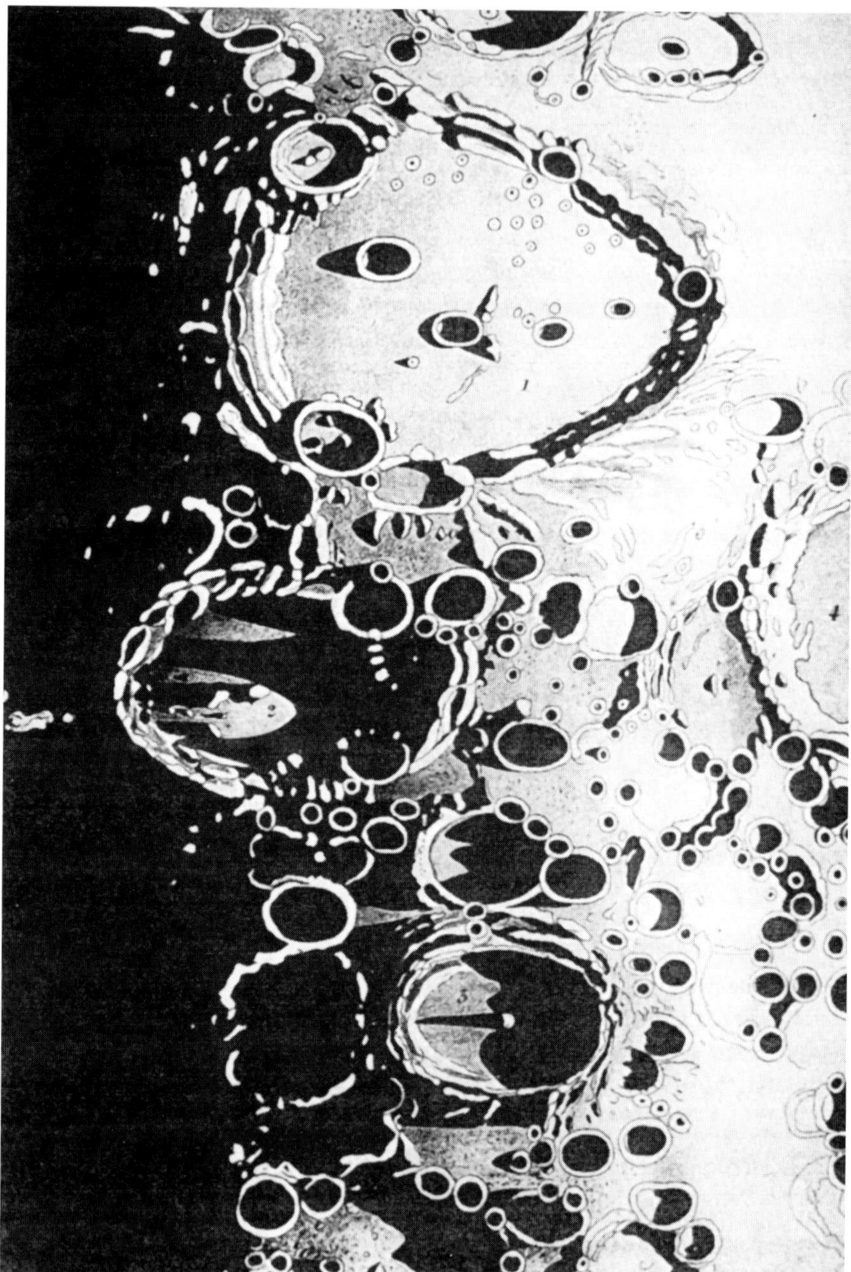

Figure 11.4 This drawing of the Moon, made by Schmidt at E. von Unkrechtsberg's observatory at Olmütz in Moravia, served as the frontispiece of his 1856 book *Der Mond*. It shows the lunar terminator near the imposing crater Clavius (top center) in the rugged southern highlands. Sheehan's collection.

Figure 11.5 The Athens Observatory, on the famed Hill of the Nymphs, as photographed from the Acropolis. Schmidt began work here in 1858. Eight years later, armed with the observatory's 6-inch refractor, he made the observations that led him to announce a dramatic change in Linné, an obscure crater in Mare Serenitatis. 1999 photograph by William Sheehan.

in a treatise entitled *Der Mond*, in which Schmidt attempted to provide a quantitative comparison of lunar and terrestrial features. Judiciously, he warned against taking too seriously the apparent similarities between the Moon and the Earth. He made observations of the Moon in March 1855 with the great refractor at Rome and again in April at the Naples Observatory. Finally, on December 2, 1858, he assumed the directorship of the Athens Observatory in Greece (Figure 11.5), where he would remain for the rest of his life.

When he set foot on Greek soil at Piraeus, Schmidt was still a comparatively young man, full of energy. Characteristically, he at once surrendered to his compulsion to collect scientific data and began to take barometric readings. Arriving at the observatory, he found it in a state of disrepair and neglect. However, within only a year he was able to restore to working order a fine 6-inch (152-mm) refractor by the Viennese optician Georg Plössl, which would serve as the main instrument for his lunar work for the next quarter of a century.

By 1865 Schmidt had assembled so many lunar observations that he began laying down his surveys of selected regions on a six-foot map. The next year he began to construct a one-meter map based on Lohrmann's observations. These had been entrusted to Schmidt by Lohrmann's publisher in the hopes that Schmidt would eventually see Lohrmann's map through the press. Schmidt was faithful to his charge, and the complete Lohrmann map finally appeared in 1878, 38 years after Lohrmann's death.

At first Schmidt planned to enter details from his own observations onto the

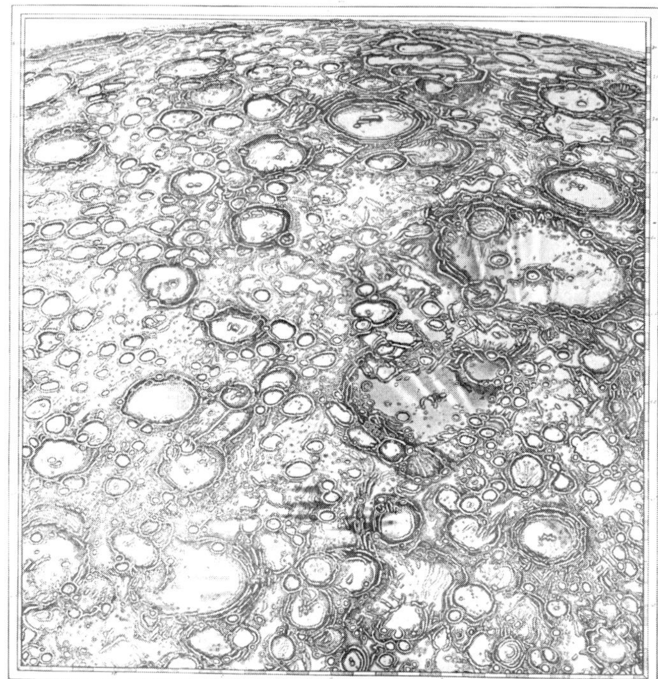

Figure 11.6 Sheet 23 of Schmidt's map. The crater Clavius is located at the upper right. Courtesy Charles A. Wood. Compare the level of detail in Schmidt's map, recorded with a 6-inch refractor, with the high-resolution photograph made with a 20-inch Cassegrain reflector by the French astrophotographer Georges Viscardy. Courtesy Georges Viscardy.

Lohrmann map. However, he soon abandoned this approach; instead he embarked on something far more ambitious—nothing less than a fresh topographic map of the Moon, on a scale of 6 feet, 6 inches to that of the Moon's diameter, to be, like Lohrmann's original design, divided into 25 sections. On this map Schmidt dreamed of rendering, insofar as possible, all the details of the lunar surface visible through his 6-inch refractor (Figures 11.6, 11.7 and 11.8).

While involved in this heroic series of observations—which he eventually realized would require "more powers of endurance and a longer lifetime than are allotted to mortals"—Schmidt made an observation which would have an almost seismic effect, and shake the world of selenographical studies to its foundation. He had, of course, been kept informed of the efforts of Birt and his associates, and actively supported the projects of the Moon Committee. He was well aware that its main objective was to establish the existence of changes on the Moon. If it is true that nature favors the prepared mind, then Schmidt's mind could hardly have been better prepared. The moment of revelation came on the evening of October 16, 1866. The Moon was eight days old, just past First Quarter, with the terminator passing through the Mare Serenitatis, one of the vast plains of the Moon and according to Mädler, a region of subtle greenish tints visible only under the most transparent atmospheric conditions. Remarkable for the uniformity of its surface, it was "like a sea of sand, and devoid of large craters."[8] In the southern part of the mare lay the prominent crater Bessel, along with several smaller craters, including a whole train of them on the ray extending across the plain from the direction of the crater Menelaus; also, to the southwest, another fine crater, Sulpicius Gallus; finally, the crater to which Mädler had bequeathed the name Linné. Since the latter was then near the terminator, it ought to have been distinctly visible under oblique illumination. Indeed, Schmidt recalled that under similar circumstances he had never before found it a difficult object.[9] It was now a mere whitish patch (Figures 11.9 and 11.10). He was convinced that the crater, hitherto so prominent, had, in fact, disappeared.

After confirming the observations at the next lunation in November, Schmidt, the German, wrote to Birt, the Englishman, in French:

> I find that a crater of the Moon is no longer visible… I have recorded this crater ever since 1841, when I found it readily apparent even at full Moon. In October and November 1866, when it should have been optimally visible—one day after the sun had risen above its local horizon—this deep crater, with a diameter of 5.6 English miles, was entirely extinct. Instead there was only a small whitish cloud in the place of Linné.

He appealed to Birt: "Would you please kindly make observations of this locality?"[10]

Schmidt maintained that until October 1866, Linné had always been recorded as a distinct crater about six miles in diameter and "very deep." Near the terminator it was "more or less overshadowed," while under high illumination it appeared as a "spot of light." His review of earlier records revealed that Schroeter had drawn it, once, in 1788, and Lohrmann had found it a "very deep" crater, the

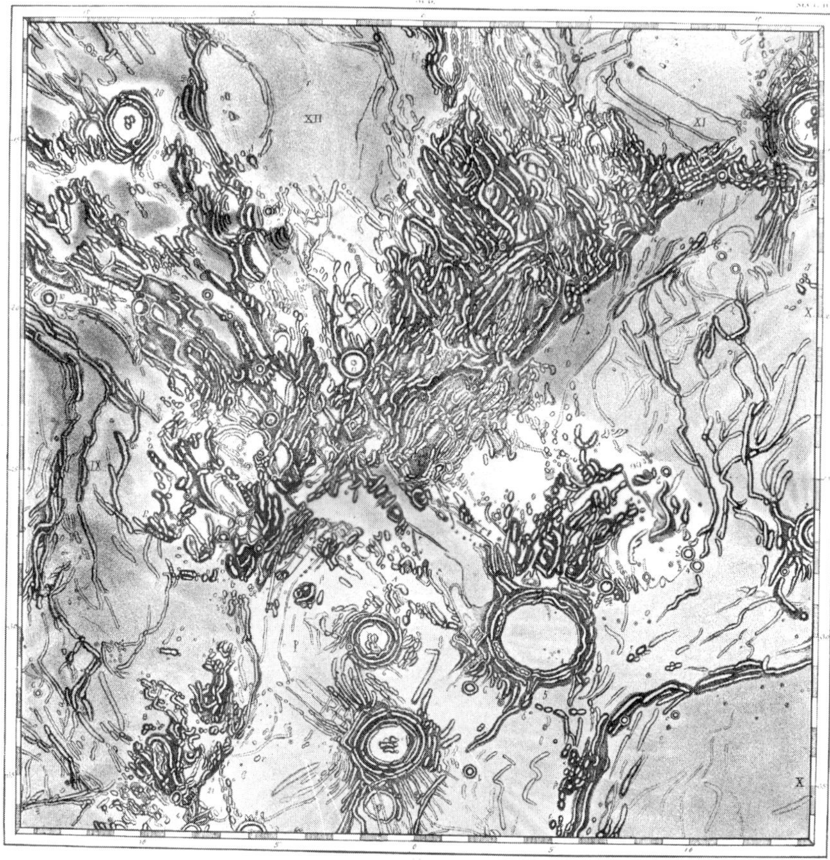

Figure 11.7 Sheet 4 of Schmidt's map, centered on the lunar Apennines. The large crater at the lower right is Archimedes; the crater Aristillus lies to its lower left. Courtesy Charles A. Wood.

third most prominent feature on the entire plain of Mare Serenitatis, in 1823. Because of its prominence under all illuminations—near Full Moon it could still be seen as a "bright spot"—Lohrmann had chosen it as a reference point for charting this part of the Moon. In *Der Mond*, Mädler had similarly described it as a "deep crater... indefinitely bounded at full Moon," and had chosen it as one of his "points of first order."

On examining his own extensive records made since 1840—more than 1,200 drawings in all—Schmidt found that he too had drawn Linné as a distinct crater on no less than five occasions between April 27, 1841 and August 17, 1843. Admittedly, he was then an inexperienced novice still in his teens, the best instrument at his disposal being a small Dollond refractor magnifying only about 20X. But the testimony of his own records tallied perfectly with those of Lohrmann and Mädler. Under the circumstances, Schmidt's conclusion seemed inescapable:

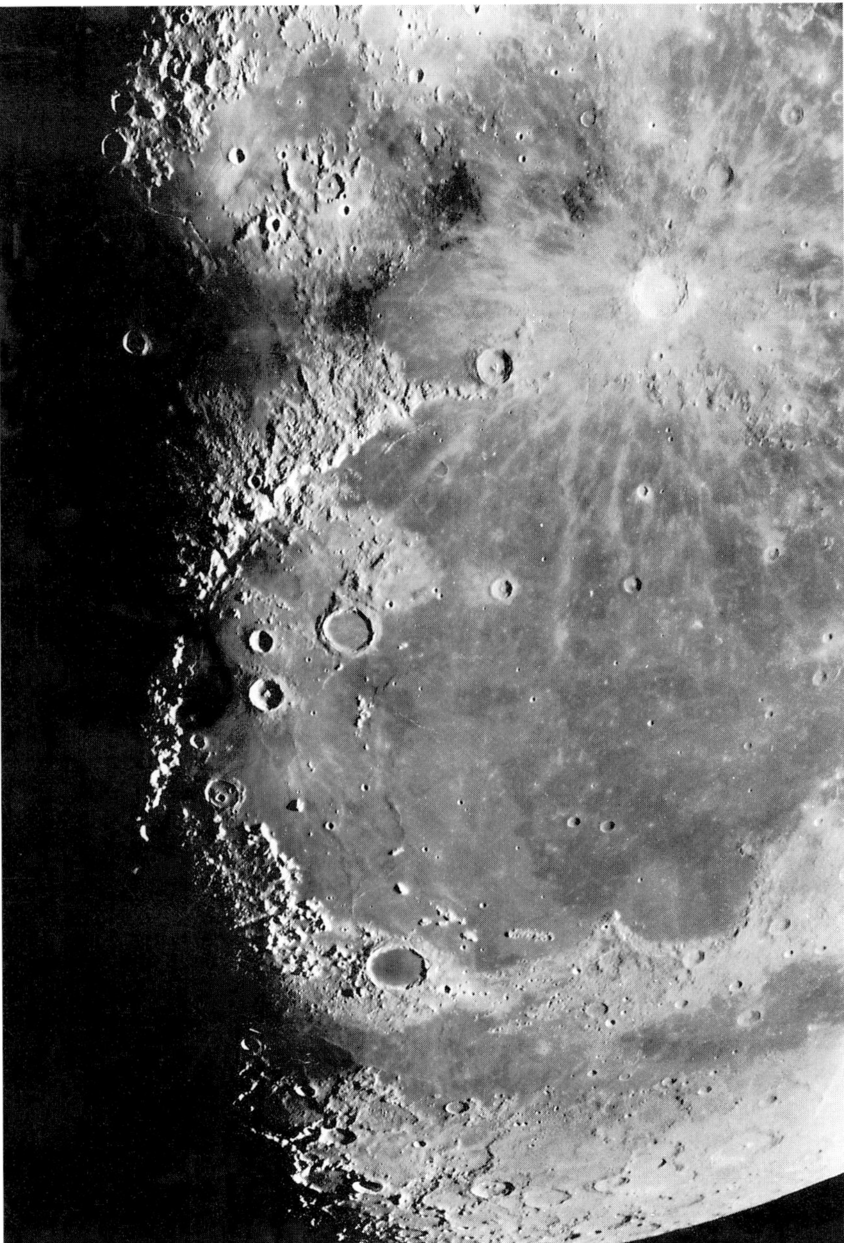

Figure 11.8 This photograph of the lunar Apennines at sunset by E. M. Carreira, S.J., using the 16-inch Zeiss refractor at the Vatican Observatory's Castel Gandolfo station, bears comparison with the corresponding section of Schmidt's map in Figure 11-7. Courtesy E. M. Carreira, S.J.

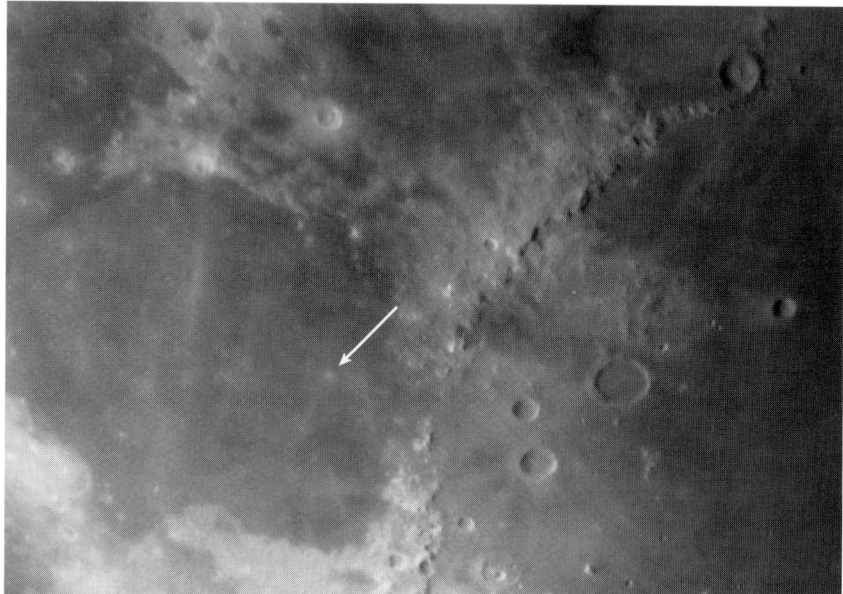

Figure 11.9 Under high lighting angles, Linné takes on the appearance of a diffuse white patch. Photograph by E. M. Carriera S.J., with the 16-inch Zeiss refractor at the Vatican Observatory's Castel Gandolfo station.

"One of the craters of the Moon exists no longer."[11]

Birt, the first in England to learn of the startling results, was only too eager to comply with the request from Athens for more observations, and immediately enjoined members of the Moon Committee to scrutinize this hitherto unremarked and unremarkable object on the plains of the Mare Serenitatis. A photograph by Buckingham taken on November 8 was examined for traces of Linné; there it was, visible but faint.

At the next lunation, when the terminator once again passed over the western boundary of Mare Serenitatis, numerous observers stood on the *qui-vive*. On December 13, Webb, with his 9.25-inch (235-mm) silver-on-glass reflector, found Linné "an ill-defined whiteness"; another observer, Talmage, using a 10-inch (254mm) refractor at Leyton, near London, recorded "a dark circular cloud." Birt and Dr. Robert Mann were prevented by thin cirrus from seeing any trace of Talmage's "cloud" on December 13, though Birt added that "our impression was that Linné ought certainly to be visible under the circumstances… No trace of Linné was visible, but… where it should be there was a faint circular cloud-like spot ('*petit nuage blanchâtre*' of Herr Schmidt)."[12] The next night, using Crossley's 7.3-inch (185-mm) refractor in London, Birt again found Linné as Schmidt had described it, "a diffuse patch of light… The *crater* is certainly obscured. No shadow visible; nor could I see into the crater."[13] Buckingham first glimpsed a tiny dark point within the larger bright cloud. Further observations by Birt showed that

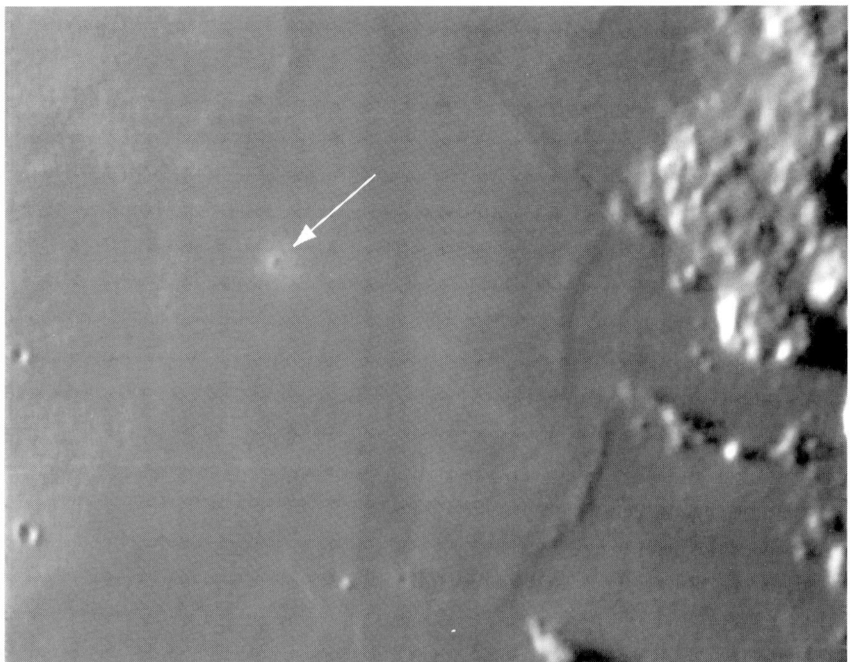

Figure 11.10 This image of Linné records the tiny crater pit first glimpsed by the British observer Buckingham using a 12-inch silver-on-glass reflector. Image by T. A. Dobbins.

the spot seemed to be of variable size, contracting but growing brighter with increased illumination.[14]

The British were not the only ones interested in what was going on in Linné. Two German professors, Förster and Tietjen, studied the Moon with the Berlin refractor on December 20, "in order," they wrote to Birt, "to convince ourselves of the disappearance of the crater Linné." Though they found the crater distinctly visible, they affirmed the reality of the change by appealing to Schmidt's authority. "An obscuration has taken place on which certainly no doubt can exist, as it is affirmed by so competent an authority as Herr Schmidt of Athens."[16]

For Birt, Schmidt's observation was the realization of the long-sought Grail of selenograpy. He was convinced that a change—and a change on an impressive scale—had at long last been identified on the surface of the Moon. "The long and indefatigable exertions of our highly honoured friend Julius Schmidt," he announced, "have... been crowned by a result for which even Mädler, although he did not resign the hope, remarked... that all the labours hitherto spent upon that object had led to no positive result."[17] No less laudatory of Schmidt's achievement were remarks by the eminent geologist Wilhelm Karl von Haidinger (1795–1871) before the Imperial Academy of Vienna: "Incontestable now is the fact that changes are taking place in the figure of the lunar surface which are not apparent, but real."[18]

As 1867 opened, telescopes everywhere were pointed to the bright cloudy patch on the greyish-green plain of Mare Serenitatis. Naturally Schmidt was in the vanguard, continuing his careful observations with the Athens refractor. As on previous occasions, he had no success in making out any signs of the former "deep crater." On two particularly steady nights, he did glimpse within the whitish spot "an extremely fine black point," which he estimated to be barely 2,000 English feet in diameter. He wrote:

> It is impossible that Lohrmann and Mädler should have drawn Linné as a sharp crater, and even have selected it as a point of the first order, if it had at the time shown its present character. In four lunations I have convinced myself, both in waning and waxing moons, *by the most careful and repeated observations, that Linné can never be seen under any illumination as a crater of the normal type.*

But if, as Schmidt alleged, Linné had indeed disappeared, what was the nature of the change? As a recent eyewitness of volcanic eruptions on the Aegean island of Santorin, Schmidt himself proposed that the lunar crater had been filled in by a similar "eruption of fluid or powdery material."[20]

By now an extensive correspondence on the "alluring subject"[21] had opened in the *Astronomical Register*, a journal to which both Webb and Birt contributed regularly. Henry Cooper Key at Stretton Rectory, Hereford, using a 12-inch (305-mm) Newtonian on January 12, 1867, reported: "The air was very tremulous... At first sight the appearance was certainly that of a whitish cloud obscuring the crater; but upon long gazing, and using averted vision, I could plainly make out a centre or nucleus, and presently afterwards a marginal ring of perhaps twice the diameter of the original Linné." Key concluded that the obliteration of the crater was "an unquestionable fact," constituting, he believed, "the first well-established instance of change on the lunar surface observed by man."[22] During the next lunation, on the evening of February 10, 1867, Angelo Secchi at Rome watched Linné emerge into the sunlight. The following evening he made out a tiny craterlet, scarcely half a mile in diameter, within the white spot. By April Secchi's craterlet had been confirmed by leading members of the Moon Committee, including the pioneer spectroscopist William Huggins (1824–1910) at his private observatory at Upper Tulse Hill near London, and elsewhere by at least two well-known professional astronomers at the Copenhagen Observatory, Heinrich d'Arrest (1822–1875) and Hans Carl Schjellerup (1827–1887).[23]

A series of highly detailed observations was reported by the Italian astronomer Lorenzo Respighi (1824–1889), who used a 4.5-inch (114-mm) refractor at the Campidoglio Observatory, atop the Capitoline Hill in Rome. On April 10, he recognized "in the white spot, and in the contiguous shadow... a small, well-defined crater." On May 9, he described Linné as having its "edge brighter and more raised on the western side, and the bottom completely dark." This profile accounted for the fact that the craterlet was scarcely visible except near the terminator, and more easily seen after sunrise than after sunset. Respighi—more cautious than most—concluded that "with regard to the notable changes which it is asserted have taken place in this lunar crater, it seems to me that the arguments are not sufficient to prove them."[24]

That May, when the lighting on Linné was most favorable, the weather was generally poor in England; even so Huggins found the craterlet distinctly visible as a black speck in the midst of the bright halo. From Paris the Moon was scrutinized by Camille Flammarion (1842–1925), an astronomer at the Bureau des Longitudes, who had recently won renown as the author of the popular book *La Pluralité des Mondes Habités*, in which he passionately advocated the existence of extraterrestrial life. On May 10, 1867, Flammarion observed Linné when the Sun had risen only a few degrees above it—observations later summarized in Flammarion's very first paper to the Académie des Sciences:

> And what was most immediately apparent was that Linné is no more a crater. It casts no exterior shadow, forms no shadow in its center. It is only a white circular cloud, or rather whitish spot; it is neither raised above nor depressed below the level of the surrounding plain.[25]

He added that if Linné had shown this aspect when Beer and Mädler had drawn up their *Mappa Selenographica*, they would certainly never have indicated it as a crater. He concluded: "One must believe that our satellite is a world not entirely dead, and that at intervals movements occur on its surface which are appreciable enough to be recorded here."[26]

So far the idea that an actual change had occurred on the Moon had encountered scant opposition. At a meeting of the Royal Astronomical Society in London, on June 14, 1867, Birt attempted to put the successive impressions of several observers into some semblance of order. His reconstruction of events indicated that "the orifice seen on December 14 by Buckingham as a black spot has become so enlarged as to present the true form of a crater." Buckingham, present at the same meeting, concurred: "Permit me to say that, having optical means second to none, I am sure the black spot has been increasing during the latter months."[27]

A benchmark for future comparisons was obtained by Huggins on July 9, 1867, when he succeeded in putting the wires of a filar micrometer on the craterlet. He measured its diameter as 1.71 arc-seconds, from which he calculated a width of 12,067 English feet, slightly more than two miles.[28] Birt noted with satisfaction that "there can be no question that the measures of Mr. Huggins deserve the highest confidence, and I apprehend we have now—that so many good observers are directing their attention to Linné—for the first time in the history of this interesting spot, *a well-determined starting point with which to compare all past and future observations*."[29]

Linné, indeed, remained the object of the most intense scrutiny during the rest of 1867 and into 1868, during which period it continued to present the aspect of a small, sharp-rimmed crater, between one and two miles across, located in the midst of a somewhat variable bright patch. Under grazing illumination just after local sunrise, the crater wall cast a small exterior shadow. As the Sun rose over the feature, the bright patch became more prominent, until by noon it became difficult to make out any trace of the crater. In the late afternoon a crescentic inner shadow formed within the bright patch; and finally, at sunset, when the crater cast its shadow westward, it resembled a very low hemispherical dome—or, as Captain William Noble put it in a memorable phrase, "it looked mammillariform."[30]

In point of fact, there was never any real doubt about the appearance of Linné *after* October 1866. Everything depended on the interpretation of the pre-1866 observations. After Respighi, the first to express skepticism about a change was apparently Warren de la Rue, who had recorded Linné as the usual bright patch on his photographs of February 1858 and October 1865, while Ernst Wilhelm Tempel (1821–1889) of the Marseilles Observatory also urged caution.

> As a whole, the Moon has been completely formed and perfect for thousands of years. Thus the only means of recognising ongoing activity consists of the perception of small changes in form of particular craters or mountains… [And yet] what wonderful changes of form do not the different degrees of illumination often show during the course of only a few hours![31]

Tempel's point was well taken. Considering the daunting complexity of lunar detail, the variable effects of shadow, foreshortening, and libration; adding to all of these the inevitable deficiencies of the selenographic record—for Beer and Madler's map was clearly inadequate for judging apparent alterations as genuine, and yet, as Tempel pointed out, "even a more beautiful and more accurate map could never lay claim to be the best and last"—the surprising fact is that Schmidt's claim of a definite change was so widely and uncritically accepted. In fact, the evidence was always rather weak and circumstantial. When Schmidt observed the crater during the years 1841–43, he was young, inexperienced, and poorly equipped. Even so, his massive archive of drawings contained two drawings showing Linné not as a deep crater but as a bright patch. He himself never called attention to them—probably, as Joseph Ashbrook suggested, for no better reason than that his ability to make observations had outstripped his ability to analyze them.[32] The descriptions of Lohrmann and Mädler had seemed unambiguous. And yet the reality was that Lohrmann's sole measure of Linné's position was made on May 28, 1823, when the Moon was in the waning gibbous phase eighteen days old. Since it was always Lohrmann's practice to sketch the larger mountains and craters under high illumination, his only examination of Linné may have been under a high Sun, when it would have been easy for him to mistake a bright spot for a bright crater floor.[33] Though Mädler's series of measures was made with Linné near the terminator of the waxing Moon, the verbal similarity of his account to Lohrmann's suggests it may not have been independent, and one must also bear in mind the small aperture he employed. Under the circumstances, it would have been only too easy for him to mistake a white spot for a crater—as indeed he did in at least two other cases, Parry B and Alpetragius D. The tiny craterlet within the bright patch, which requires moderately large apertures to be detected at all, and even then will be noticed only when one is specifically on the lookout for it, would have been quite beyond his means.

Then too there was the testimony of the oft-maligned Schroeter, who had rendered the Mare Serenitatis on two occasions. In one of these drawings, made on November 5, 1788, he recorded Linné as "a very small, round, brilliant spot."[34] This description, Schjellerup and d'Arrest remarked, indicated that 78 years earlier Linné had presented "in general about the same appearance as it does now in our large refractor."[35] Though Neison later expressed skepticism about the identity of Schroeter's bright patch and Linné,[36] Schroeter's verbal description—of which, perhaps, Neison was unaware—leaves little doubt.

A record nearly contemporary with Schroeter's was made by John Russell with a seven-foot Herschel reflector on April 12, 1788. Russell represented Linné as a small white spot without any indication of a crater-like depression—in short, just as Schroeter saw it, and in the same way that it appeared in small telescopes after 1866. Unfortunately, Russell's drawings and others made by still earlier observers did not come to light until after the controversy had largely subsided.[37]

It is interesting to note that Mädler was still very much alive when the Linné controversy erupted. Moreover, as historian Richard Baum discovered, Mädler had visited Britain in 1868 to attend the BAAS meeting at Norwich. He confided to Birt: "Respecting the crater Linné. It was a principal point of my trigonometrical network, and consequently I have observed it very often... I have never seen any real change; only optical ones."[38] Nothing could be more definite. Admittedly, his eye was no longer as keen as it once was, and he had just undergone surgery for cataract; nevertheless, as recently as May 10, 1867, he had examined the crater with one of the small telescopes at the Bonn Observatory. "I found it shaped exactly, and with the same throw of shadow, as I remember to have seen it in 1831." He concluded that "the change, of whatever nature it might have been, had passed away without leaving any observable trace by me."[39] Interestingly, Schmidt had observed the Moon that same night from Athens. The entry in his observing log speaks volumes: "L. [Linné] seems very changed... In its place is a conspicuously brighter and shadow-casting hill... Of the very bright grey cloud hardly any trace appears."[40]

In the end, the controversy never would be settled in the nineteenth century. It would linger on—sustained less by the facts, which were always open to differences of interpretation, than by the desire to believe, in Flammarion's arresting phrase, that "our satellite might not be a world entirely dead." Birt, who had so long dreamed of just such a possibility, summed up the Linné affair for the Moon Committee:

> Whatever may be the truth in the midst of the conflicting opinions which have been expressed on the state of Linné (there are no discrepancies in the evidence) it is certain that no lunar spot has received so much attention from so many observers as Linné... It augurs well for the progress of selenography that so great an interest is at present being manifested in the investigation of the physical condition of the surface of our satellite.[41]

This statement was written in 1868. For decades to come, it would continue to hold true. No spot on the Moon would be perused with keener interest or attract so much heated—and in the end inconclusive—debate.

One thing, at least, was certain: Schmidt's allegation of change had routed once and for all the Mädlerian doldrums and made 1866 as much a turning point in selenography as the Norman invasion had made 1066 a turning point in British history. Even after memories of the details of the episode had faded, the myth of a change in Linné would long afterwards haunt the imaginations of selenographers.[42] Birt was certainly galvanized by Linné. In 1870 the British Association for the Advancement of Science had withdrawn its support from the Committee for Mapping the Surface of the Moon, presumably out of dissatisfaction with the slow rate of progress. Birt, however, was more determined than ever to forge ahead. The Linné affair had underscored the inadequacy of current lunar cartography, the need for bigger and better maps. Only in terms of such an improved standard could the reality of change be confirmed.

Mädler was in complete agreement. Even supposing a change of some kind in Linné, he pointed that it was evidently only of a temporary nature, since it had left no trace by the time he re-examined the formation in May 1867. Moreover, the occurrence even of such temporary changes was obviously very rare. "In seven years' occupation with the Moon's surface," he wrote, "I never met with such before, neither have Lohrmann and other careful observers of that time. Since then no doubt can be entertained that in such a case everything depends on being able to criticize how such an object appeared formerly, so might a possibly exhaustive representation and description of all the objects visible to us on the Moon's surface be the fundamental conditions of further intellectual advance."[43]

The possibility that photography might eventually provide such an "exhaustive representation" remained a distant hope. The woefully slow emulsions of that era were able to record only the large, coarse features of the lunar surface—the greater ring-mountains, plateaus, and chains of mountains. But these undoubtedly had formed in the distant past, and changes on such a gigantic scale would not be expected to occur now. Decades would pass before the camera would be able to record features as small as those which an experienced visual observer equipped with a comparatively modest telescope could glimpse. Yet in the all-consuming question of change, Mädler stressed, "everything depends on the most exact representation of these delicate details." In the end, Baconian empiricism—the collection of masses of data, long the chief ambition of selenographers—was not enough.

No; the rich texture of data presented by the lunar surface, if it was not to be abandoned as a mere hopeless jumble, could in the final analysis only be understood by putting it into a comparative context and by reasoning from analogies. Empiricism alone was not enough. But the analogies that lurked in the minds of most of the leading nineteenth century observers were inevitably those furnished by terrestrial surface features. In particular, they remained mesmerized—then and for long afterwards—by the volcanoes of the Earth.

References

1. E. Neison, "Lunar Observations," *The Observatory*, 1 (1877), 238–242:239.

2. Neison, *Quarterly Journal of Science*, January, 1877.
3. W. R. Birt, "Morning Illumination of Hippalus (Lunar Crater)," *Monthly Notices of the Royal Astronomical Society*, 27 (1867), 92–93:92.
4. Baron Von Mädler, "Changes of the Moon's Surface," *British Association Report,* 1868, 514–518:516.
5. W. H. Pickering, *Atlas of the Moon*, introduction, p. 86.
6. It is tempting to recall Samuel Johnson's comment (in Boswell's *Life of Johnson)* on his single-handed completion of the English Dictionary in three years when the French Academy, consisting of forty members, took forty years to compile the French Dictionary. "Johnson: 'Sir, thus it is. This is the proportion. Let me see; forty times forty is sixteen hundred. As three to sixteen hundred, so is the proportion of an Englishman to a Frenchman.'"
7. The account here is based on J. F. Julius Schmidt, *Der Mond* (Leipzig,1856), J. Birmingham, "Schmidt's Lunar Map," *The Observatory*, 2 (1879), 413–415 and 3 (1880), 10–17, and J. Ashbrook, "Julius Schmidt: An Incredible Visual Observer," in *Astronomical Scrapbook*, pp. 251–258.
8. C. Flammarion, "Changement arrivé sur la Lune. Le cratére de Linné," *Comptes Rendus*; reprinted in *Astronomische Nachrichten* nr. 1646 (1867), 219.
9. See Birt, "Concealment of the Lunar Crater Linné," *Astronomical Register*, 5 (1867), 11–12; J. F. Julius Schmidt, "Über den Mondcrater 'Linné,'" *Astronomische Nachrichten*, nr. 1631 (1867), 365–366; translated by W. T. Lynn, "The Lunar Crater Linné," *Astronomical Register,* 5 (1867), 109–110.
10. W. R. Birt, "On the Obscuration of the Lunar Crater 'Linné,'" *Monthly Notices of the Royal Astronomical Society*, 27 (1867), 93–95:93.
11. Schmidt, "Disappearance of a Lunar Crater," trans. W. T. Lynn from *Cologne Journal*, March 1, 1867; in *Astronomical Register*, 5 (1867), 161–162. See also *Intellectual Observer*, 11(1867), 216.
12. Robert James Mann, M.D., quoted in *Astronomical Register*, 5 (1867), 12n.
13. Birt, "Concealment," p. 12.
14. Birt, letter, January 5, 1867; *Astronomical Register*, 5 (1867), 31–32.
15. Birt, "Obscuration," p. 94.
16. Ibid.
17. Schmidt, "Disappearance," p. 162.
18. Ibid., p. 161.
19. Schmidt, "The Lunar Crater Linné," pp. 109–110.
20. Schmidt, "Disappearance," p. 162.
21. Charles Grover, February 3, 1867; letter, *Astronomical Register*, 5 (1867), 56.
22. Henry Cooper Key, January 14, 1867; letter, *Astronomical Register*, 5 (1867), 32–33.
23. H. C. Schjellerup, "Der Crater Linné, gesehen im Kopenhagener Refractor," *Astronomische Nachrichten*, no. 1655 (1867), 367–368.
24. Anonymous, "The Lunar Crater Linné," *Astronomical Register*, 5(1867), 175–176.
25. Flammarion, "Changement," op. cit.
26. Ibid.
27. Minutes of the meeting of the Royal Astronomical Society, June 14, 1867. Buckingham's claim that the black spot was growing larger calls to mind the fact that William Herschel, who at first mistook Uranus for a comet, reported that its diameter was increasing steadily for several weeks after making his discovery. While this behavior would have been expected of an approaching comet, the apparent diameter of Uranus was actually decreasing very slightly during the interval of time involved. The history of observational astronomy is with rife with examples of expectation fathering illusion. See R. H. Austin, "Uranus Observed," *British Journal for the History of Science*, 3 (1967), 275–284.
28. W. Huggins, *Monthly Notices of the Royal Astronomical Society*, 27 (1867), 297.
29. Birt, August 10, 1867, *Astronomical Register*, 5 (1867), 237–240.

30. Quoted in Birt, "On Mapping the Surface of the Moon," p. 43. This is, insofar as the authors are aware, the only occasion on which this particular adjective has been used to describe an astronomical object.
31. Wilhelm Tempel, *Astronomische Nachrichten*, no. 1655 (1867), 365–368.
32. Ashbrook, "Linné in fact and legend," p. 275.
33. Ibid.
34. Schroeter, *Selenotopographische Fragmente*, vol. 1, p. 181.
35. Schjellerup, "Der Crater Linné," p. 218.
36. Neison, *The Moon*, pp. 187–190. It is clear that Neison wanted to believe a change had taken place.
37. Arthur A. Rambaut, "Two Drawings of the Mare Serenitatis by John Russell, R.A., affording some hitherto unpublished evidence as to the appearance of Linné in the year 1788," *Monthly Notices of the Royal Astronomical Society,* 44 (1904), pp. 156–159; S. A. Saunder, "Note on the Drawings of the Mare Serenitatis by John Russell, R.A.," *Monthly Notices of the Royal Astronomical Society*, 44 (1904), pp. 427–429. The Austrian historian Martin Stangl, who has probed even farther back in the observational record, writes: "In the astronomical literature it is repeatedly alleged that the lunar formation Linné was first depicted on Riccioli's map of 1653… While Linné does not appear on the Hevelius [full-disc] map of 1647, it was nevertheless seen by him in the early morning hours of New Year's Eve 1643, two days before Last Quarter, moreover as a white spot. This is the earliest representation of Linné. This discovery has eluded historians of astronomy, evidently because they failed to closely examine the original *Selenographia* [by Hevelius]. On Plate 27 a small white spot in the correct position is clearly engraved. It appears dull rather than brightly shining like other craters and corresponds exactly in appearance to the present Linné when seen under the same lighting. Did no change occur therefore? Has Linné always had the same appearance as today? One cannot say absolutely, for in the same engraving Hevelius also depicts the craters Ross, Maclear, and Arago—which are more than twice as large as the allegedly disappeared Linné crater—only as white spots. Probably his telescope was too poor to clearly show the shadows thrown by smaller craters." Martin Stangl, *Johannes Hevelius: der Astronom von Danzig* (Graz: Institut für Astronomie, Karl-Franzens Universität, 1994). Stangl also cites a painting of the Full Moon circa 1700 by the Italian artist Donato Creti that hangs in the Vatican Museum. Once again, Linné seems to be depicted in the form of a white spot. Stangl was unable to determine whether Creti had actually looked at the Moon through a telescope himself or, perhaps more likely, had merely consulted an existing lunar map, such as Cassini's map of 1680.
38. Anonymous, "The Lunar Crater Linné," *Astronomical Register*, 6 (1868), 213–215:214.
39. Mädler, "On Changes of the Moon's Surface," p. 241.
40. Schmidt, Julius J. F., *Charte der Gebirge des Mondes* (Berlin: Dietrich Reimer, 1878), p. 15.
41. Birt, "On Mapping the Surface of the Moon," p. 44.
42. For instance, in 1954 Richard Baum unearthed Mädler's long-forgotten comments on the subject and passed them along to a leading lunar observer, Patrick Moore. Moore replied: "To say that your letter shattered me is to put it mildly… Is it a fact that Mädler observed Linné in 1867, and said that it was unchanged since 1831? If so, it knocks the bottom out of the whole 'change' theory—in which I have believed implicitly up to now."
43. Mädler, "Changes of the Moon's Surface," p. 241.

Chapter 12:
An Evolutionary Moon

The Moon, for all the careful attention paid to it, was a subordinate detail to the Solar System at large. Inevitably, the key to its nature could emerge only as astronomers groped their way toward understanding the origin of the planetary system itself—a problem which long remained as inscrutable as metaphysics. In 1755 the Prussian philosopher Immanuel Kant speculated that the planets had emerged from a contracting nebula, but his work had little influence until it was unearthed from obscurity a century later.[1] The real progenitor of what later became known as the Nebular Hypothesis was Pierre Simon, Marquis de Laplace, the French master of celestial mechanics (Figure 12.1). Inspired by the observations of nebulae made by William Herschel, Laplace, in a 1796 footnote to his widely read *Exposition du Systéme du Monde*, made a bold attempt to account for the fact the planets all revolve around the Sun in the same direction as the Sun rotates, and do so, moreover, in paths that lie nearly in the same plane as the Earth's orbit. The known axial rotations of the planets were all in the same direction, and almost all of their satellites also revolved in the same direction in their orbits around their parent bodies. (The exceptions were the two satellites of Uranus discovered by William Herschel, which revolved in a plane almost perpendicular to its orbit).

This was the orderly world which Isaac Newton had largely framed in terms of his laws of gravitation a century before. Though Newton himself had been aware of some of these patterns, he had remained baffled about their cause. In the end he could do no better than proclaim them the (arbitrary) work of the divine hand. But Laplace—who when Napoleon asked why he had made no mention of the Creator in his work on celestial mechanics replied "Sire, I had no need for that hypothesis"—saw that the patterns of Newton's orderly world would naturally emerge if the planets had originally formed from a swirling nebula. He began with a primeval Sun glowing in the midst of the future planetary system, surrounded by a distended, hot, slowly rotating atmosphere. As the envelope began to cool, it would contract, thereby increasing its rate of rotation. At last this swirling vortex would become unstable and fragment into a series of broken fiery rings. Then the rings themselves would break apart—an example of this process could still be observed directly, in the rings of Saturn. In time, the fragments of the ring farthest from the Sun would gather together into a single spherical body, which became the outermost planet; then, indefinite ages later, the next ring would coalesce, and the next, and so on, ever inward toward the Sun. According to Laplace, then, the

Figure 12.1 A towering figure in the history of celestial mechanics, Pierre Simon, Marquis de Laplace was also politically adept. He managed to prosper during a tumultuous era that witnessed the French revolution, the Napoleonic era, and the restoration of the monarchy, changing the preface of his book *Exposition du system du monde* to suit the reigning power of the moment. Courtesy John Koester.

outer planets were the oldest, the inner planets the youngest.

Laplace envisioned the same process taking place in turn around each developing planetary center; the condensing nebulosity giving rise to systems of satellites (including the Earth's Moon) as well as—an apparently unique case—Saturn's rings, which Laplace saw as "existing proofs of the primitive extension of the atmosphere of Saturn, and of its successive condensations."[2]

At the beginning, then, was homogeneity incarnate, in the form of a dark, whirling cloud of vapor. This was the embryon Solar System in formation, perhaps figured in deep space by some of the countless nebulae being revealed by the large telescopes of William Herschel. The orderliness and symmetry of the whole process—undoubtedly pleasing to the French master mathematician—would, however, eventually become suspect. In particular, Laplace had simply glossed over the far from minor detail of why nebulous fiery rings should condense into compact globes rather than simply dissipate into space. A further embarrassment was the discovery of the anomalous motions of the satellites of Uranus and, later, of Neptune's satellite Triton.

Mathematicians would long struggle with the intricate details of the process, and in the end could do little to dispel the veil of mist behind which the Solar System's birth remained hidden. Indeed, by the 1840s, the Nebular Hypothesis seemed to be everywhere on the defensive. Observers using Lord Rosse's mammoth reflector at Parsonstown had virtually disproved it—so it was thought—by resolving a number of Herschel's nebulae into stars. However, an abrupt reversal followed in 1864 when William Huggins, one of the principals in the Linné episode, attached a spectroscope to his telescope and turned it on one of Herschel's planetary nebulae in Draco. The spectrum showed the bright emission lines characteristic of incandescent gas, demonstrating that not all nebulae were made up of stars.

Briefly out of vogue, the Nebular Hypothesis became fashionable once more as the "reigning cosmogony of Victorian science."[3] It did so in part because it was difficult to imagine a viable alternative; in part because in that post-Darwin era it became linked to a broadly evolutionary view—the recognition that not only the Earth but the entire Solar System had followed a course of progress and development. As the poet Tennyson put it:

> This world was once a fluid haze of light,
> Till toward the center set the starry tides,
> And eddied into suns, that wheeling cast
> The planets; then the monster, then the man.[4]

Instead of the stability of the Solar System, the project that had absorbed the efforts of eighteenth-century celestial mechanicians, astronomers of the nineteenth century grappled with the ways the planets and their satellites had *changed* over time—the evolution of worlds. The broad outlines seemed to be well established, and were almost intuitively obvious. As the Irish historian Agnes M. Clerke wrote: "Knowledge has transgressed many boundaries, and set at naught much ingenious theorising. How has it fared with Laplace's sketch of the origin of the world?… The groundwork of speculation on the subject is still furnished by it… Modern science attempts to supplement, but scarcely ventures to supersede it."[5]

The Nebular Hypothesis furnished a bold starting point for speculation into the nature of planetary development, while thermodynamics—another of the powerful unifying ideas of the nineteenth century—provided it with a sense of direction. For the Solar System, once and in whatever manner it got underway, must evolve in accordance with the laws of heat as the planets, each starting out as an incandescent sphere, subsequently cooled.

A source of the internal and slowly dissipating energy of the Earth became evident in the 1840s, when Julius Robert Mayer (1814–1878) and James Prescott Joule (1818–1889) announced their discovery of the equivalence of heat and motion—the grand principle of the conservation of energy. Simply put, every form of motion was convertible into heat. This process was exemplified by the spacefaring particle whose motion, impeded by the Earth's atmosphere, heated it into incandescence to produce a falling star or meteor. Out of the infinitude of meteors from which the planets had formed, Newtonian gravitation furnished sufficient heat to melt whole worlds. The hot birth of the embryonic worlds emerging from the primordial matter of the nebula fit neatly into the scheme.

Indeed, as soon as it was acknowledged that energy could be transformed, but not destroyed, the nebular cosmogony assumed a new and authoritative aspect. But it also had to be modified in detail. Daniel Kirkwood (1814–1895), an Indiana University astronomer, suggested that material separated from the rotating disc by increasing centrifugal force would be cast off as a continuum rather than in the discrete rings envisaged by Laplace. The result would have been a meteoric, not a planetary, system. For that matter, Saturn's rings, according to Laplace a solid structure condensed from the vaporous outer part of Saturn's atmosphere, were shown by the brilliant Scottish physicist James Clerk Maxwell (1831–1879) to consist of a blizzard of particles whirling around the planet in independent Keplerian orbits. As the mechanics by which worlds could be made to coalesce from nebulous rings remained elusive, it seemed more hopeful to try to extract them from streams of particles, like those Maxwell had discovered in the ring of Saturn, than from gaseous envelopes.

These, however, were details. The overall framework seemed clear enough. In whatever order the planets had formed, it was inevitable they would evolve at differing rates, as dictated by their relative sizes and consequent rates of cooling. There could be little doubt that they had started out molten and plastic, a fact made evident by their nearly spherical shapes. Lacking independent sources of heat, they must cool by conduction and radiation, since one of the fundamental principles of thermodynamics held that heat always flows from hot regions to cold ones. In the case of a planet, the direction of flow was from the hot interior to the surface; from thence, slowly but surely, it would be radiated—and hence dissipated—into the cold vacuum of space. As the outer layers of the planet cooled, they would harden and solidify, but the interior might long remain hot and liquid. Moreover, because liquids generally exhibit greater changes of volume with temperature than do solids, a gap would be left between the outermost layers of the crust and the molten interior. Periodically the overlying crust would collapse and crumple, buckling and crinkling the skin like a drying apple. These were the fissures, faults, and mountain ranges of the Earth.[6]

If it was true, however, that worlds evolved over time from hot, plastic globes to cold inert ones, then it followed that the largest planets, like Jupiter and Saturn, must still be close to their states of primordial incandescence.[7] Middling globes like the Earth would have had time to partially cool, but not yet to dissipate all the internal heat of their formation.[8] Finally, the still smaller Moon—while showing the results of its torrid youth in the vast upheavals recorded on its surface—had proceeded to the stage of being a cold, dried-up, shriveled rind of a world. Its impassive mask was that of a world nearing—if not already at—the end-point of its evolutionary career. It was equally certain that its frozen surface presaged the doom of our own world, indeed of the whole universe, which, as the Law of Dissipation of Energy or Second Law of Thermodynamics seemed to assure, must eventually end in a disordered Heat Death—*La Fin du Monde*, in the memorable phrase of Camille Flammarion.[9]

There were other conclusions that followed from the fact that the Moon had evolved more rapidly and completely than the Earth. In the near-surface rocks of the Earth, oxygen in the form of oxides, and carbonates and water in the form of hydrates, were stored in large quantities. But on the Moon, their absorption into near-surface rocks would have proceeded to near completion. "We may suppose," wrote Phillips in a valedictory address on the Moon in 1868, "the process of volcanic incineration and aqueous absorption to have gone to greater extremes in the Moon than on the Earth, and so to have reduced the original atmosphere and the original oceans to very much smaller amounts… It would seem the most probable opinion [is] that *now* at least there remains no visible trace of atmosphere or ocean."[10]

This being the case, the most violent volcanic episodes must have occurred in the remote past, when the great ring-mountains had been formed. These primordial features were less eroded on the Moon, where degrading influences of all kinds were less effective than on the Earth. Signs of such degradation in the past were suspected by some astronomers in the forms of some of the lunar features. However, Phillips, the longer he studied the Moon, became less and less assured of their presence:

> The steepness of the Moon-crater walls and slopes is much greater in general than in any… known among the volcanic regions of the Earth… In the old volcanic regions of Auvergene and Mont Dor, the Eifel and even the Phlegrean fields, which may be the fittest for comparison with the crater-covered tracts of the Moon, very steep slopes are almost unknown, except for small spaces of consolidated rocks, such as the Puy de Dôme, Monte Somma, and the Drachenfels… The great broken ridges of [lunar] mountains… the Alps, Apennines, and Riphaean[s] make themselves known as axes of upward movement… I have not [however] been able to discover in these great ridges such marks of successive stratification… The surface is, indeed, as rough and irregularly broken as that of [our] Alps and Pyrenees… Must we suppose these mountains to have undergone the same vicissitudes as the mountain-chains of our globe—great vertical displacement, many violent fractures, thousands of ages of rain and rivers, snow and glacial grinding? If so, where are the channels of rivers, the long sweeps of the valleys, the deltas, the sandbanks, the strata caused by such enormous waste? If the broad grey tracts were once seas, as analogy may lead us to expect, and we are looking upon the dried beds, ought we not to expect some further mark of the former residence of water?[11]

The lack of obvious evidence of the working of erosive forces on the Moon lent further support to the notion that the Moon's shrinkage was now nearly complete, its internal heat largely dissipated, the forces once productive of the great convulsions which had fashioned its surface long since spent. But were they *entirely* spent? Was the Moon, indeed, a frozen changeless mask? Or were there, perhaps, eruptions on a minor scale, the last gasps of a worn-out world far advanced on the road to inevitable doom?

The extinct volcanoes that dominated the lunar scene had obviously done so on a much grander scale than those of the Earth. This proved, as John W. Judd wrote in his 1888 book *Volcanoes*, that "volcanic activity [must] have been far more violent on the Moon than it is at present upon the Earth."[12] But now they were dormant, or nearly so. The fact that the Moon had once been violently convulsed and was now nearly quiescent fit perfectly the thermodynamic scheme of planetary evolution. The features of its surface suggested, moreover, the relationships of stratigraphy; they contained expressions of a narrative history—the biography of a world, with a clear beginning, middle, and end. If the rings of Saturn had been left unfinished to show us how the world began, the Moon had run headlong on its course in order to demonstrate the stark and sobering end of a world.

Figure 12.2 George Howard Darwin, son of the great naturalist Charles Darwin and author of the once-popular theory that the Moon was born by fission from a rapidly spinning primordial Earth. Courtesy Peter Hingley, librarian of the Royal Astronomical Society.

All this was evident enough. But in the end, the details of the Moon's long evolutionary career would be most clearly elucidated not by nineteenth-century students of its surface features but by mathematicians expert in celestial mechanics and concerned with the tidal interactions between the Earth and the Moon. The most notable of these was the Cambridge mathematician, George Howard Darwin (1845–1912) (Figure 12.2). The fifth son of the great evolutionist Charles Darwin, he read law at Trinity College, Cambridge, and was admitted to the bar, but never practiced. At Cambridge, he became a junior colleague of the most influential British physicist of the day, William Thomson, Lord Kelvin (1824–1907), whose calculations of a relatively brief lifespan of the Earth from considerations of its rate of cooling had loomed as an "odious spectre" for Charles Darwin's theory of biological evolution, which seemed to require hundreds of millions of years for natural selection to occur. It was at Kelvin's behest that George Darwin adopted the theory of tides as his own special domain, an important chapter in the grand story of cosmic evolution to which he devoted himself and in which he was destined to leave his mark.

Darwin first announced his theory of the tides in 1878 and published a long memoir on the subject a year later.[13] Because of the gravitational attraction of the Moon, the liquid masses of the oceans are ever so slightly heaped up on the near and far sides of the Earth relative to the Moon. In effect, the Moon holds in position a portion of the oceans. Beneath these tidal bulges the globe of the Earth rotates. Although water is a reasonably good lubricant, particularly when it is deep, it is not altogether frictionless when dragged over shallow seabeds, like the Irish and Bering seas, by the Earth's diurnal rotation. Because of this tidal drag, the Earth suffers a braking action that slows its rate of axial rotation, lengthening the day by a miniscule fraction of a second per century. Moreover, since action and reaction are equal, as the Moon pulls on the bulging oceans the oceans tug in return on the Moon, imparting energy to it and causing it to spiral slowly outward as the Earth's rotation slows.

Darwin calculated that after the lapse of indefinitely long ages, a stable configuration would be reached when the Moon revolved around the Earth in about 55 days. In that inconceivably remote future the Earth's axial rotation would also have slowed to 55 days. One could equally wind the clock backwards. Reversing the sign of his calculations, Darwin showed that in the remote past the Earth spun much more rapidly on its axis than it does now, while the Moon circled much closer to the Earth. At some point the Moon's period of revolution was equal to the Earth's period of rotation. Unfortunately, near that point the solution to the equations became unstable, Darwin noted, "in the same sense in which an egg when balanced on its point is unstable; the smallest mote of dust will upset it, and practically it cannot stay in that position."[14]

What if one attempted to wind the clock backwards still further? "It is not so easy," Darwin admitted, "to supply the missing episode. It is indeed only possible to speculate as to the preceding history." Darwin ventured a guess, however. He suggested that the Earth and Moon had once been part of a common molten mass. Due to the combined action of the tides raised by the Sun and the primordial object's rapid rotation, the object broke apart, part of it remaining behind as the Earth, the other part taking on a separate existence as the Moon (Figure 12.3). He even tried to estimate the minimum time at which the Moon had undergone this process of "fissi-partition" from the proto-Earth. "Now, although the actual time-scale is indeterminate," he wrote, "... it may be proved, in fact, that if tidal friction always operated under the conditions most favourable for producing rapid change, the sequence of events from the beginning until to-day would have occupied a period of between 50 and 60 millions of years. The actual period, of course, must have been much greater."[15]

From his belief in the "preponderating influence of the tide" in the evolution of the Earth-Moon system, Darwin found himself able to account for many of the peculiarities of that system. He also defined some of the effects that tidal evolution must have had on the body of the Moon itself:

> Once upon a time the Moon must have been molten, and the great extinct volcanoes revealed by the telescope are evidences of her primitive heat. The molten mass must have been semi-fluid, and the Earth must have raised in it enormous tides of molten lava. Doubtless the Moon once rotated rapidly on her axis, and the frictional resistance to her tides must have impeded her rotation. This cause must have added to the Moon's recession from the Earth, but as the Moon's mass is only an eightieth part of that of the Earth, the effect on the Moon's orbit must have been small. The only point to which we need now pay attention is that the rate of her rotation was reduced. She rotated then more and more slowly until the tide solidified and to the present day she has shown the same face to the Earth... Our theory, then, receives a striking confirmation from the Moon; for, having ceased to rotate relatively to us, she has actually advanced to that condition which may be foreseen as the fate of the Earth.[16]

While Darwin himself believed that the cavity left when the Moon fissioned from the Earth would have quickly closed up as the Earth's globe continued to contract, the Reverend Osmond Fisher, the rector of Harlton near Cambridge and

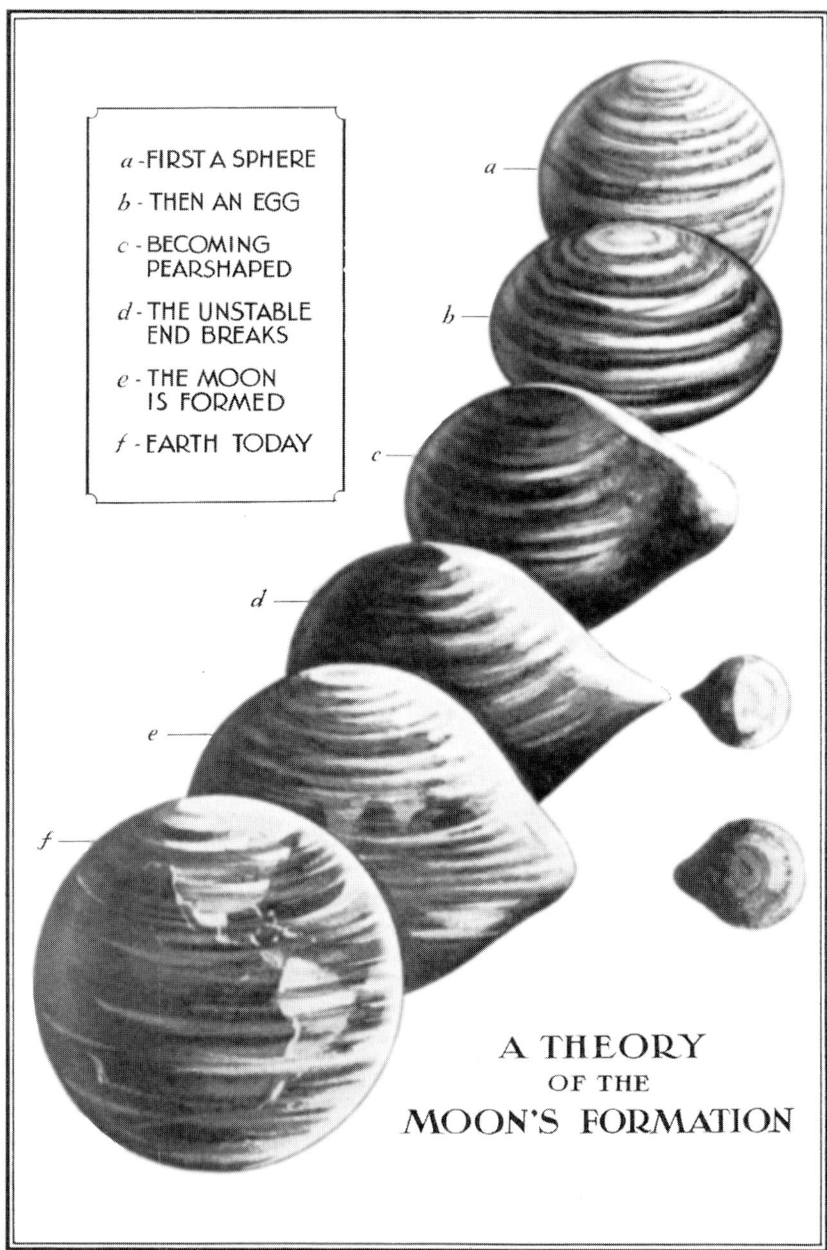

Figure 12.3 The birth of the Moon by tidal fission, as envisaged by George Howard Darwin. From H. T. Stetson, *Man and the Stars* (1930).

author of *The Physics of the Earth's Crust* (1881), disagreed. Most of the material shorn off to form the Moon would have been of the lighter continental variety, he argued, rather than the denser oceanic crust. Its departure would have left scars, including the Pacific Basin.[17] Among the most ardent early supporters of Fisher's theory was the great American geologist of the Grand Canyon, Clarence E. Dutton (1841–1912).[18]

The gripping narration of the Moon's history, an important chapter in the larger story of cosmic evolution, was already being filled in by the 1880s and 1890s. The Moon, it was imagined, had been conceived by fission from a rapidly rotating, molten proto-Earth, deformed and disrupted by the tidal force of the Sun. Since then it had been receding with almost imperceptible slowness. Once, in the indefinite past, untold millions of years ago, it had been a looming presence in the Earth's skies—a Monster Moon. Since then it had steadily retreated, grown more remote and laggard in its laps, until now it was just a small, silvery marble. At the same time, its rotation had slowed and it had cooled and contracted more quickly than the Earth, exhausting its internal fires and becoming a shrunk-shank Moon, a remote, cold, icy, tide-locked Moon. In the future, as the Moon continued retreating to ever greater distances from the Earth, it would finally come to the end of its orbital changes when its period had reached 55 days and its distance some 380,000 miles. Meanwhile the Earth, winding down like a spinning top, would also become stabilized, and the two worlds—Earth and Moon—locked in an eternal embrace like a pair of exhausted dancers, would move around each other as if connected by a rigid bar.[19]

It was a grand sequence, a sequence whose events were known with some assurance, seemingly confirmed by a brace of exact calculations made possible by the immutable laws of physics and the everlasting principles of celestial mechanics. But the exact duration of time since the formation of the Moon—the length of the ages that had passed since it had exhausted its once burning fires—remained beyond the grasp even of celestial mechanics to define. "Unfortunately, there is no mystery more inscrutable than the duration of geological time," wrote Dutton in his landmark 1882 volume on the Grand Canyon. "On this point geologists have obtained no satisfactory results in any part of the world. Whatever periods may have been assigned to the antiquity of past events have been assigned provisionally only, and the inferences are almost purely hypothetical... Here, as elsewhere, whenever we interrogate her about time other than relative, her lips are sternly closed, and her face becomes as the face of the Sphinx."[20] The same words would have applied equally well to that mask-like world beyond the world—the world in the Moon. The face of the Sphinx of the Moon was also utterly blank and inexpressive; its lips sternly closed and unyielding of the secrets of long time.

More recent methods of radiometric aging would eventually disclose the duration of an epoch which Darwin and Dutton could scarcely have imagined. But other less Sphinx-like questions, which should have been laid to rest in the nineteenth century, continued to perplex and fascinate—above all, questions about the continuing vitality of the Moon.

It should have been obvious from thermodynamic considerations, beyond a reasonable doubt, that the Moon must have long since evolved from an active fiery state to a cold inert condition where it stared back at us, blankly and changelessly and lifelessly, as from beneath a hood of bone. It surely occupied an extreme and thermodynamically advanced state in the scheme of planetary development, its inner fires long since gone cold, all traces of water or air escaped from their captivity on or within its surface into the surrounding vacuum of space. The surface of such a Moon should have frozen into a stereotyped page aeons ago; all vital changes should have ceased.

But the surface of the Moon, studied through the ever-changing interplay of light and shadow, has always been a realm of ghosts. Even at the end of the nineteenth century, revenants of change still haunted ancient cliffs, floated above chaotic jumbles, or peeked out from shattered ruins. And to those who watched, the hopeful vigil over the lunar corpse seemed to reveal, like the fog on a mirror catching the respiration of a not-yet departed soul, faint but unmistakable signs of geological life.

References

1. Immanuel Kant, *Allgemeine Naturgeschichte und Theorie des Himmels* (Königsberg: Petersen, 1755). An English translation is available: *Universal Natural History and Theory of the Heavens*, trans. Stanley L. Jaki, S.J. (Edinburgh: Scottish Academic Press, 1981).
2. Laplace, *The System of the World*, trans. H. H. Harte (London: Longmans and Green, 1830; from 5th French edition, 1824), p. 360.
3. Joe D. Burchfield, *Lord Kelvin and the Age of the Earth*, p. 50.
4. Alfred Lord Tennyson, "The Princess."
5. Clerke, *Popular History of Astronomy*, p. 376.
6. On the contraction theory in geology, see Mott T. Greene, *Geology in the Nineteenth Century: Changing Views in a Changing World* (Ithaca, New York: Cornell University Press, 1982).
7. Regarding Jupiter, Richard A. Proctor wrote in *Other Worlds than Ours:* "It seems to me that these considerations [the rapid changes in the appearance of the cloud formations] point with tolerable clearness to the conclusion that, within the orb which presents so glorious an aspect upon our skies, processes of disturbance must be at work wholly different from any taking place on our own Earth. That enormous atmospheric envelope is loaded with vaporous masses by some influence exserted from beneath its level. Those disturbances which take place so rapidly and so frequently are the evidences of the action of forces enormously exceeding those which the Sun can by any possibility exerted upon so distant a globe. And if analogy is to be our guide, and we are to judge of the condition of Jupiter according to what we know or guess of the past condition of the Earth and the present condition of the Sun, we seem led to the conclusion that Jupiter is still a glowing mass fluid probably throughout, still bubbling and seething with the intensity of the primeval fires, sending up continuously enormous masses of cloud, to be gathered into bands under the influence of the swift rotation of the giant planet."
8. Kelvin wrote, "Within a finite period of time past, the Earth must have been, and within a finite period of time to come the Earth must again be, unfit for the habitation of man as at present constituted." William Thomson (Lord Kelvin), "On a Universal Tendency in nature to the Dissipation of Mechanical Energy," *Philosophical Magazine*, Series 4, 4 (1852), 304–306.
9. C. Flammarion, *La Fin du Monde* (Paris: Flammarion, 1894).
10. John Phillips, "Notices of some Parts of the Surface of the Moon," *Philosophical Transactions of the Royal Society of London*, 158 (1868), 333–345.

11. Ibid., pp. 336–338.
12. J. W. Judd, *Volcanoes—What They Are and What They Teach* (London, Kegan Paul, Trench & Co., 4th ed. 1888), p. 367.
13. G. H. Darwin, "On the precession of a viscous spheroid," *Nature*, 18 (1878), 580–582; On the precession of a viscous spheroid, and on the remote history of the Earth, *Philosphical Transactions of the Royal Society,* 170 (1879), 447–538.
14. G. H. Darwin, *The Tides* (London: John Murray, 3rd ed. 1911), p. 286.
15. Ibid., p. 291.
16. Ibid., pp. 291–292.
17. Osmond Fisher, "On the Physical Cause of the Ocean Basins," *Nature*, 25 (1882), 243–244.
18. As noted in Greene, *Changing Views*, p. 248.
19. In fact, this is not quite the whole story. Solar tidal friction would continue to act on the system and eventually—over inconceivable ages—cause the Moon to reverse course and spiral inwards again, an eventuality unforeseen by Darwin himself.
20. Clarence Dutton, "The Physical Geology of the Grand Canyon District," in *United States Geological Survey, 2nd Annual Report, 1880–1881* (Washington: Government Printing Office, 1882), 49–166: 166.

Chapter 13:
More Changes

That the forces that had created the innumerable characters of the lunar surface had been, broadly speaking, volcanic was consistent with the idea that the worlds of the Solar System had started out hot and plastic. Indeed, even as the larger planets and the Sun were still in their fiery courses and in the process of slow but inevitable cooling, the Moon had apparently started its career as a realm of wild and vast volcanic eruptions. Sir John Herschel's comparison of the lunar landscapes to the volcanic wastes of the Phlegraean Fields and the Puy de Dôme was often quoted—in no small measure because his eminence as a geologist rivaled his eminence as an astronomer. Phillips echoed Herschel's comparison, calling the Moon a "grand Phlegraean field." Charles Piazzi Smyth (1819–1900), returning from a remarkable expedition in quest of a more transparent and quiescent atmosphere for telescopic observations on the Peak of Teneriffe (Figure 13.1), an extinct volcanic cone on the largest of the Canary Islands, was no less impressed by the similarities he noticed between that volcanic cone and the features of the lunar scenery that greeted him in the eyepiece:

> After our visit to the caldera of the Peak, with its walls of rock bleached by steam, and by acid vapours permeating them for ages, we could better understand the remarkable internal whiteness of lunar volcanoes, as shown by our telescope at Alta Vista.
>
> Some geologists have, indeed, denied that the features seen by astronomers in the Moon are to be considered as volcanoes; but we who duly noted the gentle external slope of some of those circular pits, their cliffy internal descents, their flat floors, and their central peaks—had little doubt in our minds...
>
> Could we have found in the Moon, that dynamic trace, which was so important in proving relative ages among the red and yellow lavas of Teneriffe, viz., glacier wrinkles in one, and surf-like waves in the other,—all sceptical doubts must vanish. But we failed, and this point is left to a larger telescope, more constantly employed in lunar physics, on this or some other mountain.[1]

The American geologist James Dwight Dana saw similarities between the larger lunar formations and the boiling lava-lakes of Kilauea, a still active Hawaiian volcano (Figure 13.2). The gases of Kilauea, escaping in small bubbles, created a "state of active ebullition" with little commotion, producing neither cinders nor cinder cones like Vesuvius and other more violently erupting volcanoes (Fig-

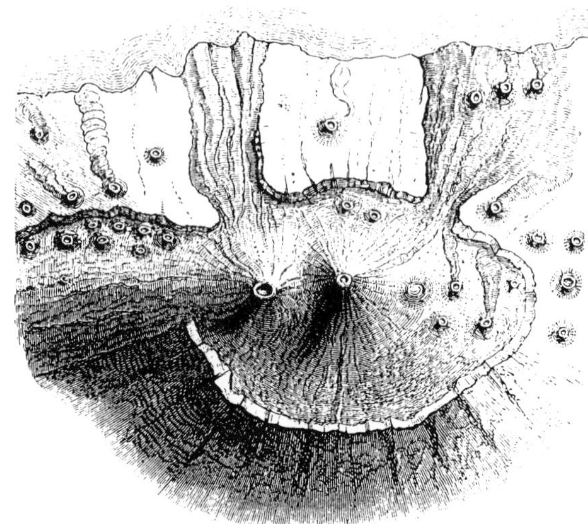

Figure 13.1 The volcanic peak of Teneriffe in the Canary Islands, as drawn by Charles Piazzi Smyth. In the quest for a more transparent and quiescent atmosphere for telescopic observations than was to be found at sea level, Smyth set up an observing station at Teneriffe in 1856. He was suitably impressed with the quality of the seeing there, and also with the similarity of the volcanic cinder cones of Teneriffe to the lunar craters he observed through his telescope. From Charles Piazzi Smyth, *Teneriffe: an Astronomer's Experiment* (1858).

Figure 13.2 As soon as the Hawaiian Islands were visited by geologists and astronomers, their volcanic features inspired new theories about the origin of lunar craters. The geologist James Dwight Dana arrived here with the Wilkes Expedition to the South Seas of 1838–1842. In addition to Hawaii, Charles Wilkes and his crew also visited the Antarctic Ocean and sailed along Antarctica in the region now known as Wilkes Land. Kilauea, shown here, suggested to Dana a possible model for the kind of volcanism that might have formed the large-scale features of the Moon's surface. Its "active ebullition," or outgassing over a large area, provided a mechanism for producing a feature of "indefinite extent," in contrast to the violently explosive release of gas occurring in a Vesuvian-style eruption. Photograph by William Sheehan, 1998.

Figure 13.3 Volcanic cinder cones as viewed from Mauna Kea, Hawaii. Photograph by William Sheehan.

ure 13.3). But if, Dana intimated, the lunar volcanoes were more kindred to Kilauea than to Vesuvius, the large scale of the lunar formations might at last be explained: "We may have boiling going on over an area of indefinite extent; for the size of a boiling lake can have no limits except such as may arise from a deficiency of heat. The size of the lunar crater is therefore no mystery."[2]

One of the most authoritative pronouncements, however, came from Britain's leading volcanologist, George Poulett Scrope (1797–1876), who for a time served with Lyell as a Secretary of the London Geological Society. Scrope wrote in his 1862 book *Volcanos* of the "remarkable resemblance presented by the surface of the Moon to some of the volcanic tracts of our Earth which are most plentifully studded with crateriform hollows surrounded by subconical banks. The analogy is so close, that it is impossible for a moment to doubt the volcanic character of the lunar enveloping crust."[3] To emphasize his point, Scrope subjoined an engraving showing the impressive similarity of the lunar crater-field near Maurolycus and the Phlegraean fields (Figure 13.4). In the rays surrounding many of the lunar craters he saw possible lava-streams or dykes extruded upwards in vertical ridges from radiating fissures. In comparison with the volcanic cones of the Earth, the lunar craters were more numerous and widely distributed; their diameters were greater, their external embankments shallower. Scrope suspected that the lunar craters derived their peculiar aspects "from the explosions of vapour that produced them breaking through the surface of soft and semi-liquid matter in successive bubbles, whose bursting would throw off all round a concentric ridge formed of repeated layers of this substance."[4] Writing on the eve of Linné, he added:

> One thing seems certain, namely, that the lunar surface is no longer eruptive—at least, that its volcanoes have been quiescent for centuries, since no change has been observed by astronomers in its mountains. It bears, indeed, the aspect of a burnt-out globe, once imbued with volcanic life and intense outward activity, probably with

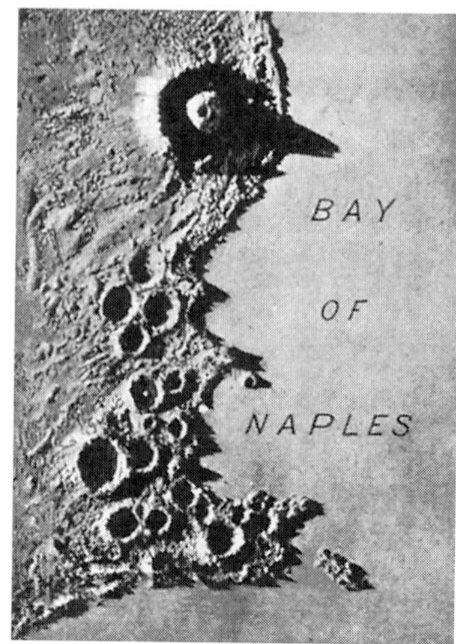

Figure 13.4 George Poulett Scrope's comparison of the Campi Phlegraei near Naples with the lunar landscape near the crater Maurolycus in his 1862 book Volcanos inspired a pair of the plaster-of-Paris models by Nasmyth and Carpenter. The Maurolycus crater field is represented on the left, the Campi Phegrai on the right. From David Peck Todd, *A New Astronomy* (1897).

seas and an atmosphere, now dried up and extinct.[5]

On Earth, Scrope observed, the explosive force of volcanoes was due to the expansive force of an "elastic aëriform fluid struggling to escape from the interior of a subterranean body of 'lava.'" Namely, water. He cited pioneering work by the French experimental physicist Henri Victor Regnault (1810–1878) which demonstrated that, under the high pressures prevailing deep beneath the Earth's surface, water was converted into steam at much higher temperatures. Under a presssure of two atmospheres water did not boil below 299° F, while an increase of pressure to twenty atmospheres raised its boiling point to 415° F. Thus, Scrope concluded, "when altogether hindered from communication with the atmosphere...[in a] closed vessel, [water] may be made red-hot without expanding into vapour. The moment, however, that an opening is made in the enclosing vessel, reducing the pressure to that of the atmosphere only, it flashes instantly into steam with an explosive violence."[6]

Water vapor powered the explosive force of volcanoes of the Earth. On the Moon, however, it was generally conceded there was little, if any, water or atmosphere. The problem of whether ongoing eruptions took place on the Moon was thus tied to that of the disappearance of former oceans and atmosphere—a problem which constituted, according to one of the leading students, England's Edmund Neison (the pseudonym of Edmund Neville Nevill; 1851–1940) "one of the most prominent difficulties in framing a consistent outline of the probable past history of our satellite."[7] Hansen had attempted to spirit away the oceans and atmosphere to the lunar far side. A more plausible idea, however, was that they had been withdrawn into the interior of the Moon—a result, it could be argued, of its smaller size and consequently more rapid cooling. As the Moon's outer crust had solidified, it had resisted further contraction. The interior continued to cool, forming a frothy, honeycombed structure with vast internal cavities, into which the lunar oceans, perhaps even the entire lunar atmosphere, might have escaped, wrote Neison, "without leaving a trace of the cause of their disappearance behind on the surface."[8]

Neison himself was one of the strongest proponents of the view that the Moon might have at least a scrap of atmosphere, sufficient to exert an influence on conditions at the surface—and published his first contribution to the Royal Astronomical Society on the subject in 1873.[9] He accepted the idea that the Moon's craters represented the results of "vast volcanic convulsions," and although he conceded that observations with small telescopes seemed to justify Beer and Mädler's view that the surface of the Moon showed an "entire dissimilarity to that of the Earth," in protracted observations with larger instruments "more points of resemblance become manifest. As [the French astronomer Jean] Chacornac found, closer examination with powerful instrumental means reveals far greater terrestrial analogy in the structures of the Moon than otherwise appears even possible; while a general analogy is often traceable between terrestrial volcanic regions and the more disturbed portions of the surface."[10] Admittedly, he partially contradicted himself by observing that through powerful telescopes the larger craters appeared "less and less like volcanic orifices or craters." Their enclosing walls lost their regularity of outline and appeared instead as confused masses of mountains broken by valleys, ravines, and depressions, "an irregularly broken surface,"[11] as Neison wrote a decade after Linné had apparently broken the Moon's inscrutable silence.

Though Neison was convinced that the celebrated change in the small crater on the plain of the Mare Serenitatis had been real, he believed that instead of a volcanic bang, it had more probably represented the comparative whimper of a landslide. "The change, if any," he wrote, "appears to have been simply the falling of the walls into the interior of the crater, filling this in great part up with *débris*, a class of occurrences of which several hundred instances could be pointed out where it has happened, and in particular the north-west wall of Gassendi, which lies in ruins on the exterior plain."[12]

After the Linné affair, many astronomers were willing to acquiesce to the possibility that the Moon was not entirely dead, that changes on a minor scale could occur there. (The change in Linné seemed to have been fairly extensive, but it was evidently exceptional; it was, morover, by no means certain that it was volcanic, and might be an ordinary landslip). Thus the stage was set for the future direction of selenographical studies: observations must focus more and more closely on minute detail. Indeed, the obsessive pursuit of detail became—with a vengeance—the preoccupation of many of the selenographers who worked the field.

The emerging generation of selenographers in Germany, France, and England produced important books: Richard Anthony Proctor's *The Moon, Her Motions, Aspect, Scenery and Physical Condition* (1873), which revived Gruithuisen's impact theory of the origin of lunar craters; Nasmyth and Carpenter's *The Moon: Consid-*

ered as a Planet, a World and a Satellite (1874); and Neison's own *The Moon* (1876), an updated—and much-revised—sourcebook modeled on Beer and Mädler, which included an amended version of their *Mappa Selenographica* and catapulted its 25 year-old author overnight to the role of most important selenographer in Britain. But it was also in some ways a lost generation, too absorbed "in the trivialities of drawing ever smaller areas of the lunar surface."[13]

It was a generation that all but abandoned the idea of any individual mastering the whole range of lunar phenomena, a generation that took for granted that progress would occur only through the cooperative endeavour of many observers. Birt's committee was, in a sense, the adaptation of a science coming to terms with its own increasingly far-reaching line of advance. Deploring the neglect of lunar studies, Neison attempted to enlist a potential army of new observers with this enticement:

> So varied and interesting, indeed, are the numerous phenomena presented by the Moon, and so strong an interest do they invariably excite in the minds of those who have once devoted themselves to their study, that it may be said once a selenographer, always a selenographer... An appeal is therefore made to those numerous observers who have not chosen a definite and entirely engrossing line of astronomical observation to turn their attention to one of the most interesting and important branches of astronomy, and help to determine the true physical condition of the surface of the Moon. In union there is strength. What any number of isolated workers may fail to accomplish, will be quickly and readily effected by half their number working in union.[14]

Many of the reports of these observers, trooping in increasing numbers to the study of the Moon, regarded possible obscurations. Blue veils had been suspected since at least Beer and Mädler's time. Schmidt pointed out that most of these had been seen with refractors, raising the justifiable doubt they were merely the spurious effects of chromatic aberration. There were reports of other possible landslips, and intimations that the great crater Tycho had at times been the subject of obscurations, since Birt sometimes found it difficult to get it sharply into focus.

But perhaps the most hotly contested field of battle on the Moon became the comparatively smooth floor of the walled-plain Plato. Smooth and uniform to a cursory examination, it betrayed to closer scrutiny—like all regions of the Moon—an ever-increasing number of minor variations in shade as well as a host of minute features. Beer and Mädler had recorded four lighter streaks extending from north to south, including three or four specks which ranked among the most elusive lunar objects visible in their small telescope. The keen-sighted Gruithuisen was able to distinguish no fewer than seven small spots and pits in 1825, and another was added by James Challis of the Cambridge University Observatory. The "eagle-eyed" amateur astronomer William Rutter Dawes later discerned that one of Gruithuisen's pits actually consisted of a pair of craterlets lying close together like a double star.

Webb, following up a hint by Gruithuisen, announced that the shadings on the floor were variable, leading Birt, in 1869, to propose that Plato be subjected to the most searching examination. The terms in which he couched his proposal expressed his motivation: "It is hardly likely that an object such as the floor of Plato should remain in *exactly* the same state during a period of eight or ten years."[15] The Plato project was embraced by several of the leading British observers of the Moon—especially Henry Pratt and Joseph Gledhill, Edward Crossley's assistant, and another newcomer to the selenographic scene who later became one of its most prominent personalities, the Liverpool engineer Thomas Gwyn Empey Elger. In 1869, 1870, and 1871 the series of observations was carried forward. Variability in the visibility of the minute features led some observers to strongly suspect that local obscurations occurred *frequently*.

The addition of new candidates to Birt's original catalog of possible changes grew from year to year until it made up a virtual "fool's paradise." As always, the need remained for a standard—thus Birt's great Moon map, the Grail he sought until the end of his life, continued to occupy most of his attentions. Only with such a map would it be possible to identify changes in the Moon, just as only with a good star catalog would it be possible to uncover the identity of a new planet. Failing a full-scale chart of the entire Moon, detailed maps of special regions were produced; many of these, for instance, recorded minor features on the floor of Plato (Figures 13.5, 13.6). Such maps satisfied, in part, the need to establish a reliable standard of comparison by which any actual change could be recognized. However, the impressions of different observers, or even of the same observer viewing a given region under different observing conditions, rendered the whole enterprise moot, since the markings themselves were delicate, fugitive, and often at the very limits of telescopic vision.

In early 1878, Birt announced in *The Observatory* the formation of a new Selenographical Society, for the furtherance of his lunar projects, including the great map. By then, his health was failing, and he was forced to relinquish an active role. Neison emerged for a time as the real leader, assuming responsibility for the society's *Selenographical Journal.*

The same issue of *The Observatory* carried exciting news from Germany. "Schmidt's Lunar Map" had appeared, published at the expense of the Prussian Government. Alone and unaided, the German selenographer, after years of work, had finally completed his great map, six and a half feet in diameter. To carry it through had undoubtedly required a single-mindedness and obsessive industry rare in any field—perhaps bordering on the pathological. Even with his industry and concentration, Schmidt had grudgingly to admit that the project had exceeded even his powers. By July 1874 he came to the realization that an exhaustive representation of all the features visible in the six-inch Athens refractor was beyond the possibility of achievement in a single human lifetime (Figure 13.7). Only then did he decided to prepare the work for publication. He presented the chart to the Berlin Observatory, where it excited admiration as a performance highly creditable "to Teutonic intellect and perseverance." Before long it was being touted as a uniquely Prussian achievement, and its publication had been sponsored and paid for by the Crown Prince of Prussia himself. Its twenty-five sections were photo-

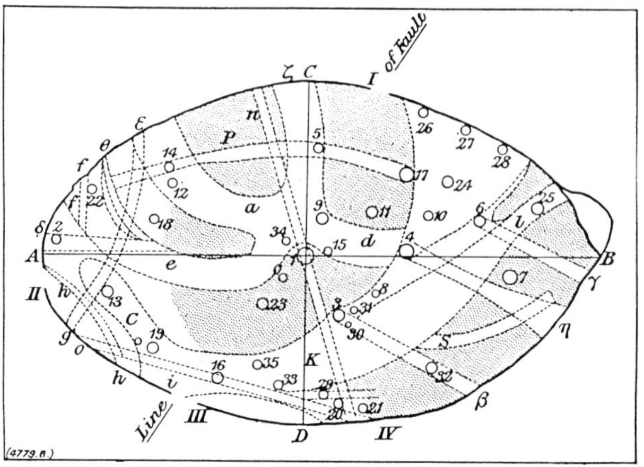

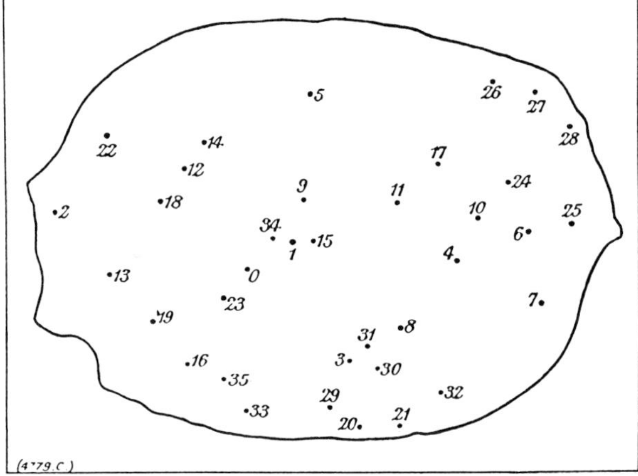

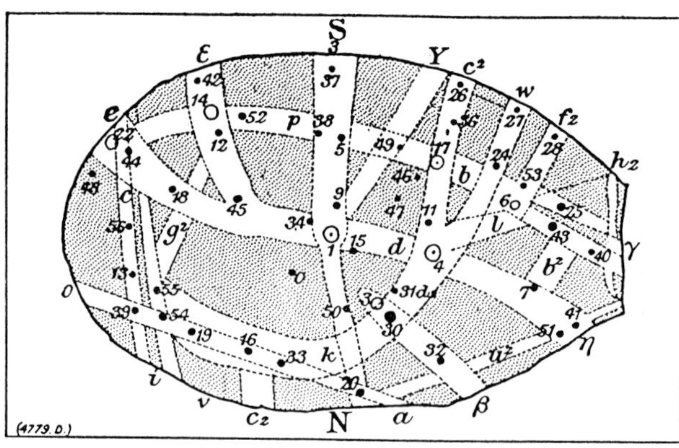

Figure 13.5 The search for the Holy Grail of lunar change focused for a time on the tantalizing features on the smooth floor of the lunar crater Plato. (Top) W. R. Birt's 1870 chart of craterlets and bright spots in Plato, attempted to assign numbers to them in order of their conspicuousness. (Middle) Another of Birt's charts, representing a compilation of his observations made during 1869, 1870, and 1871. (Bottom) A chart by another skilful amateur observer, Arthur Stanley Williams, representing a compilation of his observations made during 1882, 1883, and 1884. From "Lunar Changes," *Memoirs of the British Astronomical Association*, volume 20, part 3 (1916).

Figure 13.6 This 1992 drawing by Harold Hill accurately depicts the appearance of Plato through a 10-inch Newtonian reflector under good seeing conditions. Courtesy Harold Hill.

graphed at the General Staff Office under the direction of Count von Moltke.[16]

After the first copies of the Schmidt map appeared in 1878, the English astronomer John Birmingham commented: "In even a cursory examination of Schmidt's map its completion by a single observer must seem almost incomprehensible; but it requires protracted study to well realize the extent of the work. Any person who tries with the aid of a six-inch telescope to give a closely detailed delineation of even a small area of the Moon, will soon conclude that the period of thirty-three years was comparatively a very short one for the accomplishment of Dr. Schmidt's great task."[17] In his popular book *The Story of the Heavens*, Sir Robert Ball marveled: "To give some idea of Schmidt's amazing industry in lunar researches, it may be mentioned that in six years he made nearly 57,000 individual settings of his micrometer in the measurement of lunar altitudes. His great chart of the mountains in the Moon is based on no less than 2,731 drawings."[18] In the nineteenth-century race to the Moon, this singular German had thus outpaced Phillips, Birt, and several committees of Englishmen.

According to Schmidt's own rather compulsive analysis, Lohrmann had charted 7,177 craters, Mädler 7,735; however, his own map recorded no less than 32,856. His superiority was equally evident in his record of rilles—the 71 on Mädler's map paled in comparison with his own 348. Schmidt's effort was indeed almost monomaniacal—and yet it still left much to be desired. It was based on observations made with 13 telescopes of varying size and quality over a period of 32 years. Consequently the drawing was uneven and failed to achieve the uniformity

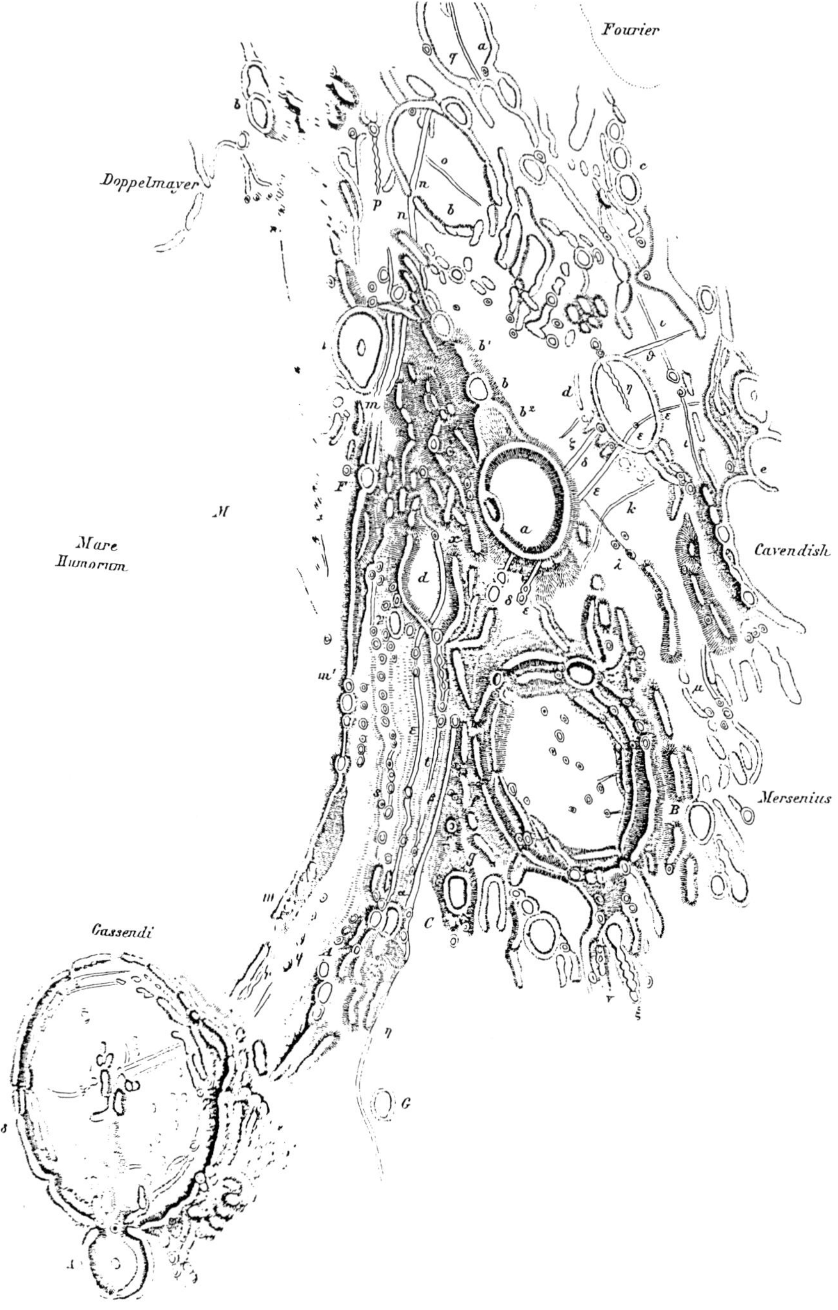

Figure 13.7 Out of his later years: Schmidt's attractive and realistic rendition of the western shore of Mare Humorum, based on a series of observations with the 6-inch refractor of the Athens Observatory. Sheehan's collection.

of detail required to serve as a standard by which to evaluate suspected changes. As Birmingham pointed out: "It would be plainly wrong to infer that a formation must necessarily be recent because it is not noticed even by Schmidt. In a work of such magnitude as his it would be manifestly senseless to imagine that even sufficiently obvious objects might not, in some instances, have escaped him."[19]

Parenthetically, 1878—the year of Schmidt's great map—was also the year when Congress organized the United States Geological Survey and assigned it the task of making large-scale topographic maps of the United States and its territories. The existing maps had been largely assembled from fragmentary reports by army engineers making railroad and geographical surveys, so it may not be too surprising that the earthward hemisphere of the Moon was depicted in greater detail and precision on Schmidt's map than many parts of the American West were depicted in existing maps of the time.

The publication of Schmidt's map took some of the wind out of the Selenographical Society's sails and weakened its resolve to carry out Birt's unwieldy and increasingly hopeless mapping project. On the other hand, interest in the detection of changes on the surface of the Moon—Birt's overarching objective and ultimately the very *raison d'être* of any map—remained as strong as ever. Indeed, it was given a new sense of urgency by the announcement in Germany of yet another possible lunar change.

Hermann Klein (1844–1914) (Figure 13.8), the director of the Cologne Observatory, was an energetic man of many talents renowned for an excellent star atlas, a map of the Milky Way, and widely-used texts on astronomy and meteorology. But above all else he was a particularly assiduous lunar observer. As a young man, he had been personally acquainted with both Mädler and Schmidt. He translated Nasmyth and Carpenter's *The Moon* into German and fostered widespread interest in selenographical work in a number of influential popular books and in the periodicals he edited—*Sirius*, *Gaea*, *Wochenschrift für Astronomie*, and the annual *Jahrbuch für Astronomie und Geophysik.* Along with Casimir Marie Gaudibert (1823–1901), a Protestant clergyman of Vaison, France, he was undoubtedly the most active student of the Moon outside of England at the time.

On May 19, 1877, Klein was observing from Cologne with a 5.5-inch (140-mm) refractor when his attention was drawn to the prominent rille running through the crater Hyginus. It is of the most fascinating regions on the Moon. The crater divides the rille into two sections of unequal lengths, which can be resolved in places into a chain of nearly confluent craterlets, as Mädler had first recognized with the Dorpat refractor. This complex region had been carefully drawn by Lohrmann, Beer and Mädler, and Schmidt. But now Klein noticed a "great black crater without a wall and full of shadow" which had not been recorded on the earlier maps (Figure 13.9). He at once concluded that it must be new.[20]

Klein delayed an announcement until March 1878, when he issued a plea for observers to examine the region at every favorable opportunity, "to prevent the matter from lapsing into the difficulties which beset the elucidation of the famous case of Linné." Unfortunately, some of Klein's drawings and descriptions were

Figure 13.8 Hermann Klein, the "German Flammarion" whose popular writings did so much to advance the cause of amateur lunar studies in Germany. This photograph of the Cologne astronomer and astronomy popularizer was taken in 1897. Courtesy Dorothy Schaumberg, Mary Lea Shane Archives of the Lick Observatory.

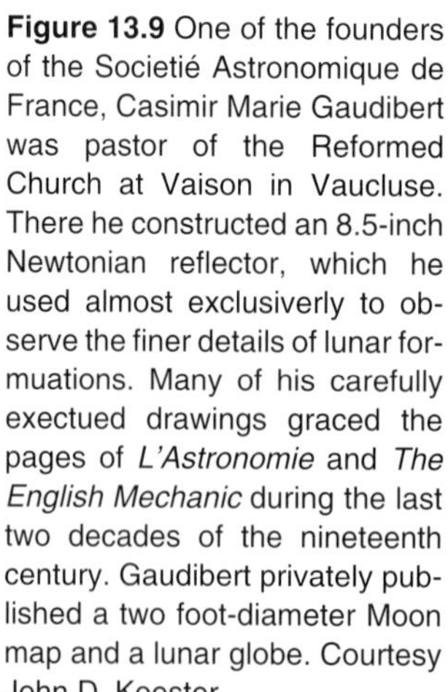

Figure 13.9 One of the founders of the Societié Astronomique de France, Casimir Marie Gaudibert was pastor of the Reformed Church at Vaison in Vaucluse. There he constructed an 8.5-inch Newtonian reflector, which he used almost exclusiverly to observe the finer details of lunar formuations. Many of his carefully exectued drawings graced the pages of *L'Astronomie* and *The English Mechanic* during the last two decades of the nineteenth century. Gaudibert privately published a two foot-diameter Moon map and a lunar globe. Courtesy John D. Koester.

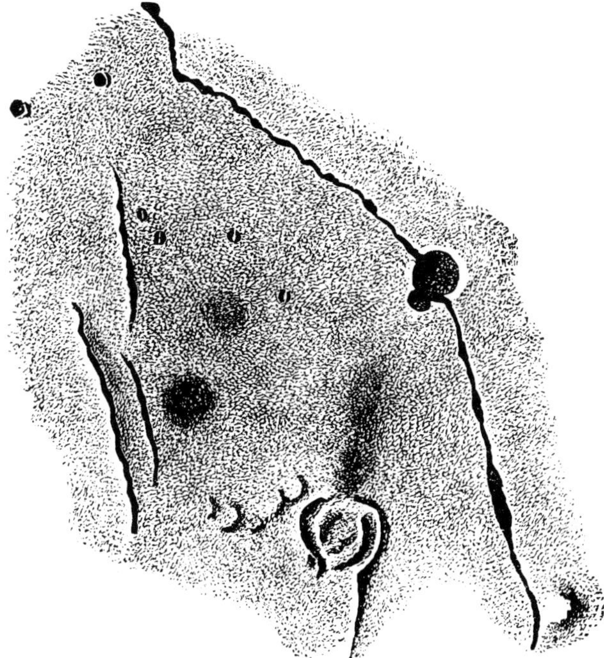

Figure 13.10 Hermann Klein's original drawing of "Hyginus N," another "new" feature which created a brief stir in the 1870s. The claim for a change in "Hyginus N" was always regarded as resting on a doubtful basis—that this feature was claimed as a novelty at all attests mainly to the preoccupation with lunar change that characterized the post-Linné era. From Hermann Klein, *Katechismus der Astronomie* (1900).

ambiguous, rendering problematical the identification by other observers of the exact feature he had in mind. Moreover, cloud and haze persistently obscured the Moon from observers in England.

Despite these difficulties, by the end of 1878 several members of the Selenographical Society had paid close attention to the region, and the presence of some sort of object, in the position indicated by Klein, had been confirmed by such leading British observers as Neison, Knott, and Sadler. The disputed feature was variously known as "Klein's crater" or "Hyginus N" (N for "Nova"), and was soon being discussed worldwide—even as far afield as Sydney, Australia, where the *Morning Herald*, on August 22, 1878, published the following account under the headline "A New Crater on the Moon":

> Having frequently observed this region during the last twelve years, Dr. Klein felt certain that no such crater existed in this region at the time of his previous observations. Dr. Klein communicated his observations to Dr. Schmidt, of Athens, the veteran selenographer, who assured him that this crater was absent from all his numerous drawings of this part of the lunar surface, neither is it shown by Schroeter, Lohrmann, nor Mädler, who carefully drew this region with the fine refractor at Dorpat… So far,

Figure 13.11 This pair of images elucidates the true nature of the supposed change in "Hyginus N." In the image at left, "Hyginus N" appears as a prominent dark spot. In the image at right, obtained under a slightly higher angle of solar illumination, it has all but faded from view. Images by T. A. Dobbins.

> the English observations of the new crater have been perfectly in accord with those of Dr. Klein, and, if the existence of this new crater be confirmed, it will form the strongest possible evidence of a real change on the surface of the Moon, a change, moreover, of a volcanic nature.[21]

Despite this enthusiastic account, it soon became evident that the appearance of Hyginus N was sensitive to slight variations in the angle of illumination, as is the case with so many lunar formations. When Klein made his first observation in May 1877, the altitude of the Sun had been only 2° and the feature had appeared large and black. In April 1878, under the same oblique conditions of lighting, he had found it "as obvious as Hyginus," i.e., deep black. But this was not its usual appearance. As the Sun rose over it, the feature faded, becoming less and less distinct until, by the time the Sun had climbed to between 10° and 20° above the local horizon, it faded to a "dull spot." Under still higher lighting it vanished altogether (Figure 13.11).

An especially important series of observations was obtained by J. Rand Capron of Guildford, using a 6-inch (152mm) Cooke refractor. Capron found that whenever the district was near the terminator, Klein's "crater" appeared as a "moderately sized, very slightly oval, cup-like cavity, of greyish-black tone, with an indistinct lighter margin or rim."[22] (Figure 13.12) The most probable explanation was that the object was not a crater after all, but only a shallow, saucer-like depression.

Whatever the exact character of the topography of this tangled region, the evidence that "Hyginus N" represented a change was, in the end, unconvincing. Pratt noted that "the probability is much against such a supposition… But whether new or not it must be confessed that the maps do not exactly represent the region in many points; and they also differ to considerable extent *inter se*."[23] Agnes M. Clerke, in her influential *Popular History of Astronomy*, was equally pessimistic. The sage Irish historian admitted that "the strongest incentive to diligence" in

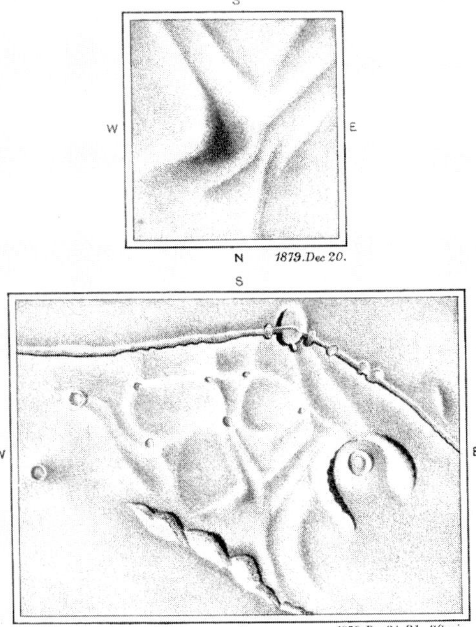

Figure 13.12 J. Rand Capron's sketches of the Hyginus region helped to dispel the notion that Klein's "Hyginus N" was a newly-formed crater. Courtesy John D. Koester.

studying the features of the Moon had always been the idea of change. However, this was obviously a case of the wish being father to the thought:

> The subject is one as to which illusion is peculiarly easy. Our view of the Moon's surface is a bird's-eye view. Its conformation reveals itself indirectly through irregularities in the distribution of light and darkness. The forms of its elevations and depressions can be inferred only from the shapes of the black, unmitigated shadows cast by them. But these shapes are in a state of perpetual and bewildering fluctuation... The novelty of Klein's observation of May 19, 1877, may have consisted simply in the detection of a hitherto unrecognised feature. The region is one of complex formation, consequently of more than ordinary liability to deceptive variations in aspect under rapid and entangled fluctuations of light and shade.[24]

The affair of "Hyginus N" was the most celebrated next only to that of Linné itself. But it seemed, even at the time, unconvincing, almost whimsical. After it boiled over, many observers seem to have become bored with the long and inconclusively debated question of "lunar changes." Perhaps not surprisingly, one detects a lapse of interest in selenography generally. The Selenographical Society proved to be just as ill-fated as the two Moon Committees of the BAAS had been, and after Birt's death in 1881 and Neison's departure for South Africa the following year to assume the directorship of the Natal Observatory at Durban, it was robbed of its very heart and soul. The Society dissolved. The great map of William Radcliffe Birt remained an unfulfilled dream—at any rate in the form he envisaged it.

Although diligent observers of the Moon remained, there was a diversion of interest to other objects of study—particularly the planet Mars, which attracted widespread attention ever since the presence of "canals" on its surface had been reported by the Italian astronomer Giovanni Schiaparelli (1835–1910). Schiaparelli had first seen the "canals" in 1877, the same year Klein made his observations of "Hyginus N." The tiny disc of the Red Planet—which at its closest approach remains 135 times more remote than the Moon and never appears larger than a modest lunar crater—seemed at the time to offer a more fertile and Earth-like field of investigation than the stereotyped page of the Moon's surface.

In retrospect, it is clear that the resources of selenography were only regrouping. The Promethean quest for lunar changes—what would be known in later parlance as "transient lunar events"—did not die with Birt anymore than it had died with Schroeter. The pursuit of veils, clouds, landslips, and eruptions had only temporarily exhausted itself. Regaining its vigor, it would remain the obsession of generations of selenographers yet unborn.

References

1. C. Piazzi Smyth, *Teneriffe, an Astronomer's Experiment: or, specialties of a residence above the clouds* (London: Lovell Reeve, 1858), pp. 317–318.
2. James Dwight Dana, "On the Volcanoes of the Moon," *American Journal of Science,* Second Series, vol. 2 (1846), 335–353.
3. George Julius Duncombe Poulett Scrope, *Volcanos: The character of their phenomena, their share in the structure and composition of the surface of the globe, and their relation to its internal forces* (London: Longman, Green, Longmans and Roberts, 1862, 2nd ed.), p. 231.
4. Ibid., p. 233.
5. Ibid.
6. Quoted in Scrope, *Volcanoes*, p. 30.
7. Neison, *The Moon*, p. 17.
8. Ibid., p. 19.
9. As noted in Patrick Moore, "E. N. N. Nevill: 'Edmund Neison,'" *Journal of the British Astronomical Association*, 75 (1965), 223–227:224.
10. Neison, *The Moon*, p. 44.
11. Ibid.
12. Ibid., p. 192. In part, Neison's reasoning was based on the recognition that a volcanic eruption on the scale of Linné ought to have produced a much more dramatic emission of gas and dust than any observed. On Earth, even the shattering explosions of Tambora in 1815 and Krakatoa in 1883 produced topographic changes that a telescopic observer on the Moon would have been very hard pressed to detect. Yet these catastrophic eruptions spewed forth dense palls of ash that circled the Earth and remained suspended in the stratosphere for many months, abruptly lowering global temperatures and producing vividly colored sunsets. Had the purported transformation of Linné been the result of a comparably violent upheaval, the accompanying clouds of smoke and ash could not have escaped the notice of a casual observer equipped with no more optical aid than a pair of field glasses!
13. Both, *A History of Lunar Studies*, p. 23.
14. Neison, "Lunar Observations," *The Observatory*, 1 (1877), 238–242. The reasons for failure of the project were also inherent in this kind of informal organization. As Allan Chapman notes, "the very nature of the informal social relationship... made it very difficult to collate effectively an extensive, detailed and sustained piece of teamwork such as was required for the production of the B.A.A.S. lunar map. In many ways, one could argue that the selenographical project only brought into focus one of the major tensions implicit within the Victorian scientific community, as the B.A.A.S. sought to direct scientific research along planned lines, whereas the independent ama-

teurs were constitutionally inclined to do their own thing. See *The Victorian Amateur Astronomer: Independent Astronomical Research in Britain 1820–1920* (New York: Wiley-Praxis, 1998), p. 82.

15. Birt, "On the Floor of Plato," *Monthly Notices of the Royal Astronomical Society,* 30 (1870), 24–26. See also, by the same author, "A few further notes on the floor of Plato," *Monthly Notices*, 30 (1870), 160–162, "Further notes on the floor of Plato," *Monthly Notices,* 31 (1871), 80–83, and "Report on the discussion of observations of spots on the floor of the lunar crater Plato, *British Association Report*, 1871, 60–97.
16. Schmidt, *Charte der Gebirge des Mondes* (Berlin: Dietrich Reimer, 1878).
17. J. Birmingham, "Schmidt's Lunar Map—II," *The Observatory*, 3 (1879), 11–17: 16.
18. R. S. Ball, *The Story of the Heavens* (London: Cassell & Co., 1908), p. 88.
19. Birmingham, "Schmidt's Lunar Map," p. 16.
20. H. Klein, *Selenographical Journal*, 1 (1878), 5–6.
21. Anonymous, "A New Crater on the Moon", *Sydney Morning Herald*, August 22, 1878.
22. Henry Pratt, "On Dr. Klein's Supposed New Crater, and the Region North of Hyginus," *The Observatory*, 3 (1879), 296–300: 296.
23. Ibid., p. 299.
24. Clerke, *Popular History of Astronomy*, p. 328. However, somewhat inconsistently, Clerke did not make the same case regarding Linné, of which she wrote: "We are… driven to accept one of two suppositions. Either Lohrmann, Mädler, and Schmidt were entirely mistaken… or a real change in its outward semblance supervened during the first half of the century, and has since passed away, perhaps again to recur. The latter hypothesis seems the more probable."

Chapter 14:
The Craters of the Moon

The volcanic theory of the origin of the lunar craters, so often cited by nineteenth-century geologists and astronomers, was given classic statement in the volcanic-fountain model, first advanced in 1874 by James Hall Nasmyth (1808–1890) (Figure 14.1) and James Carpenter (1840–1899).

Nasmyth was born in Edinburgh. The son of the painter Alexander Nasmyth and the younger brother of the landscape artist Patrick Milner Nasmyth, he showed exceptional mechanical aptitude and took an early interest in foundries and chemical laboratories. By his late teens he was making model steam-engines and even constructed a primitive automobile that he called a "steam road-carriage." Later he established a factory at Patricroft near Manchester for manufacturing machine tools, hydraulic punches, pile drivers, and steam hammers, an invention for which he bitterly disputed priority. The enterprise proved so profitable he was able to retire to the town of Penshurst, Kent at the age of 48.

Encouraged by his close friend, the Liverpool brewer and accomplished amateur astronomer William Lassell (1799–1880), Nasmyth became a telescope-maker. At Patricroft, he used a succession of ever larger Newtonian reflectors, culminating in a 13-inch (330-mm) instrument, and after removing to Kent, he erected a 20-inch (508-mm) Cassegrain reflector of innovative design (Figures 14.2 and 14.3):

> In order to avoid the personal risk and inconvenience of having to mount to the eye-piece by a ladder I furnished the telescope tube with trunnions, like a cannon, with one of the trunnions hollow, so as to admit of the eye-piece. Opposite to it a plane, diagonal mirror was placed, to transmit the image to the eye. The whole was mounted on a turn-table, having a seat opposite to the eye-piece... The observer, when seated, could direct the telescope to any part of the heavens without moving from his seat. Although this arrangment occasioned some loss of light, that objection was more than compensated for by the great convenience which it afforded.[1]

Neglected for decades, the "bent Cassegrain" or "Nasmyth" configuration would be widely adopted in the twentieth century on many large professional instruments, notably the 60-inch (1.52-meter) Mount Wilson reflector.

As an observer, Nasmyth was always keenly interested in the Moon, and became one of the first to contribute to Birt's Moon Committee. Unlike Birt, who was always on the lookout for evidences of ongoing changes in the lunar surface, Nasmyth's preoccupation became the solution of the riddle of the origin of the features which he saw scattered over its crowded surface. As a self-taught amateur, he enlisted a more learned collaborator for his work in Carpenter, a computer and later assistant astronomer at the Royal Greenwich Observatory. "A long course of reflective scrutiny of the lunar surface with the aid of telescopes of considerable power," Nasmyth and Carpenter later wrote,

> convinced us that there was yet something to be said about the Moon... Very little has been written respecting the Moon's physiography, causative phenomena of the features, broad and detailed, that the surface of our satellite presents for our study."[2]

Figure 14.1 James Nasmyth, strikingly portrayed by Paul Rajon for the frontispiece of Nasmyth's autobiography. From *James Nasmyth, Engineer: An Autobiography* (1883). After the portrait by G. Reid.

This was, in fact, the emphasis of their great book *The Moon: Considered as a Planet, a World, and a Satellite* (1874), one of the most influential works on the lunar surface ever written. Nasmyth and Carpenter accepted that the craters of the Moon had been formed by volcanic action. However, they did not attempt to gloss over the fundamental difference of form between terrestrial craters and their lunar counterparts. Terrestrial craters consisted of a hollow atop a mountain, but the interior bowls of lunar craters were almost invariably depressed below the level of the surrounding surface. And yet despite the apparent differences in form, they nevertheless declared: "... the lunar craters truly volcanic; as Sir John Herschel has said, they offer the true volcanic character *in its highest perfection*... We have some terrestrial volcanic districts that, could we view them under the same circumstances, would be identical in character with what we see by telescopic aid upon our satellite... The resemblance of which we are speaking is here so close that Professor Phillips, in his work on Vesuvius... calls the Moon a grand Phlegraeian field."[3]

Nasmyth and Carpenter endorsed a version of Laplace's cosmogony, in which particles of diffused primordial matter resulted in the formation, "by condensation and aggregation, of a spherical planetary body."[4] After formation of the planet, the next stage was the formation of a solid shell encompassing a molten core. Consistent with his empirical approach, however, Nasmyth did not take the formation of such a solid shell on faith. Instead he applied to the manager of an iron-works to verify that cold slag floated on molten slag just as solidified iron floated on molten iron, and during an 1865 visit to Vesuvius—then undergoing a "modified eruption"—he and Carpenter observed the process firsthand. They saw "white-hot lava streaming down from apertures in the sides of a central cone within the crater and forming a lake of molten lava on the plateau or bottom of the crater; on the surface of the molten lake vast cakes of the same lava which had

Figure 14.2 Nasmyth's 20-inch Cassegrain rode atop a revolving turntable. It was the first reflector to employ the convenient "Nasmyth focus" that was later emulated in such famous instruments as the 60-inch reflector of the Mt. Wilson Observatory. From *James Nasmyth, Engineer: An Autobiography* (1883).

Figure 14.3 Nasmyth's 20-inch telescope erected on the lawn of his estate at Penshurst, Kent. From *James Nasmyth, Engineer: An Autobiography* (1883).

become solidified were floating, exactly in the same manner as ice floats in water."[5]

These principles could be applied to the evolution of the Moon. Cooling from the outside in, the encasing outer shell would cause pressures to build up in the still-molten interior until it was "rent or burst open, and a portion of the molten interior ejected with more or less violence according to the circumstances."[6] As each successive shell cooled and contracted in the attempt to accomodate itself to the core beneath it, the skin would crease into alternating ridges and depressions, "like a long-kept shriveled apple."

Nasmyth and Carpenter saw everywhere on the surface of the Moon evidence of this general process. The crater Copernicus—the "Monarch of the Moon"—was evidently "the result of a vast discharge of molten matter which has been ejected at the focus or centre of disruption of an extensively upheaved portion of the lunar crust... Were we to select a comparatively limited portion of the lunar surface abounding in the most unmistakable evidence of volcanic action in every variety that can characterize its several phases, we could not choose one yielding in all respects such instructive examples as Copernicus and its immediate surroundings."[7] Tycho and its system of rays similarly represented "an instance of vast disruptive action which rent the solid crust of the Moon into radiating fissures... subsequently occupied by extruded molten matter."[8] The process that formed it was evidently analogous to the cracking of a water-filled glass globe when heat is applied, where the contents build pressure until the container is ruptured.

The main problem in understanding the origin of the lunar features remained their sheer size relative to those of the Earth. After all, the crater of Vesuvius was less than two miles across—a mere pinprick by lunar standards. In even the largest

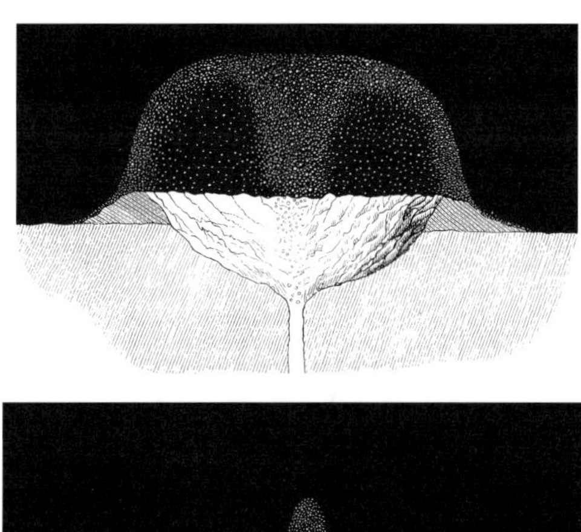

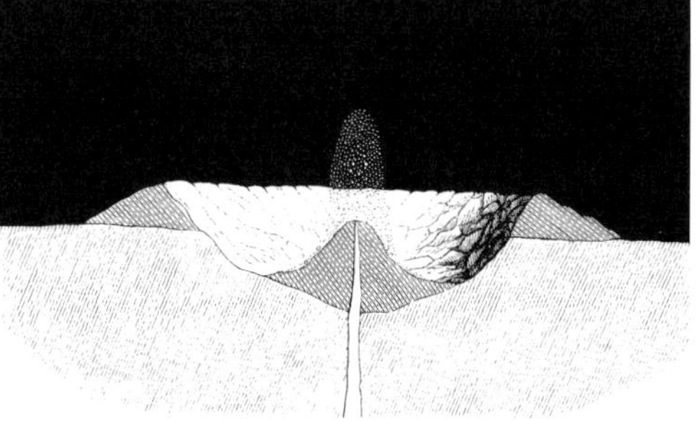

Figure 14.4 Nasmyth and Carpenter's influential "fountain model" is illustrated in these two views. (Top) During the early stages, the eruption tends to be more violent, depositing material in a distant concentric ring. (Bottom) Later, as the activity becomes more subdued, the central peak characteristic of many larger craters is formed. Nasmyth and Carpenter arrived at this process deductively; as their critics later pointed out, it had never actually been observed to take place in any volcanic formation on the Earth. From Nasmyth and Carpenter, *The Moon, Considered as a Planet, a World, and a Satellite* (1874).

terrestrial volcanic craters—Mount Asosan on Japan's Kyushu Island, the summit crater of the Bezimianniy volcano in Russia's Kamchatka Peninsula, and Lake Bombon in the Phillipines, the water-filled caldera of an extinct volcano—the craters only measured some 15 miles across. Yet on the Moon, craters like Tycho and Copernicus, with diameters of 50 and 56 miles respectively, were not even the largest specimens of their kind.

Nasmyth and Carpenter's solution was their "fountain model" (Figure 14.4). They assumed that the reduced surface gravity of the Moon, combined with the negligible resistance of its thin atmosphere, would cause lunar volcanoes to eject matter in a fountain-like fashion—as "a pyramid of ejected matter, thrown out and around an orfice." The ejected material would fall symmetrically—this, by the way, was always the chief difficulty of the theory—building up the crater walls and producing a ring structure. Inevitably, some of the material would fall back into the central structure, only to be blown out again. Once the most violent phase of activity subsided, the vent hole would at last fill in with lava to produce the crater floor, though a final resurgence of activity might give rise to the central peaks found in many of the lunar formations, which Nasmyth and Carpenter took to be cinder cones.

The central peaks found in so many craters were, then, the last gasps of the Moon's dying internal activity as the lunar crust grew thicker and more rigid and the reserves of lava dwindled. According to Nasmyth and Carpenter, such internal activity had ceased long ago, and the Moon was a dead planet—a cosmic corpse consigned forevermore to frozen oblivion.

Though the central peaks were found in most of the moderate-sized forma-

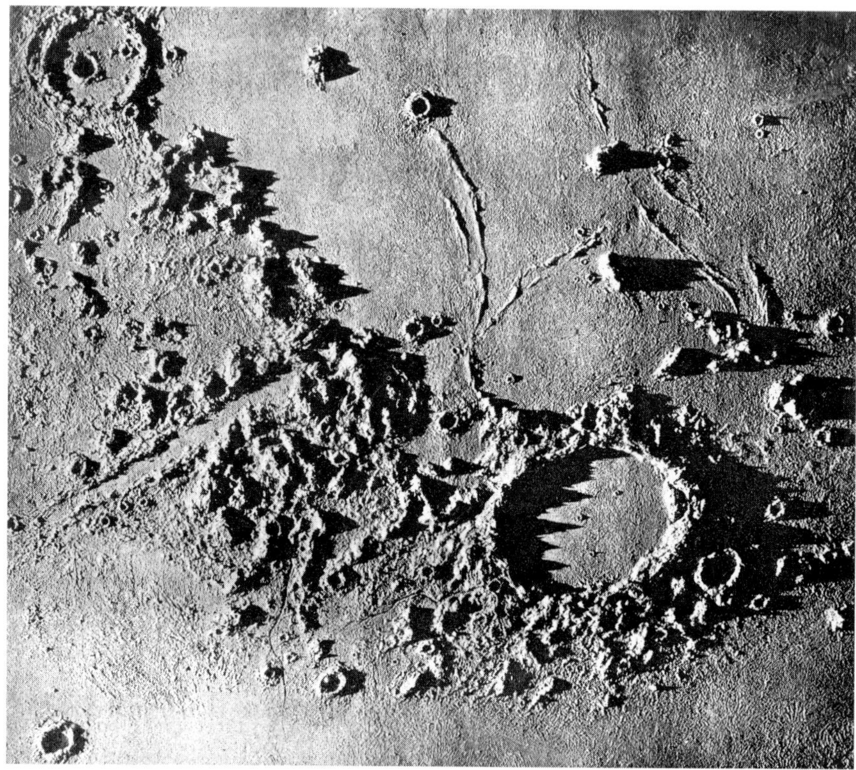

Figure 14.5 Nasmyth and Carpenter's stunning plaster-of-Paris model of the lunar Alps and the crater Plato. Photographs of these models were reproduced by the laborious woodburytype process, which requires a relief mold of the image to be made in lead. Dark areas are deep and bright areas are shallow. The mold is then filled with a suspension of ink in gelatin. The resulting prints replicate tones by the thickness or thinness of the ink and are truly grainless. From Nasmyth and Carpenter, *The Moon, Considered as a Planet, a World, and a Satellite* (1874).

tions on the Moon, they were almost entirely lacking in the largest ones, the walled plains—Ptolemaeus, Grimaldi, Schickard, and Clavius. Nasmyth and Carpenter admitted that these formations were not, therefore, "manifestly volcanic." They even regarded Mare Crisium, which measures some 270 miles from north to south by 300 miles from east to west (making it slightly larger than the state of Iowa), as "a formation not distinct from those which we have just named." The broken-bubble theory of Hooke could no longer be entertained, for there was no material capable of supporting bubbles on such an enormous scale. In the end they were at a loss to explain such gigantic features. They could do no more than recall Dana's old ebullition theory, which, however, they did not quite bring themselves to endorse.[9]

Nasmyth and Carpenter's theory was widely accepted and completely overshadowed other theories of the origin of the craters for some time. The impression made by their book was, moreover, greatly enhanced by their inclusion of a series

Figure 14.6 The spectacular Nasmyth and Carpenter model of the lunar Apennines and the nearby craters Archimedes, Aristillus, and Autolycus. From Nasmyth and Carpenter, *The Moon, Considered as a Planet, a World, and a Satellite* (1874).

of photographs of twenty four exquisite plaster-of-Paris models taken under oblique lighting. With these models the authors attempted to reproduce the appearance of "the most characteristic and instructive" lunar formations and, for purposes of comparison, the earthly Campi Phlegraei. These striking plates, which made the book almost as much a work of art as a piece of scientific literature, were widely reproduced (Figures 14.5, 14.6, and 14.7). Their rugged, shadow-sketched landscapes and sharp, craggy peaks (Figures 14.8 and 14.9)—which, though inaccurate, were dramatic, to say the least—would remain in vogue for decades as evocative illustrations in popular books on the Moon and serve as examplars faithfully followed by other lunar illustrators (Figures 14.10, 14.11).

Large drawings in black and white chalk on gray paper made at the eyepiece by Nasmyth over a period of thirty-two years served as the basis for these models, not measurements of shadows to accurately quantify elevations and slopes. Nasmyth even grossly overestimated the depth of several lunar rilles as an incredible ten miles. Consequently, the models are very deceptive; they consistently depict the Moon's surface as far more jagged than it really is.

Nasmyth detected a low, dome-like swelling on the Mare Nubium near the Straight Wall, just to the north of the crater that is now known as Birt (Figure 14.12). He suspected that this feature represented an enormous unburst bubble formed when the lunar crust was in a semi-molten, plastic state, but remarked that to his surprise he had been unable to detect other examples. In fact, Nasmyth's telescope was sufficiently powerful to reveal literally scores of these features, which are referred to simply as "domes" today.[10] Indeed, Schroeter had described

Figure 14.7 Nasmyth and Carpenter's model of the craters Theophilus, Cyrillus, and Catherina makes for an interesting comparison with Figures 9.1 and 9.8 (pages 120 and 131). From *The Moon, Considered as a Planet, a World, and a Satellite* (1874).

Figure 14.9 Nasmyth and Carpenter's "ideal" lunar landscape. From *The Moon, Considered as a Planet, a World, and a Satellite* (1874).

Figure 14.8 Nasmyth and Carpenter's depiction of the lunar mountain Pico, which stands in splendid isolation on the floor of Mare Imbrium. Though the mountain's actual profile is far more gentle, this rendering was emulated by many later artists who represented the lunar surface as uncompromisingly rugged and craggy. From *Nasmyth and Carpenter, The Moon, Considered as a Planet, a World, and a Satellite* (1874).

Figure 14.10 The craggy spires of this lunar landscape, which appeared in W. H. Warren's *Recreations in Astronomy* (1886), are vintage Nasmyth and Carpenter.

Figure 14.11 More lunar landscapes inspired by Nasmyth and Carpenter. At top, the lunar Apennines, at bottom, the southern highlands, from A. T. Arcimis, *Astronomia Popular*, published in Barcelona in 1901.

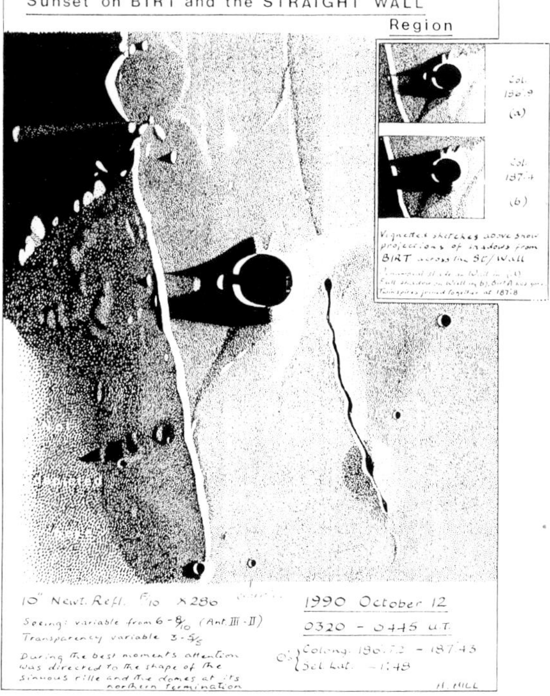

Figure 14.12 The Straight Wall at sunset, drawn by the British observer Harold Hill. Note the dome at the northern end of the rille Rima Birt at the lower right. Of the scores of domes within the grasp of his telescopes, including many known since Schroeter first recorded them, Nasmyth took notice only of this single example. Courtesy Harold Hill.

several *Beulen* ("swellings" or "boils") which he characteristically attributed to "the industrial activity of the Selenites." Nasmyth's failure to notice these features shows that he was neither an assiduous observer possessing a truly intimate knowledge of his subject nor adequately acquainted with the work of his predecessors.

In retrospect, there were many obvious problems with Nasmyth and Carpenter's theory. The most significant was the fact that terrestrial volcanoes were never observed to build up a ring-wall by fountain-like action. Successive eruptions of gradually diminishing intensity would produce a series of concentric ridges around the central fountain rather than a single ring-wall. The exterior slope or 'glacis' of a typical lunar crater is inclined only 5°, far too gentle for such an origin. By comparison, the angle of repose of ash around terrestrial volcanoes is typically 40° to 45°. While the difference in surface gravity between the Earth and the Moon *might* account for differences in the scale of analogous formations, it could never produce such systematic differences in morphology.

It was also difficult to understand why, during the filling-in phase, lava invariably stopped below the level of the surrounding wall. (The sole exception seemed to be the peculiar crater Wargentin). Similarly, Nasmyth and Carpenter could not explain why the central peaks, if they were indeed cinder cones, never grew as high as the surrounding ring-wall. These were problems that the various plutonic theorists never did succeed in overcoming—though hardly for want of trying.

At least they were right in supposing the Moon had passed through various stages in its development—an idea that was never again seriously questioned. The early efforts of Hutton and Clerk to establish a stratigraphic series for the Moon had remained buried in obscurity, but other investigators rediscovered the general relationship that had impressed them. Smaller formations were always superimposed on larger formations rather than the other way around. This was consistent with the idea of secular cooling. Great fires had obviously burned once within the Moon—seething energies had raised its great rings and mountains. Over time the power of the Moon's eruptions ebbed with the burning out of these internal fires, until the present stage was reached in which the Moon was in a state of utter paralysis—cooled throughout, and immobilized into the rigid death-mask of a frozen world. (Parenthetically, this may explain why Charles Lyell, despite having made at least a few observations of the Moon himself, never mentioned the surface of the Moon at all in the *Principles of Geology*—the history it revealed was patently at odds with the principle of uniformitarianism.)

Other observers had reached the same conclusion from the superposition of smaller features everywhere on larger ones of the same type. Thus Webb had noted: "A succession of eruptions may be constantly traced in the repeated encroachment of rings on each other, where, as Schmidt says, the ejected materials seem to have been disturbed before they had time to harden, and the largest are thus pointed out as the oldest craters, and the gradual decay of the explosive force, like that of many terrestrial volcanoes, becomes unquestionable."[11] The French astronomer

Jean Chacornac had been equally impressed by this sense of sequence: "The great continental formations [belong] to an epoch anterior to that of the production of the great plains, this again being anterior to the period of explosive energy... It would greatly assist our study if some attempt, however rough, were made to arrange a list of formations according to their supposed ages."[12]

Nasmyth and Carpenter's book attracted extensive commentary and numerous modificiations from other selenologists. The Austrian geologist Eduard Suess (1831–1914), best-known for his work *Das Antlitz der Erde* (The Countenance of the Earth), whose four volumes provided the most definitive elucidation of global contraction as an explanation of the origin of the principal features of the Earth's surface, attempted to remedy one of the deficiencies of Nasmyth and Carpenter's book by offering the following explanation of the lunar maria:

> The temperature of the large mass [meaning the Moon as a whole] is by no means completely uniform. In some areas it rises, melts down the cover of the slag once more, and from there the re-melting proceeds in all directions, for hundreds of kilometers. The shape of the molten mass is a section of a sphere, its projection a circle. Finally the process slows down, the temperature of the surface around the rim is lower, the slag is no longer melted down but is pushed outward like a moraine. Then it is all over. What is left is a large plain circular in shape and surrounded by a wall of slag... such as the enormous one which under the name of Apennines, Alps, etc., surrounds the Mare Imbrium.[13]

Admittedly, it was never clear why the great basaltic floods of Earth never formed such moraine-like walls, nor why, so late in the process when much of the Moon's internal heat had been expended, there should have remained expansive power sufficient to pile up mountainous heaps of unmelted slag.

Internal volcanic forces alone might not be sufficient, but there was another source of energy available to supplement lunar volcanoes—the tides. These had been powerfully operative long ago, before the Moon's rotation had become tidally locked into synchrony with its period of revolution. The tidal hypothesis was advanced by the French astronomer Hervé Faye (1861–1913)[14] and the German physicist Hermann Ebert (1814–1902).[15] "Let us imagine," wrote Ebert, "a shining-hot viscous planetary body that cools by radiation on all sides. Thus, on its surface, rigid 'frozen' flakes would form that float in the still liquid magma. In the case of the Moon there exists the attraction of the Earth which caused a great tidal wave of the liquid parts. The Moon rotates under this tidal wave and therefore in every part of its surface the liquid magma is alternately raised and lowered. It welled up during every tide, flooded these lakes and retreated with the coming of the ebb, and the play was repeated during the next tide."

To demonstrate the process, Ebert filled a pan with a conveniently low-melting alloy of lead, tin, bismuth, and cadmium known as Wood's Metal. The surface of the metal was allowed to solidify, while at the bottom of the pan it was kept molten by applying steam. By means of a hand pump that simulated the Earth's tidal forces, Ebert produced localized upwellings of liquified metal through small holes in the crust that spread out into circular pools. A reduction in pressure cor-

Figure 14.13 The prolific British astronomy popularizer Richard Anthony Proctor, an early champion of the impact theory of the origin of lunar craters. From Proctor's *Other Worlds Than Ours* (1892).

responding to ebb tide sucked the molten metal underneath the surface but left a solidified ring marking the confines of the emptied pool. After several repetitions of this cycle, Ebert succeeded in forming craters like the Moon's, i.e., depressed below the level of their surroundings and having terraced inner walls and central peaks.

As suggestive as the results of Ebert's experiments were, there clung to them the suspicion that they were based in a false analogy, inasmuch as it was never clear that tidal forces would really work in such a fashion on the Moon, where the solid crust must float in isostatic balance on the liquid interior, rising and falling with tide-induced waves. The German geologist Alfred Wegener (1880–1930) later wrote that "the hypothesis of Ebert is as impossible as to try to measure the tide and ebb from the floating icebergs in the North Sea."[16]

Though the volcanic theory flourished in the years after Nasmyth and Carpenter published their fountain theory, it did not go entirely unchallenged. The old impact theory was revived by Richard Anthony Proctor (Figure 14.13), who was born in Beer and Mädler's year, 1837. He received his education at King's College, London and St. John's College, Cambridge. After graduating with degrees in theology and mathematics, he was elected a Fellow of the Royal Astronomical Society and became editor of its *Monthly Notices* and—motivated in part by dire financial need following severe losses in a bank failure—he embarked on a career as one of the great popularizers of astronomy of all time.

His 1865 treatise *Saturn and Its System* was notable for providing the first detailed popular summary of James Clerk Maxwell's mathematical demonstration that the rings of Saturn, instead of being solid or liquid, consisted of, in Proctor's words, "flights of minute bodies, each travelling in its own orbit." Indeed Proctor saw the Saturnian system as a miniature and partially evolved version of the Solar System, and thus believed that it furnished important evidence of the truth of the cosmogony of Laplace. He returned to the subject of the Nebular Hypothesis in another popular book, *Other Worlds than Ours* (1870), in which he proposed modifications to Laplace's original theory. Instead of merely cooling and contracting from a nebula, as Laplace had suggested, Proctor argued that the true history of the Solar System involved the accretion of meteoric bodies into planets—a touch which, he asserted, eliminated what he regarded as the chief difficulty of the original theory, its positing the existence of planets of vastly different ages as they condensed out of successive nebular rings.

In his 1873 book *The Moon*, Proctor elaborated his view that the Solar System "had its birth, and long maintained its fires under the impact and collision of bodies gathered in outer space." He attempted to explore the implications of his theory for the "crateriferous surface of the Moon":

> According to this view, the Moon, formed at a comparatively distant epoch in the history of the Solar System, would have not merely had its heat originally generated for the most part by meteoric impact, but while still plastic would have been exposed to meteoric downfalls, compared with which all that we know, in the present day, of meteor-showers, aërolithic masses, and so on, must be regarded as altogether insignificant. It would be to such downfall mainly that the maintenance of the Moon's heat would at that time be due... It may seem, indeed, at a first view, too wild and fanciful an idea to suggest that the multitudinous craters of the Moon, and especially the smaller craters revealed in countless numbers when telescopes of high power are employed, have been caused by the plash of meteoric rain,—and I should certainly not care to maintain that as the true theory of their origin; yet it must be remembered that no plausible theory has yet been urged respecting this remarkable feature of the Moon's surface.[17]

Proctor added a few further details, such as the idea that the craters of the Moon had formed when its surface was still in a hot, plastic condition. This would explain how, even if the orbits of the infalling meteorites had been elliptical and had often struck the lunar surface at oblique angles, they still formed nearly circular impressions, since "the plastic surface would close in round the place of impact." He acknowledged that the larger formations of the lunar surface posed greater difficulties for the theory than did the smaller ones. "I am... sensible that the great craters... by no means correspond in appearance to what we should expect if they were formed by the downfall of great masses from without. The regular and battlemented aspect of some of these craters, the level floor, and the central peaks so commonly recognized, seem altogether different from what we should expect." Still, he hinted that "it is indeed just possible that under the tremendous heat generated by the downfall, a vast circular region of the Moon's surface would be rendered liquid, and that in rapidly solidifying while still traversed by the ring-waves resulting from the down-fall, something like the present condition would result."[18]

Despite its lucid style, Proctor's book did not have much immediate influence. It "stirred neither criticism nor controversy and, in fact… was largely ignored by his contemporaries in astronomy, who continued to favor the volcanic analogy in general."[19] In a note in the *Selenographical Journal,* W. R. Birt even claimed that Proctor had forsaken the meteorite-impact theory by 1878.[20] The basis of Birt's comment is uncertain, perhaps nothing more than Proctor's own remark that he "should certainly not care to maintain that as the true theory of their origin," but it is worth noting that if he did not believe in the theory, he nevertheless went to the trouble of restating it that very year with further elaborations—he pointed out, for instance, that ancient features produced by meteoric rain on the surface of the airless and moistureless Moon would survive indefinitely, while all traces of them had long since been eroded from the face of the Earth.[21]

In Germany, where the impact theory had first appeared in the early writings of the brothers Bieberstein, von Moll, and Gruithuisen, the idea was revived in 1877 by the architect Anton Meydenbauer.[22] His approach was experimental. By dropping dextrin powder from a knife onto a surface of the same material, he was able to fashion features similar to the Moon's craters, complete with central peaks. He realized, as Proctor had failed to do, that a hot and plastic condition of the surface was not required for the impact process to work. Indeed, Meydenbauer believed that the surface of the Moon was covered with dust of some sort. Perhaps, he suggested, the lunar maria were masses of formerly molten sulfur and phosphorus.

Unfortunately, there were few English astronomers who were familiar with the German literature, and *vice versa.* An exception was Nasmyth and Carpenter's book, which was widely read in Germany because of the availability of Hermann Klein's translation and which attracted the attention of Heinrich Thiersch, a professor of theology, and his son, the architect August Thiersch. They vigorously disputed Nasmyth and Carpenter's thesis, rejecting it in a forcefully written pamphlet published under the pseudonym "Asterios" in 1879, but reprinted under their own names in 1883. The Thiersches suggested that not only the smaller craters but even the great plains of the maria, including the giant Mare Imbrium, were formed by impacts. In order to account for the vast number of lunar craters, the Thiersches invoked a form of the accretion theory: "Like Saturn, the Earth could also have had a ring of moons around it… Through aeons the largest of these, our own Moon, might have attracted and assimilated those [moons]."[23]

The same conclusions were reached again, independently and far more authoritatively, by Grove Karl Gilbert (1843–1918) (Figure 14.14), one of the most respected American geologists of his generation, who began an intense and singularly productive study of the Moon in 1892.

Gilbert was born in Rochester, New York, and trained in mathematics and classical languages at the University of Rochester.[24] After a brief stint as a schoolteacher in Michigan—a gentle soul, he found himself unable to maintain class discipline—he spent five years sorting specimens as a clerk in a scientific factory in Rochester known, in true Humboldtean fashion, as Cosmos Hall. His real break

Figure 14.14 Grove Karl Gilbert of the U.S. Geological Survey made fundamental contributions to virtually every aspect of geology, including the study of glaciation, sedimentation, hydrology and geophysics. A consummate scientist, he set out for Coon Butte in northern Arizona shortly after meteoric iron was found there in an attempt to determine whether it had been formed by a "star." Later he arrived in San Francisco within hours of the devastating 1906 earthquake, making careful records of how long it took the ensuing fires to consume wooden structures of various sizes. Courtesy U.S. Geological Survey.

came when he obtained a position with John Strong Newberry's Geological Survey of Ohio. Newberry's reports, in addition to their geologic maps, stratigraphic columns, and cross sections, were organized according to the recently fashionable principle of evolution. Newberry insisted that without a knowledge of fossils, geology was gibberish. He argued before state legislators who cared little for "clams and salamanders" that his fossils were in fact practical in the highest degree; "they were the labels written by the Creator on all the fossiliferous rocks... no one can be a Geologist who has not learned their language."[25]

Through his work on Newberry's Geological Survey of Ohio, Gilbert in 1871 received an appointment to Lieutenant George M. Wheeler's wide-ranging army survey of that wonderland of geology, the American Southwest. Since Wheeler was the one who set the agenda, Gilbert's observations were often rushed, and he complained that he often felt like one obliged to read a book while its pages were quickly being turned by another. Nevertheless, he did obtain important insights. He realized, for instance, that the dried-apple theory of the Earth's contraction, though able to explain gently folded structures such as the Appalachians, could not account for sharp ridges like those of the mountains of Nevada, which he described vividly as "like an ocean wave just about to break and curl westward."[26] He was convinced that the western cordillera of Nevada could never have been formed by mere forces of horizontal compression owing to the shrinkage of the Earth. Instead it was evidence of a different kind of process involving vertical pressure from below, which was still working to elevate the land. Another important insight was that the great Bonneville basin of Utah was formed by the evaporation of a large, thousand-foot deep inland sea after the last Pleistocene ice age—the Great Salt Lake remains as its much-shrunken remnant.

Gilbert's reports of these researches already show the development of his basic approach. His "systematic geology," as his biographer Stephen J. Pyne puts it, proceeded "by an arrangement of careful, systematic contrasts, in which various geologic regions, or systems, or various geologic processes are compared with respect to their fundamental similarities and differences."[27] As Gilbert himself put it, he preferred wherever possible to make general statements rather than to draw up mere lists of facts. This approach, obviously, was in striking contrast to that being followed by most students of the Moon in the closing years of the nineteenth century.

Indeed, up to and even after the dissolution of the Selenographical Society, most of those who concerned themselves with the Moon were amateurs, still actively pursuing or recently retired from other professions and working self-consciously in the shadow of a heroic past. Their goal remained, as it had been since the days of the Moon Committee, to obtain (like Dickens's Gradgrind) "facts, facts, facts." Books on the Moon like those of Neison and Klein, and later Thomas Gwyn Empey Elger's (1837–1897) *The Moon, A Full Description and Map of its Principal Physical Features (*1895), attest more to the rapid accumulation of information than to its assimilation. They are testaments to the undampened influence of the Humboldtean vision of lunar science first elaborated by Beer and Mädler—painstaking gazetteers of walled plains and ring-mountains. The attention to detail was precise—as E. E. Both has pointed out, this generation "at times lost itself in the trivialities of drawing ever smaller areas of the lunar surface without really attempting to create a new and unified whole."[28] Occasionally it was even fanciful, as testified by following passage from Elger: "On the broken flank of Guerike is a number of little rings, all open to the N: on the [west] of these commences a linear group of lofty, isolated mountain masses extended toward the [E] side of Parry and prolonged for 30 miles or so towards the north. They are arranged in parallel rows and remind one of a Druidical avenue of gigantic monoliths viewed from above."

In 1875 Gilbert joined the U. S. Geological and Geographical Survey of the Rocky Mountain Region led by Major John Wesley Powell, the one-armed veteran of the Civil War who had achieved fame for his daring exploration of the Colorado River as far as the Grand Canyon. It was under Powell's direction that Gilbert carried out his most important investigations as a member of the the Rocky Mountain Survey, and later as a geologist of the U. S. Geological Survey. By temperament, he was different from the romantically minded Powell or his other close associate, Clarence Dutton, both of whom wrote their most important works about the Grand Canyon. Gilbert was more restrained, more controlled. His associates described him as remarkably even-keeled; always "imperturbable... his clear-eyed serenity... as distinctive a trait of his personality as his beard was to his appearance."[29] He did not indulge in flights of poetic fancy as did so many of his associates, though he did share their talent for, in Pyne's phrase, "distilling a scene into its essential lessons."[30]

In 1891, Gilbert's survey work led him to Coon Butte near Canyon Diablo in northern Arizona, and thence to his study of the Moon. Now known as "Meteor Crater" (Figure 14.15), Coon Butte consists of an arid plain whose scanty soils lie atop beds of limestone. The plain is "interrupted by a bowl-shaped or saucer-shaped hollow, a few thousand feet broad and a few hundred feet deep... In other words, there is a crater; but the crater differs from the ordinary volcanic structure

Figure 14.15 Part of the rim of Meteor Crater of northern Arizona. This view was taken from the public observing platform on the crater rim. Photograph by William Sheehan, 1982.

of that name in that it contains no volcanic rock. The circling sides of the bowl show limestone and sandstone, and the rim is wholly composed of these materials."[31]

Following the discovery of iron at the site, it was visited by a prominent mineralogist, A. E. Foote, who presented his findings at a meeting of the American Association for the Advancement of Science in Washington on August 20, 1891.[32] Gilbert, present at the meeting, heard Foote suggest the iron was of celestial origin—the remnant of a shower of meteorites. "I asked myself," he later wrote, "what would result if another small star should now be added to the Earth, and one of the consequences which had occurred to me was the formation of a crater, the suggestion springing from the many familiar instances of craters formed by collision."[33]

Eventually, Gilbert decided to investigate the question for himself. In October 1891, he and an assistant with the Geological Survey, Marcus Baker, boarded a train for the small lumbering community of Flagstaff, Arizona Territory—soon to become famous as the site of Percival Lowell's observatory founded expressly for the study of the planet Mars—located at the foot of the 12,670-foot extinct volcanoes of the San Francisco Peaks. "I am," Gilbert announced to a friend, "going to hunt a star."[34]

Gilbert's tests in the field—including his failure to detect the deflection of a magnetized needle at the crater—led him to conclude against the falling star theory. Instead, he decided the crater had been formed explosively by steam—in short, it was a volcanic feature of the type known as a *maar*. He would never discuss Coon Butte again—publicly at least—though it remained the subject of intensive study by others and would eventually and conclusively be shown to be an impact feature. But it was his study of Coon Butte which led directly to his interest in the craters of the Moon.

Indeed, almost immediately after his return to Washington, in the two months of August and December 1892, he turned the 26-inch (660-mm) Clark refractor of the U.S. Naval Observatory (Figure 14.16) and his geologically-trained eye to the Moon—thereby incurring the wrath of at least one member of Congress, critical of the Geological Survey's work, who complained: "So useless has the Survey become that one of its most distinguished members has no better way to employ his time than to sit up all night gaping at the Moon."[35] Gilbert's situation was in some ways unique. A novice in lunar observation, he readily acknowledged that previous observers had completed a survey of the visible hemisphere of the Moon—a vast area equal to that of the North American continent—so thorough that there was "no remaining space on which to write the legend 'unexplored.'"[36] However, they had not looked at it with his trained geologist's eye. The problem of the origin of the Moon's features was "largely a problem of the interpretation of form," he pointed out, which made it "not inappropriate to one who has given much thought to the origin of the forms of the terrestrial topography."[37]

As soon as he turned the telescope and his informed eye to the Moon, Gilbert found no analogy between the lunar craters, with their land-slip terraces and central peaks, and terrestrial volcanoes invoked by so many earlier writers. Terrestrial volcanoes were closely grouped around a certain maximum size, "as though constrained by a limiting condition"; the largest lunar craters—Imbrium, Serenitatis, Crisium, Humorum, Humboldtianum, Bailley, Sinus Iridum and Clavius—were widely scattered about a maximum, "like aberrant shots from the bull's eye." Even more significant were the differences in form. Craters of the Vesuvian type, which included ninety-five percent of terrestrial craters, were formed by lavas containing considerable amounts of water. As the lava rose, this water was converted into steam, and by the propulsive power of steam the lava was torn to pieces and hurled high into the air. Repeated episodes of this process—intermittent explosions followed by periods of quiescence—formed a conical mountain with a funnel-shaped cavity at its summit. Such craters, however, had little in common with those of the Moon:

> Ninety-nine times in one hundred the bottom of the lunar crater lies lower than the outer plain; ninety-nine times in a hundred the bottom of the Vesuvian crater lies higher than the outer plain... The lunar crater is sunk in the lunar plain; the Vesuvian is perched on a mountain top. The rim of the Vesuvian crater is not developed, like the lunar, into a complex wreath, but slopes outward and inward from a simple crest-line. If the Vesuvian crater has a central hill, that hill bears a crater at its summit and is a miniature reproduction of the outer cone; the central hill of the lunar crater is entire, and is distinct in topographic character from the circling rim... Thus, through ex-

Figure 14.16 At top, the dome of the 26-inch Clark refractor of the U.S. Naval Observatory. (Bottom) The telescope, built by the celebrated firm of Alvan Clark and Sons, was the largest of its kind in the world when it was commissioned in 1875. Its original location was at Foggy Bottom, a malarial swampland along the Potomac notorious for its indifferent seeing. In August of 1877, Asaph Hall made the most celebrated discovery ever made with the telescope—the two dwarf moons of Mars—fifteen years before G. K. Gilbert carried out his pioneering study of the surface of the Moon. In 1893, the telescope was moved to its present site on higher ground in the Georgetown district of northwest Washington, far from the murky lowlands along the river. Photographs by William Sheehan, 1997.

> pression of every feature the lunar crater emphatically denies kinship with the ordinary volcanoes of the Earth.[38]

Gilbert thus rejected the analogy between the lunar surface and the Phlegraean fields affirmed by Sir John Herschel, John Phillips, George Poulett Scrope, and a host of lesser writers. Next he ruled out other types of terrestrial volcanoes. Those of the Hawaiian type, produced by lavas containing little moisture, did not form by violent eruptions; instead, during the intervals between successive eruptions, their lavas pooled into lakes to form craters having inner plains like those of the Moon. "They agree with lunar craters in the possession of inner plains, and to a certain extent in the terracing of their inner walls," Gilbert admitted. "They differ in the fact that they occupy the tops of mountains; in the absence of the wreath; in the absence of the central hill, and usually in the presence of level terraces due to the formation of successive crusts. In my judgment the differences far outweigh the resemblances."[39] Finally, there were *maars*—volcanic craters created not by quiescent eruptions of lava, but by violent "point-source" steam explosions that occur when upwelling magma encounters surface or subsurface water. At best, they could account only for the smaller craters of the Moon—those less than two miles across, which even with the great Washington refractor appeared to Gilbert only as simple cups. Steam explosions could not account for the larger craters, since *maars* did not display the prominent exterior wreaths, inner terraces, inner plains, or central hills of the larger lunar features.

In short, none of the volcanic features of Earth were really at all like the lunar craters. Nor was Gilbert able to satisfy himself with tidal theories of crater origin, let alone the theory—of which more anon—of two amateur astronomers, S. E. Peal and John Ericsson, that the Moon was covered with snow or ice and that the crater rims were built up "by the quiet fall of an infinitude of ice particles or snow flakes."[40]

By a process of elimination he was left with the meteoritic theory. "If a pebble be dropped into a pool of pasty mud, if a raindrop fall upon the slimy surface of a sea marsh when the tide is low, or if any projectile be made to strike any plastic body with suitable velocity," he noted, "the scar produced has the form of a crater. This crater has a raised rim, suggestive of the wreath of the lunar craters. With proper adjustment of material, size of projectile, and velocity of impact, such a crater scar may be made to have a central hill."[41]

There were, of course, objections to be overcome. There was the sheer scale of the lunar craters. Gilbert admitted that it was "incredible that even the largest meteors of which we have direct knowledge should produce scars comparable in magnitude with even the smallest visible lunar craters." Earlier theorists had been forced to suggest that at one time these meteors were much larger than now. Again, as no evidence had been found that the Earth was subjected to a similar attack, the lunar bombardment had to be assigned to an epoch more remote than all the periods of geologic history—an epoch so remote that similar scars on Earth had been obliterated entirely by the forces of water and wind.[42]

These problems, though difficult, did not seem insurmountable. More serious, in Gilbert's view, was the circular form of the craters. By sheer chance, most meteorites would be expected to strike the Moon's surface at oblique angles. This meant elongated craters ought to be the norm—quite the opposite of what was actually observed, as nearly all lunar craters (correcting, of course, for effects of foreshortening) were circular. To get around this difficulty, Gilbert borrowed a page from the Nebular Hypothesis as modified by Proctor. He assumed meteorites had once formed a ring around the Earth—an assumption, he claimed, that rested "legitimately on the analogy of Saturn's ring."[43] Then the angle of incidence of any meteorites falling into the Moon would be nearly vertical, and the crater forms would be circular.

Gilbert supposed that the ring eventually suffered a disturbance that caused it to break up. Various centers of aggregation then formed. Smaller elements of the ring were swept up by the larger ones, and the larger ones in turn were swept into the largest of all, which became the Moon. As the Moon grew in size, the blows it received became harder and harder, and occurred with increasing frequency. Some of the heat of impact was radiated away into space at the surface, but the rate of growth of the Moon could not have been slow enough to permit all the heat to be dissipated in this way. Pockets of liquefaction near the surface formed, but never complete liquefaction of the lunar interior, since, Gilbert noted, there was no evidence of the kind of wrinkling which ought theoretically to occur if the cold crust had formed upon a cooling liquid core.

Gilbert worked out many details of the impact process. "In the production of small craters by small moonlets," he wrote, "I conceive the bodies in collision either were crushed or were subjected to plastic flow, and in either case were molded into cups in a manner readily illustrated by laboratory experiments with plastic materials. The material displaced in the formation of the cup was built into a rim partly by overflow at the edges of the cup, but chiefly by outward mass movement in all directions, resulting in the uplifting of the surrounding plain into a gentle conical slope."[44] Central peaks were formed by the recoil of material at the center of the crater. The rays—the white streaks emanating from some of the more prominent and fresher-appearing craters—were splash features, consisting of material thrown out from the impact that formed them. Indeed, what else could they be? "The ray systems resemble splashes so closely that it is difficult to understand why the idea that they really are splashes has not sooner found its way into the Moon's literature."[45]

Gilbert's most elegant piece of work was his identification of what he called "sculpture"—a pattern of parallel grooves or furrows and smoothly contoured oval hills whose trend lines converged on a point located near the middle of Mare Imbrium (Figure 14.17):

> Associated with the sculpture lines is a peculiar softening of the minute surface configuration, as though a layer of semi-liquid matter had been overspread, and such I believe to be the fact; the deposit has obliterated the smaller craters and partially filled some of the larger. These allied facts, taken together, indicate that a collision of exceptional importance occurred in the Mare Imbrium, and that one of its results was

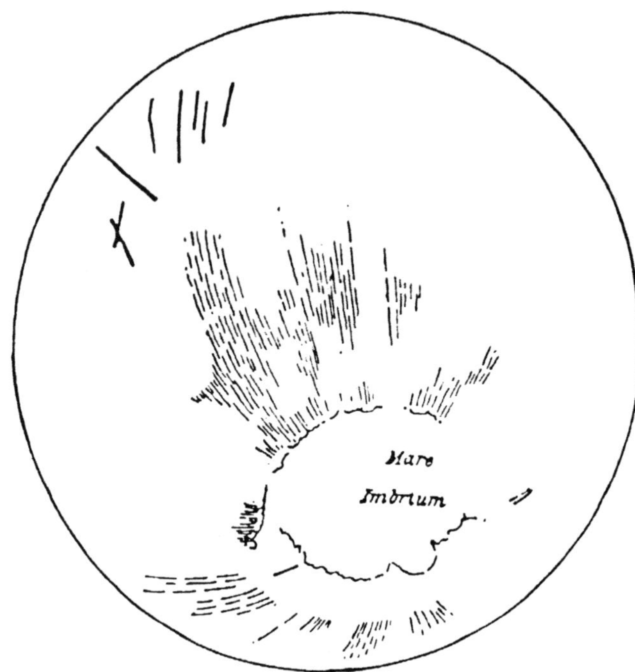

Figure 14.17 Schematic diagram by G. K. Gilbert showing the system of parallel grooves and furrows, which he called "sculpture," trending back to the middle of Mare Imbrium. He correctly cited this structure as evidence of the gigantic impact which formed the vast Imbrium basin early in the Moon's history. Courtesy U.S. Geological Survey.

> the violent dispersion in all directions of a deluge of material—solid, pasty, and liquid.[46]

The Imbrium impact threw out fragments that gouged the furrows of the sculpture, but it left unaffected more remote areas such as the closely packed craters of the southern highlands, which Gilbert regarded as representative of the general condition of the surface previous to the Imbrium event. Thus he divided the surface broadly into parts which were pre-Imbrium and post-Imbrium—antediluvial and postdiluvial, as he put it. Many of the features formerly regarded as evidence of the shrinkage of the Moon were, according to this radical reinterpretation, actually features of the sculpture. Instead of believing in a plastic and molten early history of the Moon, Gilbert reiterated his assertion that the Moon's interior had never liquified. Probably, throughout the whole period of its growth, it had been cold:

> This sketch of the life of our nearest neighbor has but little in common with the accounts of other biographers. To her has been ascribed a fiery youth, after the manner of the Sun, a middle life of dissipation, like Jupiter and Saturn, a hardening and wrinkling old age, toward which the Earth is tending, and finally, the end of change, death.

> If the record of her scarred face has now been read aright, all that remains of the old narrative is its denouement: the Moon is dead.[47]

In retrospect, Gilbert's paper "On the Face of the Moon" seems startlingly modern. Even the ring of debris around the Earth, which some have regarded as "mechanically whimsical,"[48] has its counterpart in recent conceptions of the Moon's origin from a ring of debris following a huge impact of a Mars-sized (or even larger) body with the primordial Earth. Don E. Wilhelms, a U.S. Geological Survey geologist who was a member of the Apollo science team, calls Gilbert's paper "the first in the history of lunar geoscience with a modern ring. Most earlier observers had seen the trees but not the forest: the subtypes but not the overall unity of form... Gilbert's sketches, descriptions, and interpretations could be used in a modern textbook."[49]

Gilbert, indeed, deserves to be called the Champollon of the Moon: with the insight of genius, he had presented a unified view of the Moon's incredibly diverse and hitherto largely unintelligible detail. But he was too far ahead of his time. For a long while, indeed, his work was virtually ignored. He received a polite but dissenting review from Edward Singleton Holden (1846–1914), the director of the Lick Observatory and a frequent observer of the Moon who still subscribed to the volcanic theory.[50] Gilbert's conclusions were also rejected by Proctor's literary executor, A. Cowper Ranyard,[51] and by the Harvard geologist W. M. Davis—later Gilbert's biographer—who did not find the ring of moonlets convincing.[52]

One of few later students of the Moon to give serious consideration to Gilbert's thesis was a Harvard geologist, Nathaniel Southgate Shaler (1841–1906), one of few professional geologists to concern himself to any extent with the problems of the Moon. Shaler's interest in astronomy had been sparked by a boyhood visit to the observatory founded on a river bluff in Cincinnati by the well-known astronomer Ormsby MacKnight Mitchel (1809-1862). Then and later, telescopic views of the lunar craters "greatly excited my curiosity."[53] At Harvard in the 1860s, he studied geology under the great Swiss geologist Louis Agassiz (1807–1873), famous for his theory of the action of glaciers in shaping the surface of the Earth,[54] and established his own geological reputation by an expedition to Anticosti Island in the St. Lawrence. On first experiencing the contact of the sea and land on the coastline of this island, he had the perception "that each feature—not only the details of form, but the large elements of the geography as well—not only had a history, but was in some way a record of it... [The] sense of chronicle in things."[55]

Only near the very end of his career, in 1903, did Shaler return to the subject of the craters of the Moon that had so stirred his youthful imagination. He described Gilbert's work as the "meteoric" theory without referring to its distinguishing ring of moonlets. While allowing with Gilbert that the lunar formations showed a continuous gradation from the smallest visible to those more than 100 miles in diameter, he preferred earlier views of their formation. He maintained that the craters, or "vulcanoids," had been formed by "a quiet boiling or up-welling action of lava of a very viscid character." The larger vulcanoids had no doubt formed earlier, "the size decreasing as the Moon cooled and the thickness of the crust increased."[56]

But when he came to the larger features—the great mare plains—Shaler found himself forced to abandon the tenets of the volcanic theory. The dark color of the material in the plains suggested they were of different composition from the rest of the Moon; their uniformity of level indicated they had formed from highly fluid lavas; the relative absence of large vulcanoids on their floors argued a late period of formation. All of these features seemed to require some external source of heat. Here Shaler, borrowing a page from Gilbert, proposed that late in the Moon's career, after the vulcanoids had formed, it had been struck by a few "bolides" five or ten miles in diameter. From studies of the ray systems emanating from the far side of the Moon, he concluded that the existence of mare plains were less extensive there—a finding later confirmed by the Russian Lunik 3 probe in 1959. He was also dismissive of the possibility of current volcanic activity on the Moon. Such activity must have ceased long ago, he thought, "probably before the Earth possessed even a solid crust."

A Whiggist history of lunar studies, in which one "starts with the present theory, assuming it to be correct, and asks how we got there,"[57] would essentially end with Gilbert's meteoric synthesis. But such a study would strike a false note. For it is clear that Gilbert's paper (or the qualified acceptance by Shaler of the impact origin of at least the larger mare plains) failed to produce anything like a paradigm-shift or a new consensus for late nineteenth- and early twentieth-century students of the Moon. The reality was otherwise, and more confused. It resembled the "tangled bank" of Charles Darwin, in which various competing schools continued to grow, develop, ramify, and compete. As late as 1926, Gilbert's biographer W. M. Davis summarized the controversy surrounding the origin of lunar craters with the following statement:

> It has been remarked that the majority of astronomers explain the craters of the Moon by volcanic eruption—that is, by an essentially geological process—while a considerable number of geologists are inclined to explain them by the impact of bodies falling upon the Moon—that is, by an essentially astronomical process. This suggests that each group of scientists find the craters so difficult to explain by processes with which they are professionally familiar that they have recourse to a process belonging in another field than their own, with which they are probably imperfectly acquainted, and with which they feel freer to take liberties.[58]

The features of the Moon presented, in fact, a problem in hermeneutics, a puzzle of textual interpretation. To make sense of them, one needed Shaler's "sense of chronicle in things," the prerequisite of all the historical sciences. But in the historical sciences, as historian Martin Rudwick has observed, "it should hardly be necessary to emphasize how great an interpretive gap separates the raw phenomena in the field from even the most immediate, solitary, and private record of how those 'facts' are perceived."[59]

Only the existence of this "interpretive gap" can explain how contemporaneous observers looking at the same Moon could arrive at such different interpretations of the chronicle of things presented there as did, say, Gilbert and Harvard astronomer W. H. Pickering. At the very moment Gilbert was turning the 26-inch Clark refractor to the Moon, Pickering was asking: "Are there at present active volcanoes on the Moon?" and wondering in print "whether volcanic activity upon the Moon is really dead?"[60] This remained an exciting prospect for the host of observers still electrified by the Linné episode, whose research programs had been established around the question of possible change. In contrast to Pickering's hopeful expressions, Gilbert had offered only discouragement to the quest:

> Selenographers are not yet satisfied that the condition of the lunar surface is constant, although the history of their search for changes is discouraging... If the Moon's face shall prove absolutely incommunicative of modern change, it cannot be expected to reveal the date when its expression last was varied; but, strange as it may seem, the Earth can give a partial answer, for the Earth was an actor as well as a spectator of the Moon's drama, and the record of its participation lies somewhere among the archives of its crust... Does the Earth exhibit impact craters? If not, then erosion and sedimentation have destroyed them, and the Cenozoic era did not witness the building of the Moon. Is any horizon of stratified rocks generally or widely characterized by molten disjecta? If not, then the Moon was already a finished planet in Paleozoic time.[61]

Such a cold and ancient Moon did not sit well with astronomers of the late nineteenth century, and thus Pickering, not Gilbert, would establish a "New Selenography" that became the inspiration of a new generation of lunar observers, for whom the dream of fires in the Moon had not yet been extinguished. Rather than absorbing at once the mass of insights included in Gilbert's *tour de force*, astronomers needed more time. They would have to become more gradually inured to the idea of the great antiquity—and paralysis—of the lunar surface.

References

1. J. Ashbrook, "James Nasmyth's Telescopes and Observations," in *The Astronomical Scrapbook*, pp. 144–145.
2. James Hall Nasmyth and James Carpenter, *The Moon: Considered as a Planet, a World and a Satellite* (London: John Murray, 1874), p. vii.
3. Ibid., pp. 89–90.
4. Ibid., p. 14.
5. Ibid., p. 25.
6. Ibid., p. 26.
7. Ibid., pp. 73–74.
8. Ibid., p. 82.
9. Ibid., p. 91.
10. Ironically, the domes that eluded Nasmyth are among the few lunar formations of unambiguously volcanic origin. Some are "shield volcanoes" built up by relatively gentle outpourings of fluid basaltic lavas to form low, broad, lens-like structures, often replete with summit craterlets. Others are "laccoliths" caused by intrusions of viscous magma penetrating between layers of rock which solidify in place because they cannot find a vent to the surface. The overlying layers of rock are arched up into gently sloped swellings that often exhibit a pattern of radial fissures.
11. Webb, quoted in Proctor, *The Moon*, pp. 174–175.

12. Quoted in Birt, "Lunar Physics," *The Observatory*, 4 (1881), 47–51; 108–111.
13. Eduard Suess, "Einige Bemerkungen Über den Mond," *Sitzungsberichte d. k. Akademie d. Wissenschaften zu Wien*, 104 (1895), 21–54.
14. H. Faye, "Les volcanoes de la lune," *Revue scientifique*, Series 3, 1 (1881), 130–138.
15. H. Ebert, "Ein Vorlesungsversuch aus dem Gebiete der physikalischen Geographie" (Bildung der Schlammvulkane und der Mondringgebirge), *Ann. d. Phys. u. Chem. N. F.* 41 (1890), 351–363.
16. A. Wegener, *Die Enstefehung der Mondkrater* (1921), trans. by A. M. Celal Sengor, in *The Moon*, 14 (1975), 211–236: 214.
17. Proctor, *The Moon*, pp. 258, 262.
18. Ibid., pp. 261–262.
19. W. G. Hoyt, *Coon Mountain Controversies* (Tucson: University of Arizona Press, 1987), p. 23.
20. [W. R. Birt], *Selenographical Journal*, 1 (1878), 35.
21. R. A. Proctor, "The Moon's Myriad Small Craters," *Belgravia*, 36 (1878), 153–171. It is argued that further evidence of Proctor's change of heart is found in his last work, *Old and New Astronomy* (London, 1895), where he simply states there were "mighty volcanic eruptions, during which the great craters were formed." However, since this work was incomplete at the time of Proctor's death in 1888 and was later completed and published by A. Cowper Ranyard, a known partisan of the volcanic theory, it is equally likely to record Ranyard's view of the matter as Proctor's.
22. A. Meydenbauer, "Über die Bildung der Mondoberfläche," *Sirius*, 10 (N.F. 5), 1877, 180; "Die Gebilde der Mondoberfläche," *Sirius* 15 (N.F. 10), 1882, 59–64.
23. Heinrich W. J. Thiersch and August Thiersch, *Die Physiognomie des Mondes* (Augsburg, 1883).
24. For detailed information on Gilbert, see the excellent biography by Stephen J. Pyne, *Grove Karl Gilbert: A Great Engine of Research* (Austin, Texas, the University of Texas Press, 1980).
25. John Strong Newberry, "Report of Progress in 1870," *Geological Survey of Ohio* (Columbia, Ohio: Columbia Printing Co., 1871), pp. 32–33.
26. Pyne, *Grove Karl Gilbert*, p. 39.
27. Ibid., p. 52.
28. Both, *A History of Lunar Studies*, p. 23.
29. Pyne, *Grove Karl Gilbert*, p. 73.
30. Ibid.
31. G. K. Gilbert, "The Origin of Hypotheses, Illustrated by the Discussion of a Topographic Problem," *Science*, New Series 3 (1896), 1–11:3.
32. A. E. Foote, "A New Locality for Meteoric Iron with a Preliminary Notice of the Discovery of Diamonds in the Iron," *Proceedings of the American Association for the Advancement of Science*, 40 (1892), 279–283.
33. Gilbert, "Origin of Hypotheses," p. 4.
34. W. M. Davis, "Grove Karl Gilbert," *Biographical Memoirs of the National Academy of Sciences*, 21 (1926), no. 2, p. 183.
35. Quoted in Don E. Wilhelms, *To a Rocky Moon* (Tucson: University of Arizona Press, 1993), p. 8.
36. Gilbert, "The Moon's Face: A Study of the Origin of its Features," address as retiring president to the National Academy of Science, November 1892; *Bulletin of the Philosophical Society of Washington*, 12 (1893), 241–292: 241.
37. Ibid., p. 241.
38. Ibid., p. 250.
39. Ibid., p. 251.
40. John Ericsson, *Nature*, 34 (1886), 248; S. E. Peal, *Nature*, 35 (1886), 100. If formed by snowflakes, Gilbert argued, the configuration should then be smooth and regular instead of exhibiting the roughness of outline actually observed.
41. Gilbert, "The Moon's Face," p. 256.
42. Ibid., pp. 257–258.

43. Ibid., p. 270.
44. Ibid., p. 271.
45. Ibid., p. 285.
46. Ibid., pp. 275–276.
47. Ibid., p. 290.
48. W. G. Hoyt, "G. K. Gilbert's Contribution to Selenology," *Journal for the History of Astronomy,* 13 (1982), 155.
49. Don E. Wilhelms, *To a Rocky Moon*, p. 8.
50. E. S. Holden, editor's note, *Publications of the Astronomical Society of the Pacific*, 4 (1892), 263–264.
51. A. C. Ranyard, "The great lunar crater Tycho," *Knowledge*, August 1, 1892, 149–152.
52. W. M. Davis, "Lunar craters," *The Nation*, 41 (1893), 342–343.
53. Nathaniel Southgate Shaler, *Autobiography* (Boston, Houghton Mifflin, 1909), p. 56.
54. Ibid., p. 103.
55. Ibid., p.166.
56. N. S. Shaler, "A Comparison of the Features of the Earth and Moon," *Smithsonian Institution Contributions to Knowledge*, vol. 34, no. 1438 (1903).
57. Brush, *Nebulous Earth*, p. 4; referencing the work of historian Herbert Butterfield.
58. W. M. Davis, "Grove Karl Gilbert," *Biographical Memoirs of the National Academy of Sciences,* 21 (1926), no. 2, p. 183.
59. Rudwick, *The Great Devonian Controversy*, p. 431.
60. W. H. Pickering, "Are There at Present Active Volcanoes on the Moon?" *The Observatory*, 15 (1892), 250–254.
61. Gilbert, "The Moon's Face," p. 291.

Chapter 15:
The Madman of Mandeville

The name William Henry Pickering (Figure 15.1) evokes both admiration and ridicule. He was a wide-ranging man of ideas and a diligent observer who is now remembered chiefly for his discovery of Phoebe, Saturn's ninth satellite, in 1898. But Pickering was also so prone to rash and unconventional interpretations of his data that most astronomers of his day and historians since have had a hard time taking him seriously. He was a man endowed with many gifts—"skill and perseverance as an observer, mathematical ability, and literary flair... almost a recipe for greatness," in the words of Joseph Ashbrook.[1] In the end he seemed to lack only one virtue, though ultimately the one most essential: sound judgment.

Pickering was born in 1858 in the affluent Beacon Hill district of Boston, the younger brother of Edward Charles Pickering (1846–1919), who became one of the most accomplished and respected astronomers of the age, a pioneer in stellar spectroscopy and photometry, the essential elements of the emerging discipline of "astro-physics." Grandsons of George Washington's first Indian Commissioner and nephews of naturalist Charles Pickering (chief zoologist of the Wilkes expedition to the Pacific, notable for making the first recognition of Antarctica as a separate continent), the brothers grew up on the same Mount Vernon Street where another Brahmin family destined for astronomical notoriety, the Lowells, lived for several years. William graduated from the Massachusetts Institute of Technology, served there briefly as an instructor of physics and photography, and in 1883 went to the Harvard College Observatory, where Edward had risen to the post of Director. For the next eight years William assisted Edward's early efforts in photometry and the spectral classification of stars.

Inevitably, the brothers' paths began to diverge—in part because of William's growing resentment at his brother's domineering personality, in part because of Edward's failure to sympathize with William's all-consuming passion for planetary astronomy. When the Boston engineer Uriah Boyden bequeathed a sum of more than $230,000 to the Harvard College Observatory for the purpose of establishing an observing station "at such elevation as to be free, as far as practicable, from the impediments to accurate observations which occur in the observatories now existing, owing to atmospheric influences," a site high in the Andes at the foot of the dormant volcano El Misti (Figure 15.2), near the city of Arequipa, Peru, was selected.[2] Edward appointed William as its first director.

Arriving in Peru in January 1891, William was charged with his brother's fa-

Figure 15.1 An 1896 portrait of William Henry Pickering. By this time his reputation and career were already in decline. Courtesy Dorothy Schaumberg, Mary Lea Shane Archives of the Lick Observatory.

vored programs of securing photographs suitable for making new charts of the southern skies and obtaining large numbers of stellar spectra. However, he at once began to devote most of his time to making visual observations of the Moon and planets. In many ways it is hard to blame him. He had a keen eye, the superb Boyden telescope—a 13-inch (330-mm) refractor made by the renowned firm of Alvan Clark and Sons—and a good observing site. Located at an altitude of 8,100 feet in the high desert, the Boyden Station routinely enjoyed not only clear, dry air but also exquisitely steady "seeing," which was all too rare back in Massachusetts.

The Pickering brothers had pioneered the use of the dry plate that was revolutionizing the photography of stars, nebulae, and comets. In the older wet-collodion process, each plate had to be individually prepared just prior to use by coating a pane of glass with a freshly prepared solution of collodion (cellulose nitrate or "guncotton") containing potassium bromide. As soon as this mixture had dried, the plate was immersed in a solution of silver nitrate to form *in situ* a myriad of tiny, light-sensitive crystals of silver bromide. Once sensitized, the wet plate had to be immediately placed in the camera and exposed, for its sensitivity rapidly diminished as it dried. This failing imposed a practical limit on the length of exposures of only several minutes.

The advent of the dry plate enabled astronomers to make exposures many hours in duration, accumulating the feeble light of faint stars and nebulae far beyond the grasp of any visual observer. Harvard College Observatory played a leading role in the revolution that ensued. Laborious visual searches for variables and asteroids—indeed all visual methods of conducting astrometric, photometric, and spectroscopic stellar astronomy—were rendered obsolete in the span of a few years.

Capturing lunar and planetary features, however, was quite a different matter. To record even these comparatively bright subjects with the slow, grainy emulsions of the late nineteenth century required exposures several seconds or even tens of seconds in duration, so atmospheric turbulence at even the best sites

Figure 15.2 This striking 1945 aerial photograph of Arequipa, Peru, nestled at the foot of the quiescent volcano El Misti, captures the stark, almost lunar appearance of the arid Peruvian Andes, which played an important role in shaping the selenological views of W. H. Pickering. Courtesy E. M. Carreira, S.J.

hopelessly blurred the finest details. Lunar and planetary observing has been compared to watching a motion picture in which the projector is out of focus except for occasional sharp frames thrown in at random.[3] While the eye-brain combination is not quite instantaneous, it requires only about one-fifteenth of a second to register an image, so while visual observers had been hopelessly outclassed by the camera when it came to the study of faint subjects, they continued to enjoy an insuperable advantage over the photographer when it came to recording delicate lunar and planetary markings. A century would pass before the widespread availability of the "charge-coupled device" (CCD), an electronic sensor that replaces the chemical silver bromide grains of the photographic emulsion with a far more efficient chip of silicon, would finally bring to a close the long era of the supremacy of the eye.[4]

In the closing years of the nineteenth century, a handful of professional astronomers did manage to capture on film lunar details that would challenge a skilled visual observer equipped with a six-inch telescope, but the feat required huge instruments far beyond the means of even the wealthiest amateur. Following its commissioning in 1888, the 36-inch (91-cm) refractor at the Lick Observatory at the 4,250-foot summit of California's Mount Hamilton was almost immediately put to use by Sherburne Wesley Burnham (1838–1921) and Edward Singleton Holden (1846–1914) to secure nineteen plates that served as the basis of a photographic lunar atlas on the same scale as Beer and Mädler's 1837 map. Meanwhile, at the Paris Observatory, Moritz Loewy (1833–1907) and Pierre Puiseux (1855–

Figure 15.3 Constructed in 1882, the Paris Observatory's 23.6-inch coudé refractor was employed by Loewy and Puiseux to secure scores of outstanding lunar photographs over a span of more than two decades. The pioneering instrument's 60-foot light path was jack-knifed by means of two plane mirrors to a stationary eyepiece located in a heated room, providing an unprecedented level of comfort for the observer. Though the telescope was dismantled in 1933, the objective lens soldiered on as the heart of a more conventional telescope installed at the Pic du Midi, a site high in the Pyrenees of southern France renowned for its steady seeing. It was used during the 1940s, '50s, and '60s to take hundreds of additional lunar photographs, often achieving a three-fold improvement in resolution over the plates by Lowey and Puiseux. From C. A. Young, *General Astronomy* (1903).

1928) were systematically photographing the Moon using a refractor of 23.6 inches (60 cm) aperture with a focal length of 60 feet (18.2 meters) (Figure 15.3). The image of the Moon measured almost seven inches across at the focal plane of this enormous instrument. Exposures scarcely one second long were required, so on steady nights resolution was limited largely by the granularity of the emulsion. Soon scores of these beautiful plates, enlarged to a scale of 80 to 90 inches to the Moon's diameter, were widely reproduced and distributed for study.

Yet William preferred to study the Moon and planets by traditional visual methods. In fact, he was entirely justified; despite their excellence, the Lick and Paris photographs recorded only a fraction of the detail that he could see through the eyepiece of the Boyden refractor in the steady Andean air. The objective lens of this state-of-the-art instrument was a "photo-visual" doublet. The curves of the glass had been laboriously calculated so that by removing and inverting the forward crown element, the color correction of the objective could be converted from "visual" correction in the yellow-green region of the spectrum, where the human eye is most sensitive, to "photographic" correction in the blue-violet region of the spectrum, where the emulsions of that era were most sensitive. With Edward almost half a world away, William assembled the objective in its visual configuration and only

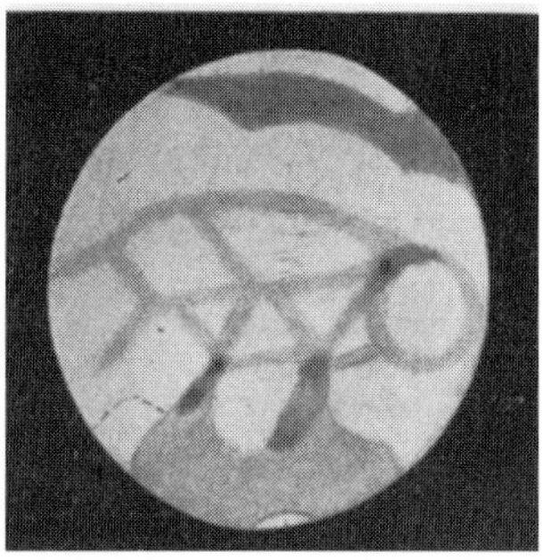

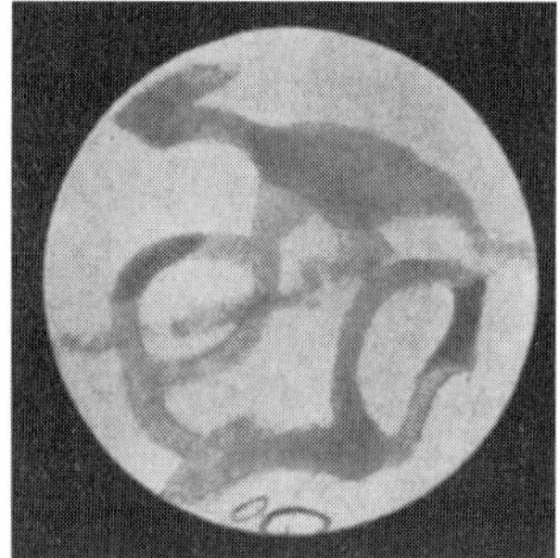

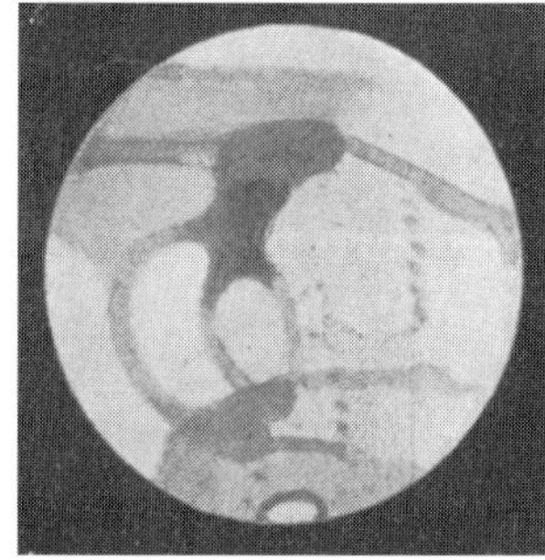

Figure 15.4 These drawings of Mars by W. H. Pickering showing a handful of broad "canals" are typical of his depictions of the planet, and far less artificial looking than Percival Lowell's intricate rectilinear network of gossamer lines. From T. E. R. Phillips and W. H. Steavenson, *Splendour of the Heavens* (1925).

made sporadic, half-hearted attempts to use the instrument photographically.

Indeed, after nine months passed without a single photograph from Arequipa, an exasperated Edward tersely cabled William a direct order: "Photograph with the thirteen inch." Gentler, almost pleading letters followed, to no avail. William's attention during his second year at Arequipa was largely devoted to Mars, which in August 1892 came to an eagerly awaited favorable opposition that brought the planet to within a distance of only 35 million miles. Because of its southerly declination, Mars was much better placed for observers in the southern hemisphere than in the northern, and William observed the planet on every night but one between July 9 and September 24, managing to make out many canals (Figure 15.4), though they appeared to him less narrow, uniform, and geometrically regular than those depicted by their discoverer, Giovanni Schiaparelli. At the intersections of the canals he detected over forty dark spots which he incautiously referred to as "lakes" in a series of reports telegraphed to the *New York Herald*.[5] Edward immediately fired off a rejoinder to William:

> In my own case I should have restricted myself more distinctly to the facts in this as in other cases. You would have rendered yourself less liable to criticism if you had stated that your interpretations were probable instead of implying that they were certain.[6]

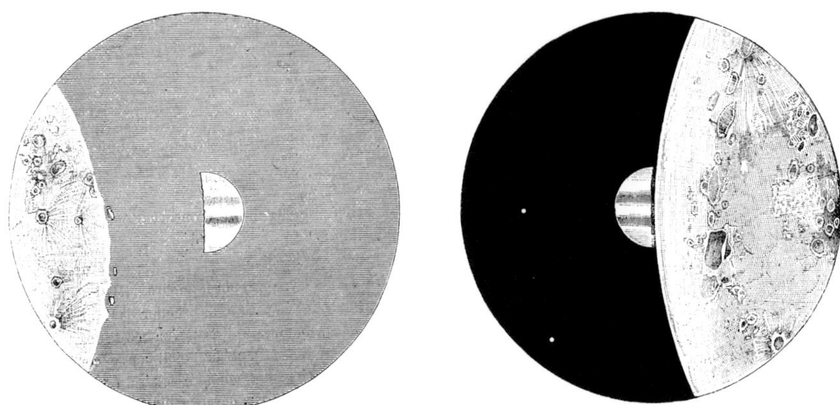

Figure 15.5 Pickering was not the first observer to report a dusky band parallel to the bright lunar limb during an occultation of Jupiter—an appearance which he cited as evidence of a lunar atmosphere perhaps eight or ten thousand times more tenuous than that of the Earth. These drawings by the French astronomer and popular writer Camille Flammarion depict the disappearance of Jupiter behind the unilluminated limb of the Moon (left) and its re-appearance from behind the bright limb (right) during an 1889 occultation. From A. T. Arcimis, *Astronomia Popular* (1901).

At the same time that his sensational Mars observations were capturing widespread attention, William Pickering began another study—more pertinent here—destined to become a lifelong obsession. What he later called the "new selenography" began with his observation of a relatively rare but mundane celestial event—an occultation of Jupiter by the Moon (Figure 15.5) on the night of August 12, 1892. As the Moon passed between Jupiter and the Earth, he detected a diffuse, dusky band parallel to the limb of the Moon and roughly perpendicular to the characteristic cloud-belts of the planet. This curious appearance, when seen at earlier occultations of Jupiter, had been dismissed as a result of contrast—in effect, as an optical illusion—since it was evident only when Jupiter was adjacent to the bright sunlit limb of the Moon and not when the planet was in contact with the dark lunar limb.[7]

Oddly enough, Pickering was convinced that this optical illusion constituted nothing less than "ocular proof that the Moon has an atmosphere." The absence of the band at the dark limb he simply took as an indication that "the absorbing medium, whatever it is, is condensed to a solid by the intense cold that must prevail during the lunar night." Although he had managed to photograph the occultation and claimed that the obscuring band had been recorded, and could even be measured on his negatives, he immediately clouded the issue with an admission that it was too delicate a feature to be printed satisfactorily. Moreover, the photograph that he did publish showing the presumed band was an accidental double exposure.

This whole affair was vintage Pickering. What was in question was not Pick-

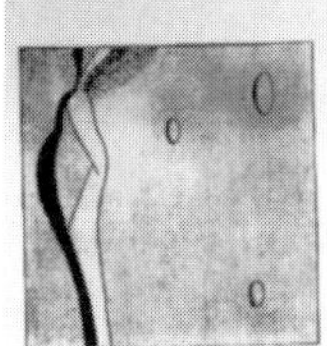
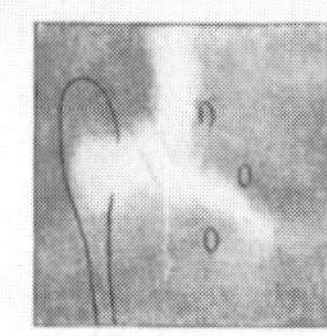
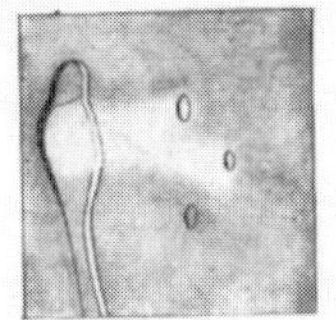
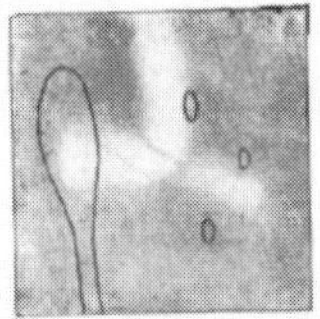

Figure 15.6 Pickering's "fumarole" at the mouth of Schroeter's valley is depicted in these 1892 drawings. From W. H. Pickering, *The Moon* (1904).

ering's descriptions—which were always meticulously accurate—but his seemingly fanciful intepretations of them. Critics countered that no dusky band is seen around the mountainous profile of the Moon during the partial phases of solar eclipses, when the delicate granulation of the Sun's photosphere remains visible right up to the crisply silhouetted lunar limb. Pickering responded that on his photographs the disk of Jupiter, when in contact with the lunar limb, appeared flattened in a direction radial to the Moon by perhaps 0.5 arc-seconds (a mere 1⁄80 of its apparent diameter). This infinitesimal distortion he did not hesitate to attribute to refraction by the Moon's atmosphere.[8] Years later he would implausibly claim that a painstaking series of filar micrometer measurements of widely spaced double stars revealed a minute relative shift in position of 0.2 to 0.4 arc-seconds when the stars were placed near the lunar limb. Although the magnitude of the alleged displacement was at the very limit placed on resolution both by the aperture of his telescope and the effects of atmospheric turbulence, Pickering advanced these pedantically precise measurements as evidence for refraction by a lunar atmosphere some eight to ten thousand times more rarefied than the Earth's. The existence of such a rarefied lunar atmosphere would be debated inconclusively for the next fifty years.

Though corresponding in density to our own air at an altitude of over 40 miles, Pickering claimed that this thin lunar atmosphere was "a factor in selenography by no means negligible." Indeed, he argued, if present it must imply the existence of ongoing volcanic activity on the Moon. Drawing upon the kinetic theory of gases, first published in 1860 by James Clerk Maxwell, according to which the Moon's feeble gravity must have permitted the escape of any primordial air or water in the distant past, Pickering surmised: "The chief gases of our air—oxygen and nitrogen—would escape from the Moon's atmosphere about as readily as hydrogen does from that of the Earth. Carbonic acid [carbon dioxide] would be retained with somewhat greater facility, but in general it is likely that any gas that was not constantly renewed from the Moon's interior would have practically disappeared from its surface long ago."[9]

Carbon dioxide and water vapor are the principal constituents of the mixtures of gases emitted by terrestrial volcanoes. There was no reason to suppose it would be otherwise on the Moon. Both of these gases would freeze out during the frigid lunar nights—carbon dioxide as "dry ice," water vapor as hoarfrost—which provided a neat explanation for the absence of a dusky band adjacent to the dark

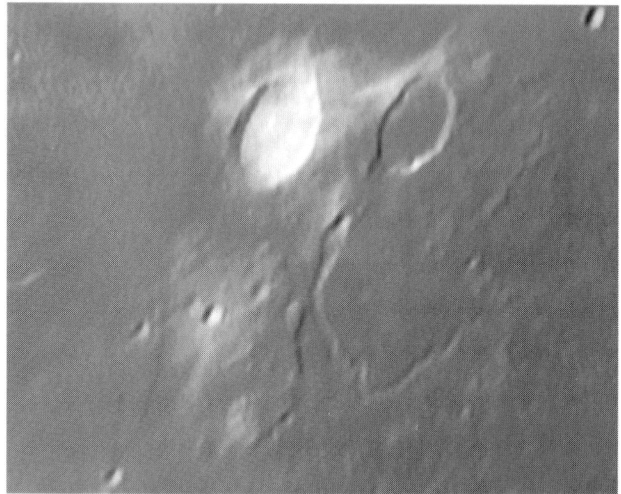

Figure 15.7 This CCD image of Aristarchus, Herodotus, and Schroeter's Valley records Pickering's "vapor column" as nothing more than a bright, amorphous patch of lunar soil. One of the most geologically diverse areas of the Moon, this hilly jumbled terrain, containing numerous domes, rilles, and secondary impact craters, has also furnished a favorite hunting ground for observers in search of "transient lunar phenomena." Image by T. A. Dobbins.

lunar limb during occultations of Jupiter. The gases had frozen out onto the surface. These solids would then rapidly sublime (i.e. pass directly from a solid to a gaseous state) in the heating that followed local sunrise. Having deduced that the tenuous lunar atmosphere he had identified must be continuously replenished by gases and steam seeping from volcanic vents, Pickering promptly set out to find such features. He soon detected what he took to be an erupting fumarole or geyser in the feature later named the "Cobra Head" (Figures 15.6, 15.7, 15.8), a pear-shaped depression at the mouth of the prominent Schroeter's Valley near the crater Herodotus in the Oceanus Procellarum:

> Dense clouds of white vapor were apparently arising from its bottom and pouring over its southeastern wall in the direction of Herodotus. So striking, indeed, was this appearance, that, notwithstanding the fact that the supposition was distinctly opposed to what was at that time generally believed regarding the condition of the surface of our satellite, I determined to make a series of careful drawings of the apparent vapor column, in order to determine whether any variations in its outline might be detected from time to time, or whether, like a stain, it was immovably attached to the lunar surface... A casual examination of the sketches shows the great changes that are from time to time undergone by the vapor column, as we shall for convenience call it—changes that are readily detected by a six inch telescope under ordinary atmospheric conditions...
>
> The most marked of these changes depend for their existence upon the altitude of the Sun, for apparently no volcanic activity whatever is exhibited until about one day after sunrise. The activity then increases to a maximum, diminishes, and finally ceases a few days before sunset.[10]

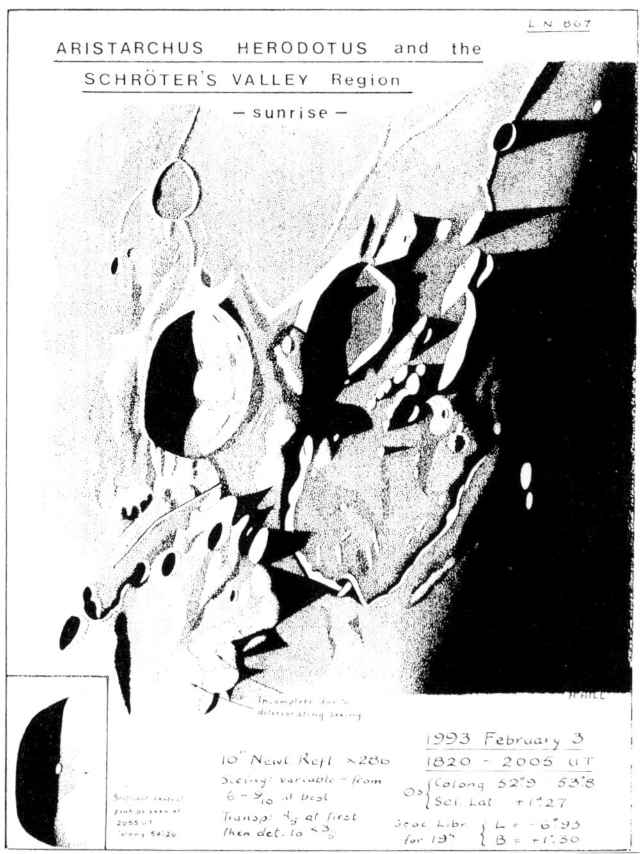

Figure 15.8 The Aristarchus, Herodotus, and Schroeter's Valley region at sunrise. Note the dusky bands running down the terraced inner wall of Aristarchus. These features were formed by landslips. This remarkably detailed drawing was made by Harold Hill using a 10-inch Newtonian reflector. Courtesy Harold Hill.

There was, of course, a less forced explanation for these changes than light-dependent volcanism. In fact, Pickering's "vapor column" is simply an amorphous patch of soil with a higher reflectivity than its surroundings, presumably due to differences in chemical composition. Under high Sun it undergoes more pronounced brightening than the terrain around it. Small changes in the angle of illumination produce a rather convincing illusion of growth and motion, while differences in libration also contribute to the deception by slightly modifying its outline from lunation to lunation.[11]

Pickering made scores of observations of his "geyser" during the coming years, but was never able to decide "if these shifting white objects are due to streams of gas issuing from the craters and carrying with them white crystals of ice, forming real clouds upon the Moon, or whether the crystals are deposited as soon as formed, or whether they are formed only on the lunar surface itself like hoarfrost... That irregularly occurring lunar changes of some sort are in progress is all that we are sure of at the present time."[12] Of course none of his proposed explanations is correct.

Another object of Pickering's obsessive scrutiny at Arequipa was the floor of the crater Plato, one of the favorite objects of Birt and his circle. Collectively, the British observers had mapped 36 craterlets and a delicate filigree of dusky bands and pale streaks; some had disappeared only to reappear months or even years later, then to be reported as new discoveries. Birt had strongly suspected the occasional presence of obscuring mists.[13]

With the 13-inch (330-mm) Boyden refractor in the clear Andean air, Pickering valiantly charted no fewer than 42 craterlets and 29 white spots, which he described as "small volcanic cones." (Admittedly, he was never able to see more than 39 of these features at one time.) Methodically comparing the relative conspicuousness of the craterlets and white spots and the outlines of the streaks in his series of drawings to the map compiled by Birt and his co-workers, Pickering noted a number of subtle differences. All but ignoring the subjectivity inherent in depictions of these slight nuances of light and shadow, and heedless of the fact that such exceedingly delicate objects can be easily effaced by slight atmospheric tremors, he attributed the discrepancies to actual physical changes resulting from volcanic activity of a quiescent sort; more like the emissions of steam in Yellowstone—or in the near-dormant summit crater of El Misti itself (Figure 15.9)—than eruptions of Vesuvian or Pelean violence. "When we speak of boiling springs," he reminded his readers, "we must remember that on the Moon [because of the extremely low atmospheric pressure] water would boil at the freezing point."[14]

To Pickering, the rays surrounding many craters, invisible under oblique illumination and appearing as if from nowhere under a high sun, argued for the presence of frost or loosely-packed snow lying at the bottom of small hollows and shallow depressions. He believed the most extensive ray system on the Moon, that of the crater Tycho, might consist of a low-lying mist of ice crystals resembling terrestrial "mare's tail" cirrus clouds, and not unlike the white veils often seen in the polar regions of Mars.[15] The configuration of divergent radial streaks he attributed to a host of minute, steam-puffing "active craterlets" lying along "lines of weakness or cracks in the surface" too narrow to be seen.

In the harsh glare and nearly vertical lighting of the Full Moon, the ramparts of many craters are seen as brilliant white circles. The most conspicuous examples of these features, which Pickering dubbed "snow-craters," are an eye-catching pair of small craters in Mare Foecunditatis, Messier and Messier A (Figure 15.10). Messier A comprises the "head" of a striking "Comet Tail," consisting of two straight, narrow rays that extend 120 kilometers across the surrounding plain. Schroeter had been the first to draw attention to the curious variations in apparent shape and relative size of these craters under changing illumination, variations that were thoroughly confirmed by Gruithuisen and had long occupied the attentions of Webb, Neison, and Elger. (Mädler alone had departed from the consensus view,

Figure 15.9 Steam often billows from the summit crater of El Misti, as shown here. Was the power of suggestion at work when Pickering imagined that vapors emanate from the mouth of Schroeter's Valley on the Moon? Courtesy of E. M. Carreira, S.J.

having described them as "alike as two peas in a pod.") Pickering's description of the monthly cycle is characteristically precise:

> Sometimes one of these craters is the larger and sometimes the other. Sometimes they are triangular and sometimes elliptical in shape. When elliptical their major axes are sometimes parallel and sometimes perpendicular to one another. When the sun first rises on them they are of the same brilliance as the mare upon which they are situated, but three days later they both suddenly turn white, and remain so until the end of the lunation. When first seen the white areas are comparatively large, especially that surrounding Messier itself, but it gradually diminishes in size under the Sun's rays.[16]

Pickering could think of no explanation for these pronounced alterations in appearance more convincing than the alternate evaporation and deposition of frost. Although most of his contemporaries remained convinced that differences in lighting combined with peculiar topography were sufficient to explain them, they remained intriguing enough for the formation to be singled out for detailed scrutiny by the Lunar Orbiter and Apollo missions three-quarters of a century later.[17]

Not surprisingly, Pickering was a firm believer in the variability of Linné. His own observations showed that the diffuse bright halo surrounding the minute central craterlet was "subject to change in size dependent on the altitude of the Sun," a phenomenon, he suggested, analogous to "the changing size of the polar caps of Mars and of our own Earth," not only in superficial appearance but in its very nature.[18] "What reason is there to believe that there is ice upon the Moon?"

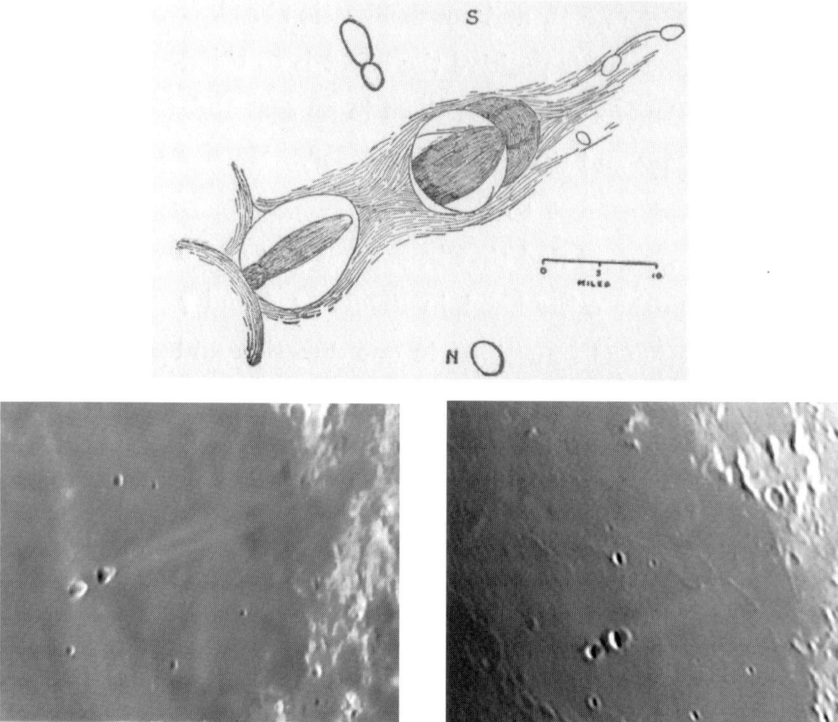

Figure 15.10 (Top) A detailed drawing by W. H. Pickering of the "twin" craters Messier and Messier A (formerly known as W. H. Pickering!), made at Arequipa using the 13-inch Boyden refractor. From W. H. Pickering, *The Moon* (1904). (Below) A pair of CCD images of Messier and Messier A under different lighting conditions. At left, under a high Sun, the comet-like ray is prominent; at right, with the craters lying close to the terminator, the ray is subdued and the appearance of the craters changes markedly. Images by T. A. Dobbins.

he asked. "The answer is: For the same reason that we believe there is ice upon Mars, because the phenomena observed can be more readily explained that way than any other."[19] The dazzling interior of Aristarchus—where William Herschel had seen one of his supposed "eruptions" in 1787—was another example of Pickering's "snow fields."

While Pickering's notions of frosts and snows on the Moon may strike the modern reader as outlandish or even grotesque, it must be remembered that at the turn of the last century many astronomers believed that even midday temperatures on the Moon were quite frigid. According to a standard textbook by the respected Princeton astronomer Charles Augustus Young (1834–1908):

> As to the temperature of the Moon's surface, it is difficult to affirm much with certainty. On one hand, the lunar rocks are exposed to the Sun's rays in a cloudless sky for fourteen days at a time, so if they were blanketed by air like our own rocks they would certainly become intensely heated. Some years ago, Lord Rosse inferred from his own observations that the temperature of the lunar surface rose at its maximum

> (about three days after full Moon) far above that of boiling water. But his own later investigations and those of [Samuel Pierpont] Langley throw great doubt on this conclusion. There is no air-blanket at the Moon's surface to prevent it from losing heat; and it now seems rather more probable that the temperature never rises above the freezing point of water, as is the case on the highest of our mountains, where there is perpetual ice. So far as we can judge, the condition of things on the Moon's surface must correspond to an elevation many times higher than any mountain on the Earth; for no terrestrial mountain is so high that the density of the air at its summit is even nearly so low as that of the densest supposable lunar atmosphere.[20]

In this climate of opinion, Pickering's views appear both reasonable and moderate:

> What the temperature of the day side may be under a vertical Sun is very uncertain. Sir John Herschel and Lord Rosse thought it might exceed that of boiling water. Ericsson concluded that it was far below zero, while Professor Langley considers it very uncertain, but probably not far from the freezing point... All we can say for the present is that when the Sun is in the zenith the temperature on the Moon must lie somewhere between the freezing and boiling points of water, and not very near either of them.[21]

Although his claims of frost and snow met with widespread skepticism, Pickering presented more persuasive evidence of the presence of moisture (if only in the distant past) in the form of various features which appeared to have been carved by running water. One of the principal objections to volcanic theories of the origin of lunar craters had always been: volcanoes emit water; there is no evidence of water on the Moon; therefore the lunar craters cannot be volcanic. Before Pickering, hundreds of rilles had been charted, which included clefts composed of relatively straight segments like the spindly cracks near the crater Triesnecker, or concentric, gently curved "arcuate" fractures paralleling the borders of the maria, like the wide, shallow trenches that span the eastern shores of the Mare Humorum. The very favorable atmospheric conditions at Arequipa enabled Pickering to catalog no fewer than 35 examples of a third, much more delicate kind of rille, "undoubtedly, taken as a whole, the most difficult class of objects to be found upon the Moon." Some of these features had been recorded by earlier observers such as Julius Schmidt—the delicate Rima Hadley (Figure 15.11), for instance, which meanders through the foothills of the Apennine range and to the verge of which the astronauts of Apollo 15 would one day range with their lunar rover. But Pickering added others—including the delicate rille that courses through the Alpine Valley—and he was the first to accurately describe and depict them, referring to them as "riverbeds" because they were "composed almost entirely of curves of very short radius, giving them a zigzag, winding appearance exactly resembling a terrestrial river as drawn on a map.[22] To Pickering they represented "the only strong evidence that water in the liquid state ever existed upon the surface of the Moon."[23]

While acknowledging that under current conditions "no water in the liquid form can exist upon the Moon," Pickering speculated that a denser lunar atmosphere—probably never exceeding one-fifteenth the density of the Earth's—had

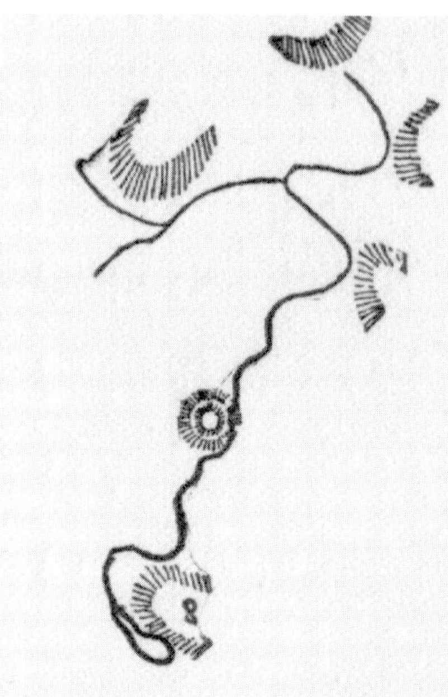

Figure 15.11 (Top) Pickering's sketch of the meandering Hadley Rille, made with the 13-inch Boyden refractor at Arequipa. He regarded this feature as a dried-up riverbed. From W. H. Pickering, *The Moon* (1904). (Bottom) Hadley Rille as photographed by the Lunar Orbiter 5 spaceprobe in 1967, testimony to the basic accuracy of Pickering's sketch. Courtesy NASA.

existed in some distant epoch. "That there was never very much water on the Moon's surface is rendered certain by the lack of extensive areas of erosion," he noted. "Nothing is seen at all comparable to what an outside observer would notice in inspecting the Earth under similar circumstances."[24]

Today Pickering's riverbeds are known as "sinuous rilles." A grouping of five located on slopes of the Harbinger Mountains north of the lava-flooded crater Prinz looks for all the world like parched streambeds carved by runoff from the adjacent heights. Originating in pear-shaped craterlets or elongated depressions that Pickering suggested were drained lakes, they gradually taper and grow ever shallower before petering out on the floor of the Oceanus Procellarum—just the sort of behavior that would be expected of a stream of water under conditions of extremely rapid evaporation. Pickering himself drew comparisons to rivers, creeks, and wadis in South America and the southwestern United States that diminish and finally disappear after entering desert regions.[25]

But if the Moon possessed an atmosphere—no matter how rarefied—containing water vapor, the stage was set for speculation about organic growth upon its surface, including vegetation. Conversely, evidence of such growth could be used to buttress his evidence of the existence of water vapor and hoarfrost.[26] Thus Pickering's fascination with the "variable spots" and "pseudo-shadows" in the Moon's tropical and temperate latitudes, to account for which he returned to the ideas of old Gruithuisen, who had given comparably detailed descriptions of these features and resorted to the same class of explanations. Inconspicuous or even invisible under low Sun, during each lunation they darken rapidly as the Sun rises over them, only to suddenly blanch again toward sunset. Always found in close proximity to "small craterlets or deep, narrow clefts," to Pickering as to Gruithuisen they were evidence of some form of "organic life resembling vegetation" sustained by carbon dioxide and water vapor seeping from volcanic vents. Pickering—like Gruithuisen before him—could find no variable spots beyond latitudes 55° North and 60° South, surmising that it was too cold for even the hardiest organisms to survive nearer the poles. Their apparent cycles of growth and decay suggested an analogy between the two week-long lunar day and the seasons of the terrestrial year, with morning corresponding to spring, midday to high summer, evening to autumn, and night to winter.

The most prominent variable spots were three irregularly-shaped patches on the floor of the 120 kilometer-diameter crater Alphonsus (Figure 15.12). They were located just within the formation's inner walls amid a tangle of delicate crater-studded clefts. Pallid under early morning and late evening illumination, under high Sun they took on the appearance of charcoal smudges. Other examples included the variable spots in the crater Atlas (Figure 15.13), which Elger and Arthur Stanley Williams (1861–1938) had studied in the heyday of the Selenographical Society. Williams had gone so far as to construct a light curve characterizing their changing aspects under various angles of illumination between 1883 and 1887. From Arequipa, Pickering was able to detect delicate craterlets and clefts within the spots, which he regarded as probable sources of life-sustaining

Figure 15.12 A high sun image of the long-debated dark spots on the floor of Alphonsus. Both Gruithuisen and Pickering had no doubt that these seemingly variable spots were tracts of vegetation. Image by T. A. Dobbins.

gases and moisture.

However, the most intriguing dark patches were those in and around the great crater Eratosthenes (Figure 15.14). Favorably placed for observation at the southern end of the Apennine range not far from the center of the lunar disc, Eratosthenes is a majestic sight under a low Sun, with finely terraced inner walls and a complex of central peaks. Every month as the Sun rises higher and the lunar shadows disappear, its familiar outlines are replaced by a confused array of streaks and splotches with little obvious relationship to the formation's topographic relief. Pickering drew a veritable maze of dusky linear features intersecting in dark spots "practically identical in appearance with those seen upon the planet Mars." Borrowing from Martian nomenclature, he even called them—injudiciously, one might add—"canals" and "lakes."[27] Pickering mapped the changes in these "pseudo-shadows" in obsessive detail, changes which, he found, occurred with predictable regularity. The initial small darkening near the center of the crater proceeded to spread outward toward the crater's periphery, advanced up the inner slopes and spilled over the ramparts, then rapidly faded out again with the approach of evening shadows. Seemingly capricious minor changes in this cycle of development in what he called the "Gardens of Eratosthenes" would occupy much of his attention in coming years."[28]

Sternly upbraided by his brother because of his lunar and planetary philandering and extravagant financial mismanagement, William was recalled to Boston late in 1893. After an unsuccessful attempt to change Edward's mind about his fitness for the post, William briefly joined forces with businessman turned Oriental-

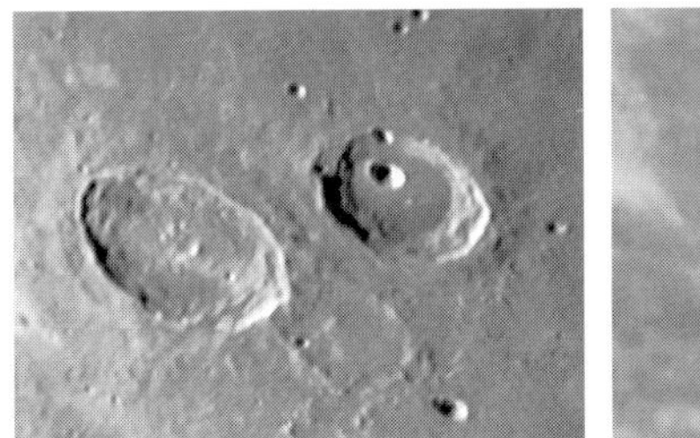
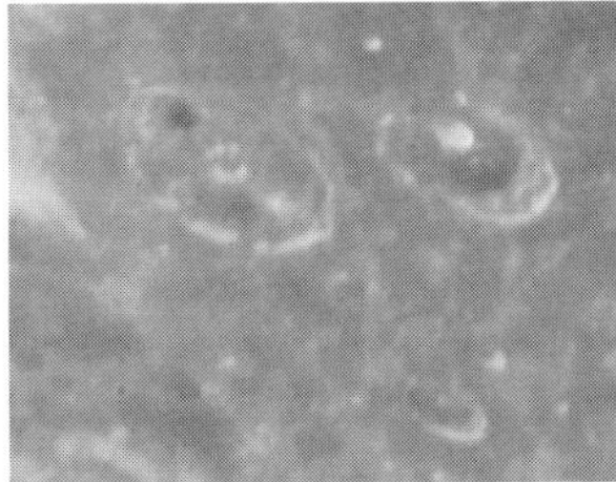

Figure 15.13 The crater pair Atlas and Hercules, another subject studied obsessively by W. H. Pickering and others because of the marked apparent changes under different conditions of lighting. Under late morning illumination (Left), several delicate rilles can be made out on the floor of the crater Atlas; though difficult, they are within reach of only a six-inch telescope. By midday (Right), this maze of delicate floor detail vanishes and is replaced by the dusky patches that Pickering took to be vegetation. Images by T. A. Dobbins.

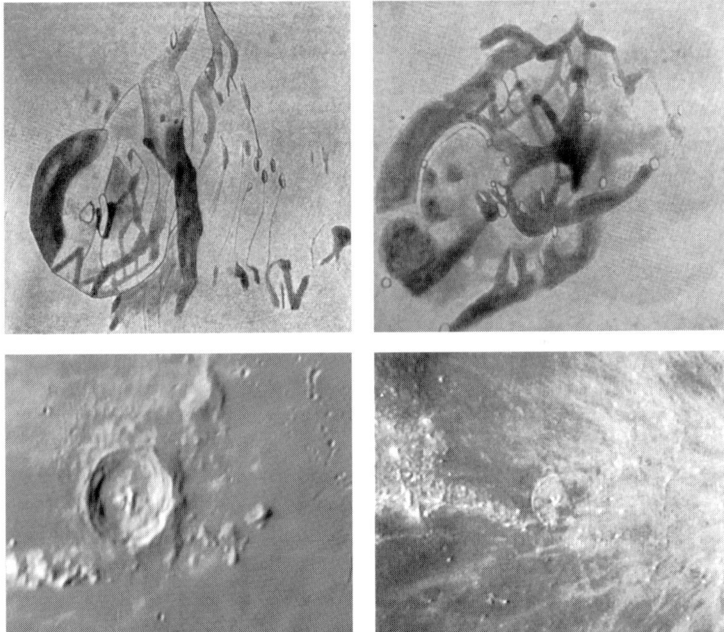

Figure 15.14 The "Gardens of Eratosthenes." (Top) Two drawings by Pickering showing the various "pseudo-shadows" and "canals" in and around the crater. Initially he regarded these features as tracts of vegetation, but as his thinking about them evolved, he came to regard them as swarms of insects scuttling across the lunar landscape. From W. H. Pickering, *The Moon* (1904). (Bottom) Eratosthenes as it looks under conditions of morning lighting and under the high Sun illumination of Full Moon, when only its ramparts can be recognized as a bright oval outline. Note the dramatic change in the appearance of the Copernican ray system in the right part of these frames, from nearly invisible under morning lighting to magnificently displayed in the high Sun image. Variegated material thrown off by the Copernicus impact that blanketed the floor of older Eratosthenes accounts for the features that so intrigued Pickering. Images by T. A. Dobbins.

ist turned Martian explorer Percival Lowell (1855–1916), who in January 1894 met with Pickering and his assistant A. E. Douglass (1867–1962) in Boston to lay plans for a Martian observing expedition to the Arizona desert the following autumn. In Arizona William hoped to find atmospheric conditions as good or even better than those at Arequipa for lunar and planetary work. Not everyone endorsed Lowell's judgment in choosing Pickering as his astronomical mentor, however. Boston astronomical computer Seth Carlo Chandler, Jr., wrote to Lick Observatory director Edward S. Holden, "Lowell, I am sorry to say, has not selected the right kind of companions for his astronomical picnic."[29]

Pickering did accompany Lowell and helped him to establish his observatory in Flagstaff—among other things, he designed a pre-fabricated dome which housed a borrowed 18-inch refractor—but Lowell, who had no interest in being anyone's understudy for long, severed the relationship immediately after the opposition of Mars. Pickering then returned to Cambridge, where Edward attempted to persuade him to return to Flagstaff—despite his rather low opinion of what was being done there, he pointed out to William that "your skill seems to lie especially in such work." Once again, William failed to heed Edward's advice. Instead, he diverted himself with laboratory experiments in which he used hot paraffin to represent the partially molten surface of the Moon and a piston applied beneath the surface of the wax to represent the tidal forces of the Earth. He persuaded himself that the lunar maria were created by enormous fissure eruptions like those that created the Deccan Plateau in India, which is covered with an enormous sheet of basalt.[30] Today this is the prevailing view among planetary geologists. Despite their shortcomings, Pickering's ideas had considerably greater merit than the volcanic theories being advanced by many of his contemporaries. But his restless mind soon turned to other subjects.

In August 1898, while inspecting plates taken with a new photographic telescope at Arequipa, he made his most notable discovery, Phoebe, the outermost satellite of Saturn,[31] a feat for which he was awarded the prestigious Lalande Medal of the French Academy of Sciences. In the summer of 1899, while vacationing in Jamaica, he became convinced that the quality of the island's "seeing" surpassed even that of the arid mountain stations at Arequipa and Flagstaff. He managed to persuade his brother to dip into the Boyden Fund to establish a temporary observatory on a rented estate called Woodlawn, the site of a former coffee plantation at an elevation of 2,000 feet in the hills near the town of Mandeville. From two anonymous donors Edward soon secured the funds to purchase a doublet objective lens of 12 inches (305 mm) aperture with a focal length of 135 feet (41 meters) from the firm of Alvan Clark and Sons, which would serve as the basis of William's next telescope (Figure 15.15). Such an extraordinarily long focal ratio all but eliminated the lens's chromatic aberration and provided a generous image scale ideal for precise astrometric work, but it precluded mounting the instrument in a conventional manner. A man of considerable mechanical aptitude, William cleverly selected a hillside with a gentle slope equal to the modest 18-degree latitude of his Jamaican site. On a north-south axis he erected a rectangular wooden

Figure 15.15 The Harvard Moon Atlas telescope. The mounting of a conventional instrument of a focal length sufficient to give the requisite plate scale was out of the question, so Pickering constructed a "tube" in the form of a cloth-enclosed tunnel running up a hillside. The 12-inch objective lens and 18-inch flat mirror were placed at the bottom of the tunnel; the plateholder was situated in a shed at the top. Unfortunately, the resulting photographs fell far short of expectations due to the faulty figure of the flat mirror, which required stopping down the objective to an aperture of only six inches. From W. H. Pickering, *The Moon* (1904).

framework covered with wire mesh overlaid in turn with cotton cloth, the whole affair supported by a series of posts driven into the ground. The lens was mounted at the lower end of this rudimentary tube, facing an 18-inch (457-mm) diameter flat mirror that was mounted separately on an equatorial steel fork driven by a pair of electric motors which served both for remote pointing and to compensate for the Earth's rotation. By adjusting the orientation of the mirror, which had an unobstructed horizon-to-horizon view from east to west, the light from celestial objects could be reflected through the axis of the objective lens to the focal plane at the upper end of the tube. There the image rotated as the mirror turned to track celestial objects, so a third motor imparted a corresponding rotation to the photographic plateholder, keeping the image fixed in both position and orientation. An enclosure complete with a darkroom was constructed where Pickering could work in comfort at the instrument's stationary focal position, sheltered from insects and the heavy dewfall of humid tropical nights.

Edward had hoped against hope that William would use the telescope to photograph the Earth-crossing asteroid Eros as part of an international effort to refine the value of the Earth-to-Sun distance (the so-called "astronomical unit"), but William disappointed yet again. Due to unforeseen delays, he missed the opportunity to secure a single photograph suitable for measurement. Instead, he casually informed his brother, he intended to take advantage of Jamaica's favorable atmospheric conditions and the new telescope's enormous image scale—it provided a lunar image no less than 15 inches (381-mm) in diameter—to pursue a long-contemplated project of producing a new photographic atlas of the Moon.[32]

Unfortunately, the results left much to be desired, in part because of the poor optical quality of the flat mirror. Pickering's lunar images were markedly inferior to Loewy and Puiseux's as well as to those taken by the meticulous George Willis Ritchey (1864–1945) with the recently commissioned 40-inch (1,010-mm) Yerkes refractor at Williams Bay, Wisconsin. His work was not entirely without merit, however, since in his series of photographs, taken over a period of five months, he managed to record the entire face of the Moon five times over under different conditions of local lighting—sunrise, two days after sunrise, noon, two days before sunset, and sunset. The series served to illustrate the dramatic effects of changing illumination on the appearance of many formations, the first atlas of its kind to do so in a systematic fashion.

Although Edward made no secret of the fact that he thought that William's photographs were second-rate, he did throw in $500 from his own pocket to help defray the costs of publishing the work, which appeared as *The Moon: A Summary of the Existing Knowledge of Our Satellite, With a Complete Photographic Atlas* (now generally referred to simply as "the Harvard Atlas"). The accompanying text, which combined a series of popular articles that had appeared during the preceding two years in *The Century* magazine with abridged, watered-down versions of material published in 1903 in *The Annals of the Harvard College Observatory,* promoted what Pickering referred to as "the new selenography—the selenography which consists, not in the mapping of cold dead rocks and isolated craters, but in a study of the daily alterations which take place in small, selected regions, where we find real, living changes—changes that cannot be explained by shifting shadows or varying librations of the lunar surface."

The reaction of professional astronomers—those who bothered to take notice of the publication at all—was sharply hostile. A review by Edward Emerson Barnard (1857–1923), the foremost American observational astronomer of that era, was typical:

> [Pickering] claims to have found evidences, not only of present volcanic life, but of snow and ice, clouds and vegetation. Perhaps if these discoveries had come slowly and one at a time, with long intervals between, they might have been received with better grace, but they have been turned out by wholesale, and almost any place on the Moon would seem to be conspicuously productive of one or more of the above phenomena... These observations are so startling that until they are fully verified one hesitates to accept them as real.[33]

But the popular imagination was more receptive. During the closing years of the Victorian age the English science teacher turned journalist turned novelist

Figure 15.16 The insect-like Selenites who inhabited the interior of the Moon are shown carrying off Bedford, the hero of H. G. Wells' *The First Men in the Moon* (1901). In general, Wells derived many of his ideas about lunar conditions from Pickering. Courtesy of the University of Minnesota Library.

H. G. Wells (1866–1946) penned two of the immortal classics of science fiction. Published in 1898, *The War of the Worlds* is the tale of an invasion by a murderous race of Martians that drew its inspiration largely from the theories of Percival Lowell. Pickering's equally sensational 1890s "discoveries" served as the basis of the strange lunar world that greeted Wells's intrepid explorers in *The First Men in the Moon*, published in 1901 (Figure 15.16). The description of the lunar sunrise that unfolded as Wells protagonists Cavor and Bedford peered through the window of their Moon-faring Cavorite sphere was vintage Pickering:

> The distant cliff seemed to shift and quiver, and at the touch of the dawn a reek of grey vapour poured upward from the crater floor, whirls and puffs and drifting wraiths of grey, thicker and broader and denser, until at last the whole westward plain was steaming like a wet handkerchief held before the fire...[34]

> Imagine it! Imagine that dawn! The resurrection of the frozen air, the stirring and quickening of the soil, and then this silent uprising of vegetation, this unearthy ascent of fleshiness and spikes. Conceive it all lit by a blaze that would make the intensest sunlight of earth seem watery and weak. And still around the stirring jungle, wherever there was shadow, lingered banks of bluish snow.[35]

In 1904, Pickering visited the Mauna Loa and Kilauea volcanoes in Hawaii that had so impressed James Dwight Dana, an experience that strongly reinforced his own theories of the nature and origin of lunar formations. "There seems, indeed, to be no feature found upon the Moon which is not presented by these Hawaiian volcanoes," he wrote, "and there is no feature of the volcanoes that does not also have its counterpart upon the Moon."[36] Indeed—perhaps motivated by desire to elude the overbearing supervision of his older brother—Pickering began to travel extensively. In addition to Hawaii (part of what were then known as the Sandwich Islands), he visited volcanic landscapes of the Azores, Central America, Alaska, and western Canada. But these carefree days of globetrotting came to an abrupt end in 1911.

"The easiest way to make a small fortune," runs an old saw, "is to start with a large one." For years Pickering had toyed with developing a "mathematical formula for beating the stock market." His early experiments resulted in the loss of $14,000 in the spring of 1901, a sum that represented almost six years' salary. A loan of $1,000 from his brother followed only a month later by an inheritance of $45,000 upon the sudden death of their mother brought temporary relief. But in 1911 applying a revised version of the formula resulted in a total, devastating loss. As ever, his older brother was on hand with a remedy. Edward proposed that William return to Jamaica, where living expenses were modest, and try to get his life back in order. Using his prerogative as director, Edward agreed to equip William with Harvard's Henry Draper Memorial telescope, an 11-inch (279-mm) Clark refractor, on the condition that he employ the instrument photographically to supplement the photometric and spectroscopic investigations being conducted at Arequipa. Predictably, William once more defied his brother's directive and chose to concentrate on visual observations of the Moon and planets. The ensuing correspondence with Edward revisited the disputes of twenty years earlier that had resulted in his dismissal and recall from Arequipa.

William and his family settled in at the Woodlawn estate near Mandeville that had been the site of the first Jamaican expedition. The Draper refractor was erected on a patio next to the plantation house, protected by a roll-off shed in inclement weather. William quickly grew to love the island's delightful climate and colonial splendor, and he emerged as a local celebrity known simply as "The Professor." A succession of government officials, scientists, and admirals of the British and American navies were entertained at Woodlawn, where views of the Moon and planets through the telescope were customary after-dinner fare.

Having found a measure of independence, status, and contentment in this tropical redoubt, William somehow managed to persuade Edward to dip one last time into the Boyden Fund to formally establish Mandeville as a permanent station of the Harvard College Observatory. But in the end, this self-imposed exile led to more than just physical isolation from his peers. The interests and attentions of the professional astronomical community were by now increasingly directed beyond the confines of the Solar System to the problems of the stellar and nebular universe, and the Moon and planets—William's favorite haunts—became almost quaintly provincial. In 1914, Edward urged his brother to publish lunar and planetary observations in the Harvard College Observatory *Annals* in order to ease his task as director of justifying the expense of maintaining the Mandeville station to the college's administrators and trustees. William replied bitterly: "One might as well publish in a botanical journal... The *Annals* are read only by astronomers (professionals), and very few professionals care a rap about either Mars or the

Moon."[37] Indeed, William had by now largely foregone submitting his work to professional journals and had instead become a frequent contributor to the magazine *Popular Astronomy*, published by W. W. Payne at Carleton College in Northfield, Minnesota. He repeatedly appealed to its readership of amateur astronomers to verify and expand upon his work. In these circles his influence would be deep and enduring.

Estranged from professional astronomers, it was with amateurs that Pickering now felt a spirit of kinship. He still spent countless hours at the eyepiece, unlike most professionals who seldom looked through telescopes except to guide long-exposure photographs or to measure double stars. Haunting the ruins of his career, he launched blistering attacks on the professional astronomical community that by now was beginning to regard him as the "madman of Mandeville":

> Those sidereal astronomers with no planetary experience of their own whatever, rush into print to say that there are no canals and no changes, seasonal or otherwise, to be found either upon the Moon or Mars, and do it in spite of the lifelong careful studies and measures of the planetary astronomers, are not only foolish, but as far as their reputations count for anything, they utterly misrepresent the extent of their own information, or lack of it, to the public who trust in them for guidance.[38]

While Pickering's theories about some subjects—notably the Martian canals and the calculated positions of various planets he predicted beyond Neptune, in sometimes bitter rivalry with Percival Lowell—were almost capriciously changeable, showing a mental flexibility bordering on fickleness, his thoughts about the Moon demonstrated what G. K. Chesterton once called "the clean, well-lit prison of one idea." Strange, given his conviction that "the body most resembling the Moon with which we are acquainted is the planet Mars."[39] Stubbornly loyal to the tenets of his "new selenography," he continued to buttress his theories with additional observations—telling proof of E. E. Barnard's dictum that "the eye easily pleases and satisfies the minds of those who have firmly rooted and preconceived ideas of things."[40]

From Mandeville Pickering reported that he could distinguish unmistakable evidence of glacial erosion on the central peaks of many craters—U-shaped valleys and the linear deposits of rocky debris geologists call moraines.[41] He detected drifting clouds and even recurring "snow storms" over the central peaks of the craters Theophilus, Plinius, Alphonsus, and Bullialdus, as well as over several isolated mountains such as Pico and the Straight Range. He regarded the central peaks as volcanic cones; the exhalations of steam that precipitated as snow were the last vestiges of activity.[42] He began to write of a "meteorology of the Moon."[43] It was, as a wit might have been justified in observing (and as has been said of theology and of exobiology) a science in search of a subject.

Pickering spent many balmy tropical nights in Mandeville continuing his studies of the curious high Sun markings of Eratosthenes. While he had satisfied himself years before that tracts of vegetation grew on the floor of this crater, he remained puzzled by subtle "non-seasonal" differences in the location of certain dusky patches from lunation to lunation. In the final installment of a collection of six articles on Eratosthenes that appeared in the pages of *Popular Astronomy*, he suggested that from a lunar vantage point, a telescopic observer in the mid-nineteenth century would have seen similar moving patches on the plains of North America—vast herds of buffalo. Although the shifting "pseudo-shadows" in Eratosthenes were of comparable size, he calculated that they moved at a rate of only a few feet per minute, leading him to conclude that they were probably composed of creatures considerably smaller than buffalo. Most likely they were swarms of insects, though he refused to rule out lumbering larger animals: "While this suggestion of a round of lunar life may seem a little fanciful, and the evidence on which it is founded frail, yet it is based strictly on the analogy of the migration of the fur-bearing seals of the Pribiloff Islands."[44]

Now well beyond the pale of sober science, the publication of these speculations destroyed the last vestige of respect for Pickering harbored by his peers in the professional community. The prolific British author Valdemar Axel Firsoff (1910–1982) (Figure 15.17), who continued to give at least qualified support to Pickering's views of the Moon as late as the 1960s, admitted that the insects of Eratosthenes "provided handy ammunition to those who wished to see him as a crank not to be taken seriously."[45]

Meanwhile, Edward had died in 1919, leaving William to the mercy of his successor, Harlow Shapley (1885–1972), who quickly insisted on the return of the Draper refractor. In order to continue his observations, William purchased a 12-inch (305-mm) Newtonian reflector by British instrument-maker George Calver. With this amateur-sized telescope, he bravely soldiered on. Walter Haas, a young Ohioan who visited Pickering in Jamaica in the summer of 1935, recalled that by then he could "sense a certain disappointment that his achievements had gone unappreciated... By then, the professor's astronomy was perhaps behind the times by a decade or two."[46]

Pickering was then tired and worn out, with little interest in observing, a man in the disappointed twilight of his career. He died in 1938. However, right up to the end of his life he remained confident that his "new selenography" would one day be vindicated:

> The astronomers who are not selenographers declare that there is no atmosphere or water upon the Moon, and that, therefore, changes are impossible. The selenographers' reply is simply that they have seen changes take place. The argument reminds one somewhat of that which occurred before the general acceptance of the Copernican system. The majority of the scientific world at that time, and the great weight of authority, insisted that the Earth did not move, because it was impossible. The reformers claimed that their observations and computations proved it.
>
> The question of lunar changes is one of much less consequence to the world at large than the other, but we can hardly doubt that the final outcome of the controversy will in the end be the same—the observer will triumph over the man who depends exclusively on his reasoning faculty... But this is of minor consequence; the Moon is always with us, and these discoveries may well be left for the next generation of astronomers.[47]

Figure 15.17 The British selenographer Valdemar Axel Firsoff, shown here at a 1980 meeting of the Terrestrial Planets Section of the British Astronomical Association at Chester, England. Firsoff continued to defend many of Pickering's interpretations of the lunar surface and its supposed changes as late as the 1960s. Though Firsoff threw quite a number of errant horseshoes, he did manage an occasional ringer, such as his notion that torrid Mercury might harbor deposits of water ice in the shadowed floors of craters near its poles. Although the idea seemed bizarre when Firsoff proposed it in the early 1970s, it has recently been confirmed by radio astronomers. Courtesy Alan Heath.

Though not, of course, vindicated, Pickering certainly had a profound influence on lunar observers of the next generation. He was the patron saint of many amateurs who continued to search for changes on the Moon—including Haas himself, who in 1942 published an influential article entitled "Does Anything Ever Happen on the Moon?"[48] The quest for what became known as "transient lunar phenomena"—carried on mainly by amateurs, but joined even by some professionals during the 1950s and 1960s—was inspired in no small measure by Pickering's work.

For all that, Pickering's influence on lunar studies has never been fully appreciated. He himself noted somewhat bitterly in his 1904 book on the Moon: "The names of the chief craters are generally those of men who have done little or nothing for Selenography." For example, he pointed out, the crater named Newton was so near the limb it might as well have been on the other side of the Moon; Schroeter was commemorated "by a broken-down crater of which so little is left that it is often only recognisable with difficulty"; a crater only six miles in diameter represented Schmidt, who "perhaps devoted more of his life than any other man to the study of the Moon."[49] Ironically, poor William Pickering would fare even worse, for he has no crater named after him at all. One of the formations that he studied intensively, the crater Messier A, was long unofficially designated "W. H. Pickering," but the name was dropped by the Lunar Nomenclature Committee of the International Astronomical Union at its meeting in Dublin in 1955—not, according to Ewen Whitaker, who was one of the delegates there, out of any animus against him, but simply because there was already a Pickering on the Moon—E. C. Pickering. Even on the Moon, his special province, William has been overshadowed by his older and more respectable brother.

References

1. Joseph Ashbrook, "W. H. Pickering and the Satellites of Jupiter," *Sky & Telescope*, 27, 6 (1963), 335.
2. A fascinating account of the founding and early years of the Boyden Station can be found in Donald Fernie, *The Whisper and the Vision* (Toronto: Clarke, Irwin, and Co., 1976).
3. William Sheehan, *Planets and Perception*, (Tuscon: University of Arizona Press, 1988), p. 99.
4. See Thomas Dobbins, "Recording the Moon and Planets with a Video Camera," *Journal of the British Astronomical Association*, 106 (1996), 309–314 and Dobbins, "Shoot the Moon and Planets!" *Sky & Telescope*, 91, 6 (1996), 94–97.
5. Harold Plotkin, "William H. Pickering in Jamaica: The Founding of Woodlawn Observatory and Studies of Mars," *Journal for the History of Astronomy*, 24 (1993), 103.
6. B. Z. Jones and L. G. Boyd, *The Harvard College Observatory: The First Four Directorships, 1839–1919* (Cambridge, Mass.: Harvard University Press, 1971), p. 307.
7. Dark bands like the one reported by Pickering are often seen when two unequally bright surfaces are brought into contact with one another. They are known as "Mach bands" after the Austrian physicst Ernst Mach, who gave the first satisfactory explanation of them; see Floyd Ratliff, *Mach Bands: Quantitative Studies on Neural Networks in the Retina* (San Francisco: Holden-Day, 1965). The effect can be demonstrated by positioning a 40-watt frosted-envelope incandescent light bulb in one's line of sight so it appears to lie behind a 100-watt counterpart. An illusory band tangent to the edge of the brighter bulb will be readily apparent, perfectly mimicking Pickering's observation. The bright halo that has often been reported around the jet-black disk of Mercury when that airless planet transits the Sun is a similar effect that can be duplicated in a parlor experiment using a light bulb and a coin.
8. W. H. Pickering, "The Lunar Atmosphere," *Popular Astronomy*, 32 (1924), 1–4.
9. Pickering, *The Moon: A Summary of the Existing Knowledge of Our Satellite, With a Complete Photographic Atlas* (New York: Doubleday, Page, and Co, 1904), p. 18.
10. Ibid., pp. 41–42.
11. Other experienced observers have fallen victim to the very same illusion, notably Francis H. Thornton, co-founder of *The Moon*, the periodical of the Lunar Section of the British Astronomical Association. Observing with an 18-inch (457mm) reflector on February 10, 1949, he reported "what seemed to be a diffuse patch of smoke or vapour, apparently originating from the east side of the Valley near the Cobra-Head, where the landslip is, and spread over the edge on to the plain for a short distance." See H. Percy Wilkins and Patrick Moore, *The Moon* (London: Faber and Faber, 1955), p. 263.
12. Pickering, *The Moon*, p. 42.
13. Jackson T. Carle, "Three Riddles of Plato", *Sky & Telescope 14*, 6 (1955), 221–224.
14. W. H. Pickering, "The Lunar Atmosphere", *Popular Astronomy*, 32 (1924), 4.
15. W. H. Pickering, "Meteorology of the Moon", *Popular Astronomy*, 23 (1915), 12.
16. W. H. Pickering, "Changes Upon the Surface of the Moon", *Nature*, 71 (1905), 226
17. For a superb illustrated discussion of the nature and causes of the sequence of changes in appearance exhibited by Messier and Messier A based on an unprecedented series of observations spanning some four decades, see: Harold Hill, *A Portfolio of Lunar Drawings* (Cambridge: Cambridge University Press, 1991), pp. 210–214. Messier A is itself a compound structure consisting of a comparatively recent impact not quite superimposed on an older crater. Under morning illumination the older component is prominent, giving Messier A its triangular aspect, but under higher lighting it fades and the formation takes on the appearance of a longitudinally-oriented oval.
18. Pickering, *The Moon*, p. 42.
19. Pickering, "Meteorology of the Moon", *Popular Astronomy*, 24 (1916), 12.
20. C. A. Young, *A Text-book of General Astronomy for Colleges and Scientific Schools* (Boston: Ginn & Co., 1900), p. 173.
21. Pickering, *The Moon*, p. 20.

22. Ibid., p. 42.
23. Pickering, "Changes Upon the Moon's Surface," p. 227. It was by no means an unreasonable hypothesis. Indeed, the mystery of the nature and origin of Pickering's riverbeds would endure until men set foot on the Moon. The highly detailed images of sinuous rilles transmitted by the Lunar Orbiter spaceprobes during the mid-1960's revealed an absence of tributaries and deltas of sediment deposits, showing that Pickering's supposed analogy with terrestrial arroyos and canyons was less than perfect. Nevertheless, theories of lunar watercourses died hard. Several geologists conjectured that these features might represent channels cut by glacial meltwater that had flowed beneath pressurizing canopies of ice—evidence perhaps of an extensive subsurface layer of lunar permafrost.
24. Pickering, "Evidences of Erosion on the Moon," *Popular Astronomy*, 24 (1916).
25. R. E. Lingenfelter et al, "Lunar Rivers," *Science,* 161 (1968), 266.
26. Pickering, "Lunar Changes," *Memoirs of the British Astronomical Association*, 20 (1916), 111.
27. Pickering did attempt to defend his terminology. In his article "The Canals in the Moon," *The Century Magazine*, 64 (1902), p. 189, he argued: "The word 'canal' as used in astronomy is applied to a dark, narrow, straight, or smoothly curved surface marking. The term does not necessarily imply the presence of water." Indeed, the dark lanes of dust in the Andromeda galaxy, discovered by G. P. Bond with the 15-inch Harvard refractor in 1847, were referred to as "canals" until long-exposure photographs revealed the object's true spiral form shortly before the turn of the century. See J. Ashbrook, "The Discovery of Spiral Structure in Galaxies" in *The Astronomical Scrapbook,* pp. 402–403.
28. According to Harold Hill, *Portfolio*, p. 48: "Pickering's conclusions were based upon his interpretations of small-scale details of so subjective a character that another observer might draw and interpret them quite differently." Val Axel Firsoff wrote in 1959: "In British conditions and with a less powerful instrument I have been unable to trace the minor 'canals' drawn by Pickering and am generally inclined to suspect that he fell into the common error of exaggerating the linearity of lunar markings, which often appear to me as irregular curves." Firsoff, *Strange World of the Moon* (New York: Basic Books, 1959), pp. 95–96.
29. S. C. Chandler, Jr. to E. S. Holden, September 4, 1894; Mary Lea Shane archives of the Lick Observatory.
30. W. H. Pickering, "Evidences of Erosion on the Moon," *Popular Astronomy,* 24 (1916).
31. A. F. O'D. Alexander, *The Planet Saturn* (London: Faber and Faber, 1962), pp. 263–264.
32. In the article "Investigations in Astronomical Photography," *Annals of the Harvard College Observatory,* 32 (1895), p. 110, Pickering had written that a "polar siderostat" refractor of extremely long focal length—exactly like the one later placed at his disposal—would be the ideal instrument for lunar photography.
33. E. E. Barnard, "Review of Pickering's 'The Moon,'" *The Astrophysical Journal,* 20 (1904), 359–364.
34. H. G. Wells, *First Men in the Moon*, ed. by David Lake (New York: Oxford University Press, 1995), p. 52. See also Thomas Dobbins and Richard Baum, "Observing a Fictional Moon," *Sky & Telescope*, 95, 6 (1998), 105–109.
35. Ibid., p. 59.
36. Pickering, *The Moon*, p. 42.
37. W. H. Pickering to E. C. Pickering, 20 October, 1914; cited in Plotkin, "William H. Pickering in Jamaica," p. 112.
38. W. H. Pickering, "Report on Mars No. 36", *Popular Astronomy*, 34 (1926), 363.
39. Pickering, "Lunar Changes," *Memoirs of the British Astronomical Association*, 110–111.
40. E. E. Barnard, "The Milky Way and the Great Nebula of Andromeda", *American Annual of Photography*, 8 (1898), 1–8.
41. Pickering, "Evidences of Erosion on the Moon," *Popular Astronomy,* 24 (1916).
42. Pickering, "The Snow Peaks of Theophilus", *Popular Astronomy*, 25 (1917).

43. Pickering, "Meteorology of the Moon".
44. Pickering, "Eratosthenes—VI, A Study for the Amateur", *Popular Astronomy,* 33 (1925).
45. Firsoff, *Strange World of the Moon*, p. 160.
46. Walter Haas to William Sheehan, personal correspondence, May 31, 1994.
47. Pickering, "Lunar Changes," pp. 110–111.
48. Walter H. Haas, "Does Anything Ever Happen on the Moon?", *Journal of the Royal Astronomical Society of Canada*, 36 (1942).
49. Pickering, *The Moon*, p. 86.

Chapter 16:
Fauth's Frozen Moon

Even as the "new selenography" was taking shape and Pickering was invoking Gruithuisen with his full-blown claims of a lunar meteorology and migrating swarms of insects, the inevitable reaction set in. In many ways the most outspoken counter-revolutionary among active observers of the Moon would be a German school teacher, Philipp Fauth, whose abilities as an observer and draftsman might have lent great authority to his insightful, carefully reasoned arguments against the tantalizing claims of changes and volcanic activity on the Moon. He was Parmenides to Pickering's Heraclitus, opposing to the perennially fashionable idea that all was in flux on the Moon the view that nothing there changes—in fact, Fauth was convinced that the Moon was frozen, covered with a rind of ice. But this quarrelsome man's stubborn advocacy of outlandish theories tarnished his reputation so badly that his work has been relegated to undeserved obscurity, particularly in the English-speaking world.

Philipp Johann Heinrich Fauth was born in 1867 at Bad Dürkheim, a village nestled among the picturesque rolling hills and vineyards of the Rhineland. The oldest of three children born into a long-established family of pottery-makers, his interest in astronomy was kindled at the age of seven when he was awakened by his father and carried outside to see Comet Coggia gleaming in the predawn sky.

Like William Herschel, Fauth was a musical prodigy, having taken up the violin at the age of five. While music would remain a life-long passion, Fauth chose to become a teacher. Few professions were more highly esteemed in late nineteenth-century Germany. A portrait of him taken in 1890, the year he embarked on his career in a public high school, captures a self-confident professorial face adorned with pince-nez eyeglasses and a stylish moustache with upturned waxed tips.

Like so many amateur astronomers before and since, Fauth began work with a humble refractor of just under 3 inches aperture (72 mm), which he purchased in 1887. Three years later he graduated to a better instrument of 6.4 inches (162 mm), its doublet lens the handiwork of Professor Max Pauly (1849–1917), who six years later would become the founding Director of the *Astroabteilung* ("Astronomical Division") of the famous Carl Zeiss optical works at Jena.[1] At the time it was unusual to find such an instrument in the hands of an amateur; in fact Fauth was as well equipped as the nearby university observatories at Karlsruhe, Heidelberg, and Cologne.

To house his telescope Fauth erected a rudimentary structure of stone, brick, and timbers surmounted by a cylindrical turret with hinged shutters. He located it on a grass-covered knoll, the Lämmchesberg, on the southern outskirts of Kaiserslautern. He immediately set to work, devoting himself chiefly to studies of the Moon and Jupiter.

In 1892 Fauth was transferred to a post in the remote village of Oberarnbach. Suddenly the use of his observatory required a three-mile tramp through the hills to a train station, followed by a ten-mile trip by train, and finally a half-mile climb up the slopes of the Lämmchesberg. This journey had to be repeated in reverse each morning, when another day's work in the classroom awaited him at 8 o'clock. Fauth later recalled those days of hardship:

> I earned my spurs in those years and, I believe, bitterly earned them. In the summers I more than once saw the Sun set as I began to observe, and rise as I closed up the observatory... Deep snow and angry cold made for many a burden... I was at that time 25 years old and had a strong constitution. I don't believe that there has ever been another astronomer who has endured so many bodily and, in despairing moments, spiritual pains as I did in this voluntary work, again and again goaded on by successes that I would never have dared to dream.[2]

So unflagging were Fauth's self-discipline and zeal for observing that he managed to keep up this grueling routine for four years, during which he established his reputation as a keen-eyed observer and a superb draughtsman (Figure 16.1). In 1893 and again in 1896 he issued impressive monographs, *Astronomische Beobachtungen und Resultate aus den Jahren 1890 und 1891 auf der Privatsternwarte zu Kaiserslautern* (Astronomical Observations and Results from the Years 1890 and 1891 at the Private Observatory at Kaiserlautern) and *Neue Beiträge zur Begründung einer Modernen Selenographie und Selenologie* (New Contributions Toward Establishing a Modern Selenography and Selenology). The latter contained topographic charts of 25 selected regions of the Moon masterfully executed in the hachure technique employed by all the leading German selenographers since Lohrmann, and an announcement that the author intended to eventually produce a new lunar map on a scale of 1:1,000,000—more than three and a half times larger than Lohrmann and Mädler's maps and almost twice as large as Schmidt's—that would be based on outlines derived from photographs, with finer details inserted from visual observations. Articles by Fauth began to appear in the leading German astronomical journals, *Astronomische Nachrichten* and *Sirius*, and he began to correspond regularly with kindred spirits such as Hermann Klein in Cologne and Johann Nepomuk Krieger in Gern-Nymphenberg (Figure 16.2), an emerging rival Fauth would quickly come to regard as a "bitter mortal enemy.[3]

Only two years older than Fauth, Krieger was, like Hevelius, the son of a master brewer. He was little more than a boy when he started to observe the Moon with a small refractor from the sleepy mountain hamlet of Unterwiesenbach in Bavaria, where his formal education ended at the age of fifteen. Six years later he travelled to Cologne to visit Klein. The professor not only warmly encouraged Krieger to make selenography his life's work, but assumed the role of his mentor,

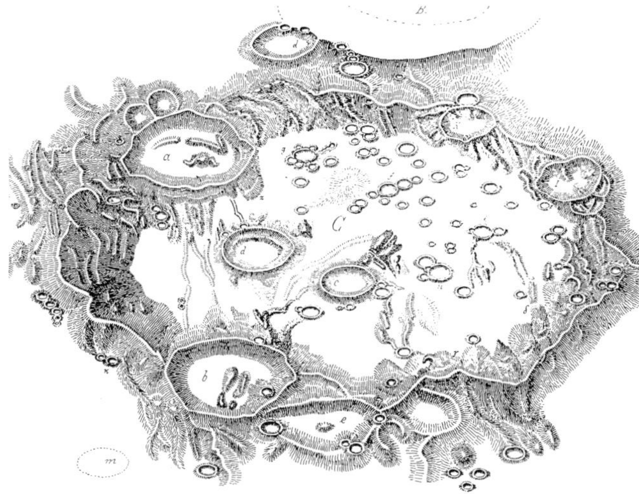

Figure 16.1 Fauth's topographic chart of the crater Clavius is typical of his early work. From Philipp Fauth, *Neue Beiträge zur Begründung einer Modernen Selenographie und Selenologie* (1896).

Figure 16.2 The Bavarian amateur Johann Nepomuk Krieger, whose premature death at the age of 37—at least partly due to overwork—ended a career of rare promise. This portrait appeared as the frontispiece of Krieger's *Mond-Atlas* (1912). Sheehan's Collection.

directing the young man to study mathematics, physics, photography, and the graphic arts.

Krieger's ensuing academic career soon faltered, for he lacked the mathematical aptitude required for the rigorous curriculum at the University of Munich. Undeterred, he spent his inheritance to establish a private observatory in the Munich suburb of Gern-Nymphenburg, equipped with a fine 10.6-inch (270-mm) refractor. Like Fauth, he announced his intention to produce a definitive lunar atlas.

Klein provided Krieger with photographs made from the best Lick and Paris negatives, enlarged to a scale of almost 12 feet to the Moon's diameter. These grainy, low-contrast prints served as the basis for Krieger's drawings, ensuring an exceptional level of positional accuracy and proper proportion. At the eyepiece Krieger used different colored pencils on successive nights to sketch in the finest

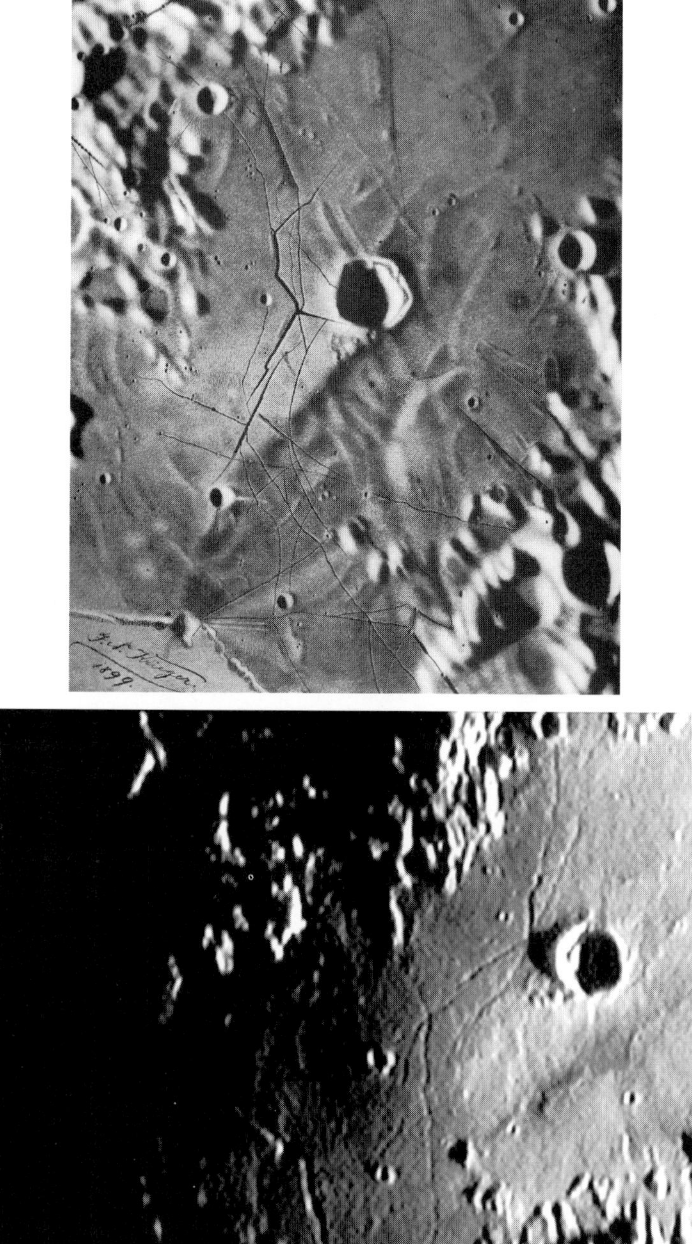

Figure 16.3 (Top) Krieger's drawing of the spidery network of rilles near the crater Triesnecker, from his *Mond-Atlas* (1912). Sheehan's Collection. Although in general the level of detail is excellent, the rectilinear appearance of the clefts is somewhat exaggerated, as is evident from the comparison CCD image of the Triesnecker rilles shown at grazing sunset illumination (bottom). Image by T. A. Dobbins.

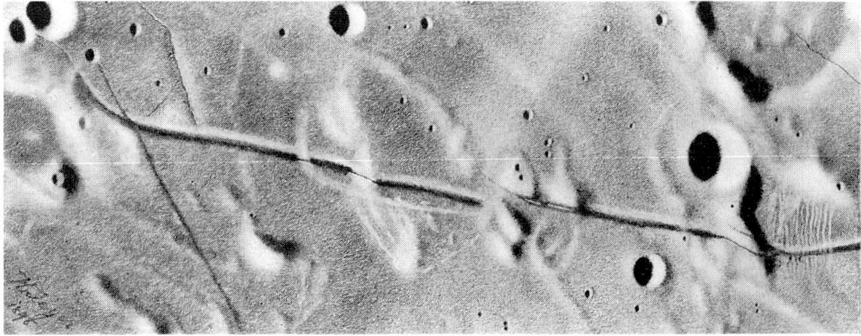

Figure 16.4 (Top) Krieger's drawing of the Ariadaeus Rille, a striking example of a "graben" —an area of dropped terrain lying between two parallel faults. The drawing shows that both ends of the rille are curiously forked, and also clearly represents the displacement of the rille by its own width at one point—the clearest example of what geologists call a strike-slip fault on the entire Moon. From his *Mond-Atlas* (1912). (Bottom) Comparison image of Ariadaeus Rille by T. A. Dobbins.

details glimpsed in fleeting moments of steady seeing—delicate features that were far beyond the capability of photography to record. These sketches then served as the basis for magnificent shaded drawings executed with India ink, graphite pencil, charcoal, and paper stumps that were almost univerally recognized as "startlingly superior… in meticulous accuracy, aesthetic appeal, and legibility"[4] to any that had been produced up to that time (Figures 16.3, 16.4).

Not, however, by Fauth. The German schoolmaster accused Krieger of "arrogance" and "superficiality," and boasted that with his smaller telescope he routinely saw more than Krieger (Figures 16.5, 16.6). Perhaps Fauth secretly dreaded that his own praiseworthy efforts would be overshadowed by Krieger's work, but

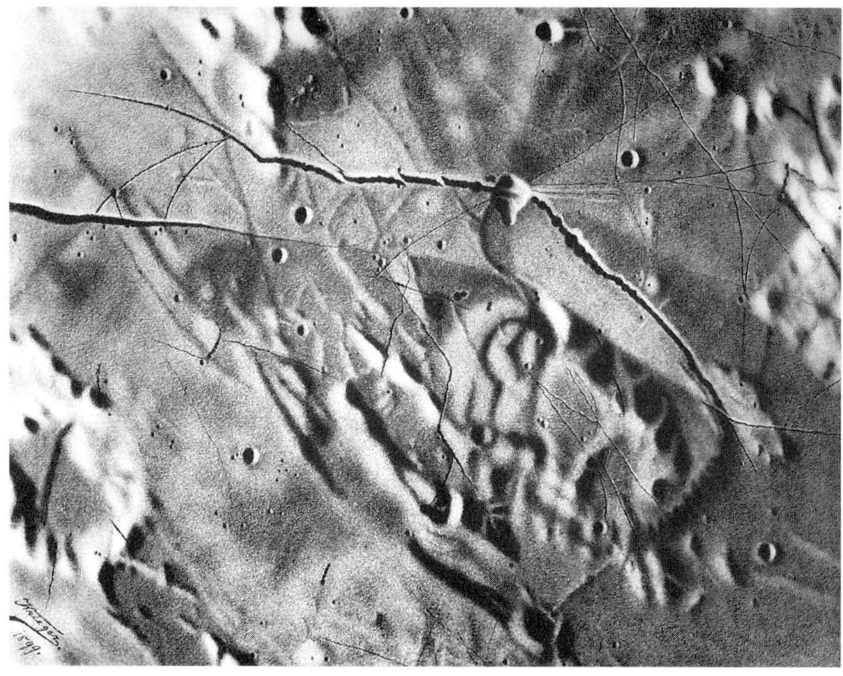

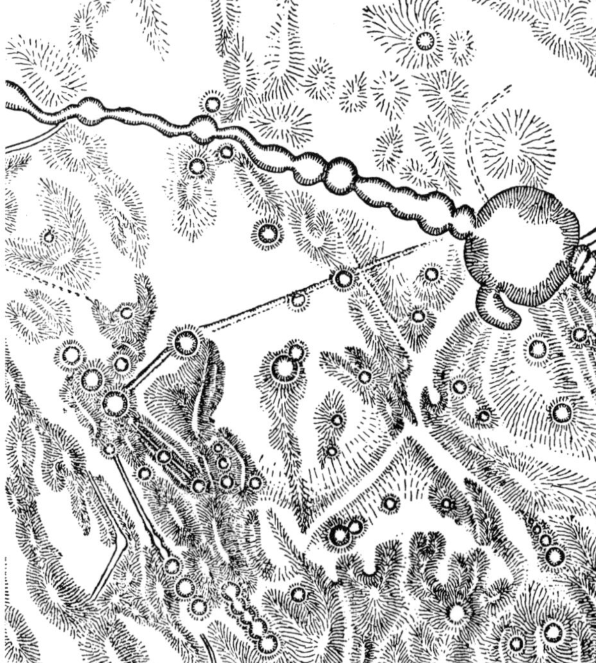

Figure 16.5 Krieger vs. Fauth. (Top) Krieger's drawing of the Hyginus Rille, which contains many rimless craterlets strung out along its floor like pearls on a string. They probably represent localized areas of subsidence along the fault that created the rille itself. From his *Mond-Atlas* (1912). (Bottom) Fauth's more schematic rendering of the eastern arm (astronautical convention) of the Hyginus rille, with elevations denoted by the method of hachures. From Fauth's *Der Mond und Hoerbigers Welteislehre* (1925).

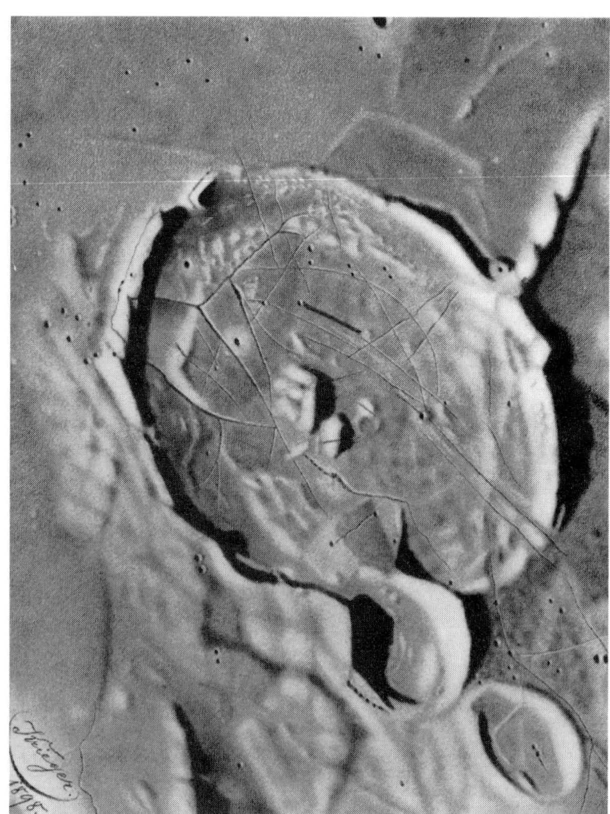

Figure 16.6 Krieger vs. Fauth. (Top) Krieger's drawing of the 68-mile (110-kilometer) diameter crater Gassendi, which dominates the northern shore of Mare Humorum. A delicate network of rilles criss-crosses the eastern half of Gassendi's floor, which exhibits a rough, jumbled texture. The southern ramparts of Gassendi were breached by lavas from the adjacent mare, flooding a small crescent on the crater's floor. From Kriegers' *Mond-Atlas* (1912). (Bottom) Fauth's Gassendi, an example of his early work. The level of detail is remarkable considering the modest 6.4-inch aperture he employed. From Fauth, *Neue Beiträge zur Begründung einer Modernen Selenographie und Selenologie* (1896).

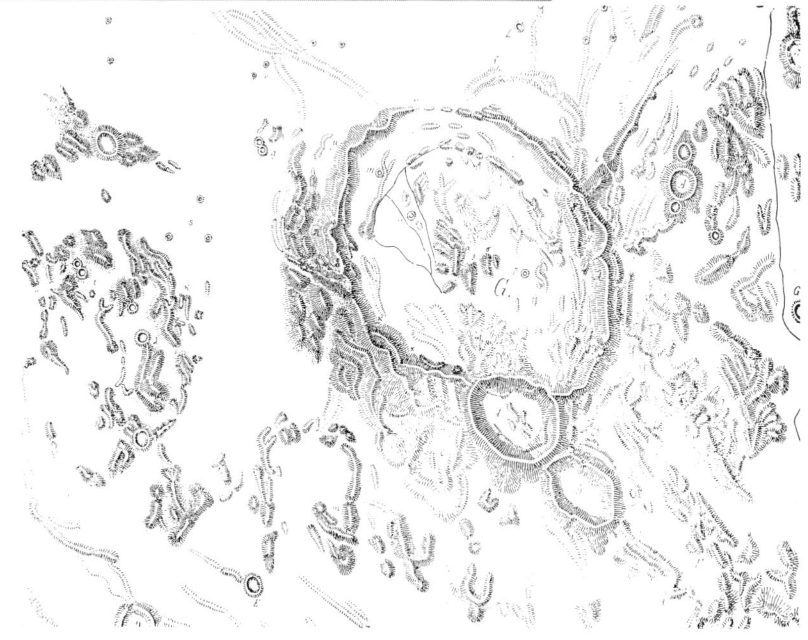

Figure 16.7 The vast majority of Krieger's drawings were still incomplete at the time of his premature death. This depiction of the partially flooded crater Fracastorius is typical. The blurry, over-enlarged photographic substrate is evident beneath the outlines of craterlets and clefts that he penciled in at the telescope. From Krieger's *Mond-Atlas* (1912).

if so, his fears were uncalled for. The frantic, monomaniacal pace at which Krieger labored quickly took its toll. In only a few short years Krieger's health utterly collapsed. He died in February 1902, a martyr to selenography, having completed less than a third of the plates for his atlas (Figure 16.7). This fraction would be published, in a rough and fragmentary form, only ten years after his death.[5] It was only by default that Fauth emerged as the leading German selenographer of the era.

In 1895 Fauth was transferred again, this time to a school in nearby Landstuhl. Aided by a subsidy from the Royal Prussian Academy of Sciences, he constructed his second observatory at an elevation of 1,300 feet above sea level. It consisted of a dome that he made by stretching and sewing sailcloth over a framework of metal rods, supported by a 26-foot tall sandstone tower that rose above the treetops on the forested summit of the Kirchberg, a 450-foot hill overlooking the town.

As Fauth's reputation grew, his 6.4-inch achromat was supplemented by a state-of-the-art triplet apochromat refractor of seven inches (178-mm) aperture, on loan from the Zeiss Institute during the years 1896 to 1903. The two instruments shared the same mounting, with the Zeiss telescope riding atop its smaller

Figure 16.8 Fauth's domed observatory atop the Kirchberg near Landstuhl, which housed 6.4-inch and 7-inch refractors that rode together on an overtaxed mounting. From Fauth, *The Moon in Modern Astronomy* (1907).

counterpart in "over-and-under" shotgun fashion (Figure 16.8).

Despite the modest aperture of these telescopes, Fauth was extravagant in his claims: "I have penetrated into the features of the Moon's face to an extent that no other eye has yet done," he boasted. "I have an eye that is remarkably adapted in structure for such research, and has improved more and more with use in the power of perceiving tiny features.[6] Unwilling to trade crisp definition for increased image scale, Fauth preferred to use surprisingly low magnifications, advising that "for observation of the Moon clarity is worth more than size." He was most successful in detecting minute details with a power of only 160X on the 6.4-inch refractor and 176X on the 7-inch (Figures. 16.9, 16.10). Even on the steadiest nights, he very seldom resorted to magnifications in excess of 210X on the 6.4-inch or 233X on the seven-inch.[7] Here was proof, if any were needed, of the arbitrariness of W. H. Pickering's oft-repeated dictum that 400X is the minimum magnification required for original work on the Moon and planets.[8]

There can be no doubt of Fauth's extraordinarily acute vision. Like Gruithuisen, he consistently recorded features that eluded experienced observers equipped with considerably larger instruments. By 1899 he had charted 2,532 previously undiscovered craterlets and rilles, and in another three years of work he had more than doubled this number. His innate artistic skill allowed him to record what he saw with a high degree of precision and aesthetic appeal. A ruthless perfectionist, many of his drawings are marvels of accuracy in both proportion and position. "Fauth was the last of the great visual observers, and the very high standards he set for himself were never approached by his contemporaries," according to E. E. Both.[9]

But instead of letting his achievements speak for themselves, Fauth exhibit-

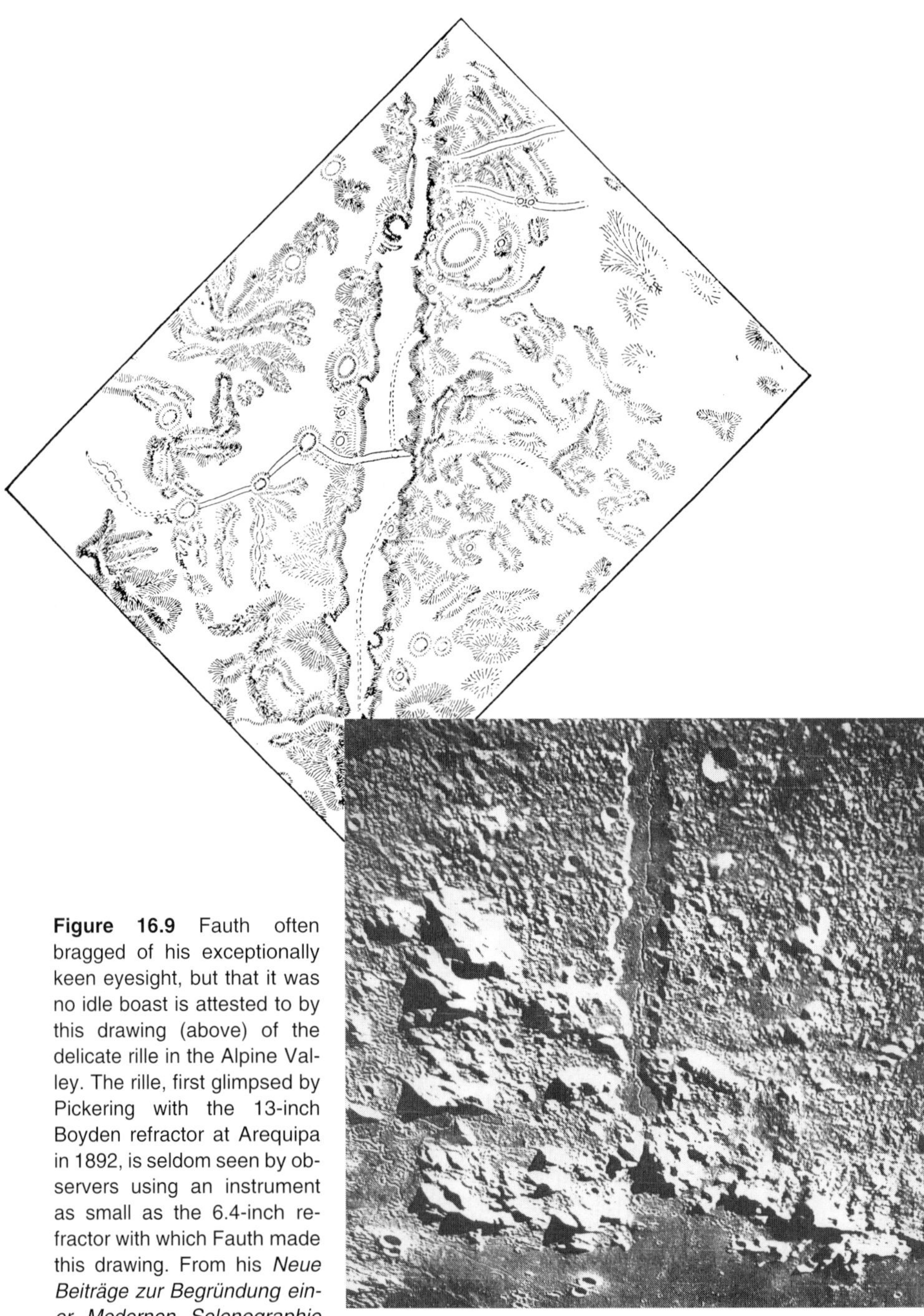

Figure 16.9 Fauth often bragged of his exceptionally keen eyesight, but that it was no idle boast is attested to by this drawing (above) of the delicate rille in the Alpine Valley. The rille, first glimpsed by Pickering with the 13-inch Boyden refractor at Arequipa in 1892, is seldom seen by observers using an instrument as small as the 6.4-inch refractor with which Fauth made this drawing. From his *Neue Beiträge zur Begründung einer Modernen Selenographie und Selenologie* (1896). (Right) The same feature is prominent in this photograph, taken from a more favorable viewing angle—overhead—by the Lunar Orbiter 5 spacecraft in 1967. The central rille is 90 miles long and up to 5 miles wide. The Alpine Valley is a lava-flooded swath of terrain that dropped between two parallel faults. Courtesy NASA.

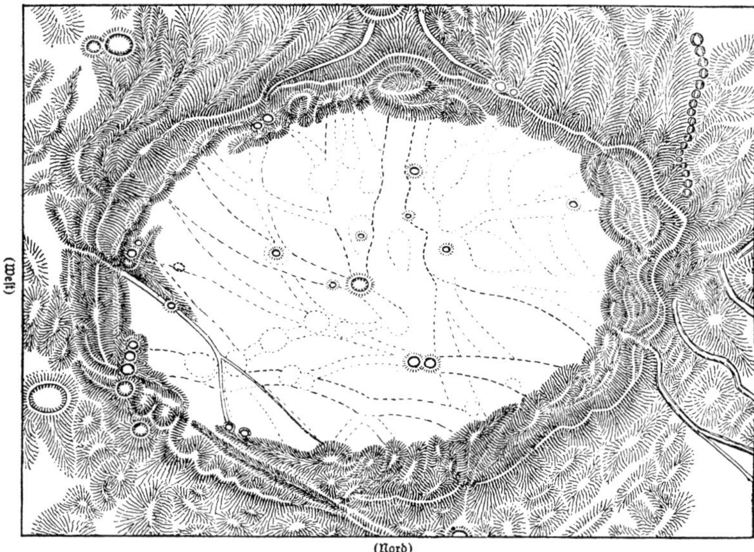

Figure 16.10 Fauth's depiction of Plato, made with his 6.4-inch refractor, with albedo features denoted by dashed lines and topographic features by hachures. Few observers can make out more than four or five craterlets, even with much larger instruments under excellent seeing conditions. From *Neue Beiträge zur Begründung einer Modernen Selenographie und Selenologie* (1896).

ed not only an intense egotism but a relentless proclivity to denigrate the work of others. His printed attacks on the respected Czech selenographer Ladislas Weinek (1848–1913) were particularly bitter and unrestrained. The director of the Prague Observatory, Weinek had scrutinized copies of the Lick Observatory lunar plates through a microscope and announced the "discovery" of a host of minute craterlets and rilles that were later proved, to his embarrassment, to be no more than grains of silver and bubbles and veins in the gelatin of the emulsion.

Though these traits of Fauth soon alienated him from many influential figures, including Klein, he was drawn, through a kind of mutual attraction, to one of the strangest characters in a field rife with eccentric personalities. "Leo Brenner" was the pseudonym adopted by Spiridion Gopcevic (1855–1936?) (Figure 16.11), a Serbian journalist, novelist, playwright, and dabbler in the ever tumultuous politics of the Balkans. After marrying into wealth, Brenner took up astronomy at the age of 39. In 1894 he established the "Manora Observatory" on the Dalmatian island of Lussin, located in the northern Adriatic Sea off the coast of present-day Croatia, then an outpost of the Austro-Hungarian Empire. Equipped with a fine 7-inch (178-mm) refractor and aided by excellent seeing that resulted from the modest diurnal temperature variation characteristic of the island's delightful climate, Brenner soon established himself as an enthusiastic and perhaps a gifted amateur, issuing a torrent of observational reports that won him a measure of respect among lunar and planetary specialists, including Percival Lowell, who visited

Figure 16.11 (Above) Fauth's friend, the Serbian astronomer Spiridion Gopcevic ("Leo Brenner"), one of the strangest characters in the history of astronomy. Courtesy Martin Stangl. (Right) Brenner's 7-inch refractor, manufactured by the Munich firm Reinfelder und Hertel, was equipped with a filar micometer, spectroscope, camera, and a host of other accessories.

Lussin during the winter of 1895–96.

In April 1896 Fauth visited Lussin to "observe under the same conditions, in the same location, during the same hours, through the same refractor together with Brenner. Our side-by-side checks of the Moon, Jupiter, Saturn, Venus, and other objects brought complete clarity to our interpretations and I experienced wonderful impressions."[10] The two got along famously, and Fauth cherished memories of the visit for the rest of his life.

But while Fauth's star would continue to rise, Brenner's burned out quickly. According to the Austrian historian and Brenner biographer Martin Stangl, Brenner was driven by "a nearly pathological craving for fame and recognition" combined with an "overestimation of the possible."[11] For example, he imagined he could glimpse oceans through transient clearings in the cloud canopy of Venus and published a laughably precise axial rotation period of 23 hours 57 minutes 36.27728 seconds—rather like a geologist estimating the age of the Earth to the nearest minute. (The true rotation period is 247 days, retrograde). Equally spurious rotation periods for Mercury and Uranus followed, while Brenner's renderings of Mars featured a canal network even more spidery and intricate than Percival Lowell's. When these observations began to be greeted with scepticism, he retaliated by making scurrilous *ad hominem* attacks on his critics. Several influential figures became targets, notably the popular French astronomer Camille Flammarion and his highly respected assistant E. M. Antoniadi. Eventually Brenner even fell out with Lowell, though he reserved his most scathing abuse for the staff and equipment of the Vienna Observatory.

Brenner's reputation was all but destroyed by this conduct, and he came to be regarded as a pariah by the astronomical establishment. As his claims grew ever more incredible, many even began to suspect that his observations were outright forgeries. (It remains an open question to this day whether he was a pathological liar or simply delusional.) In 1898 Heinrich Kreutz, editor of the *Astronomische Nachrichten*, banned Brenner's submissions from the publication. Brenner responded by adopting Gruithuisen's strategy: he promptly established his own journal. His monthly *Astronomische Rundschau* served as a vehicle for self-promotion and allowed him to conduct personal vendettas against the growing ranks of astronomers who dared to disagree with him. It occasionally featured counterfeit endorsements of Brenner's work by various luminaries.[12] Many of the articles, authored by such well-known figures as Simon Newcomb, T. J. J. See, and E. E. Barnard, were simply pirated from other journals. Few contributions were actually submitted. Of these few, the only one of any consequence was by Fauth—his "Linné und Mondveränderungen" ("Linné and Lunar Changes"), which appeared in the issue of March 1901. This masterly polemic, based on an intimate knowledge of the Moon, effectively demolished the claim that Linné had somehow been transformed from a well-defined crater into a diffuse bright spot.[13] Fauth cautioned that to equate mere discrepancies between the maps of Lohrmann, Mädler, and Schmidt with real physical change was foolhardy. Since Mädler had mistakenly recorded diffuse white spots as shadow-filled cavities when he charted the features that he named Parry B and Alpetragius D, it struck him as strange that Schmidt "only expended his zeal on Linné and not these other quite analogous cases." He later summed up his arguments:

> The progress of lunar research has taught us that we cannot swear to any single one of the thousands of details on the early maps. Even Schmidt's map, which was completed fifty years later [after Mädler's], has serious defects here and there. The present author himself has had to strike out a 12 mile-wide crater ("Melloni"), which does not exist at all. And on testing the two older maps, we found the following surprising results: On half the surface of the map (not counting the chaotic mountainous districts, for instance, to the south, and without the border [limb] districts) Mädler has omitted 17 craters of such dimensions that they ought not to have escaped him; and on the other hand, he gives 337 pits and small crater-structures that are not found on the Moon at all... Lohrmann also has not seen many craters that were certainly within the range of his telescope, and gives 95 craters that did not exist. These instances warn us not to draw general conclusions from the small crater-forms given on these two maps.[14]

Unfortunately, given the obscurity of Brenner's journal, Fauth's comments were little noticed at the time, and the case for a change in Linné continued to be debated for decades. It was settled once and for all only when the astronauts of Apollo 15 imaged it from lunar orbit and showed it to be to be a fresh, albeit otherwise quite ordinary, impact crater (Figure 16.12).

Figure 16.12 It is hard to believe that this ordinary, if relatively fresh, impact crater on the floor of Mare Serenitatis was once the subject of one of the most heated debates in lunar astronomy. Linné, as photographed by the astronauts of Apollo 15. Courtesy NASA.

In the March 1909 issue of the *Astronomische Rundschau*, Brenner dramatically unmasked himself. He announced not only his real identity but also revealed his intention to abandon the world of astronomy that had failed to properly recognize his "achievements." This man, to whom Fauth had been attracted as Narcissus to his reflection, sold his telescope and magnificent astronomical library, left his wife, and turned his attentions once again to political and literary pursuits. For a while Fauth considered purchasing Brenner's observatory and taking up residence on Lussin.

Krieger had named an inconspicuous, eroded crater near the southeastern limb of the Moon after his discredited friend, who had long since faded into obscurity. (Even the exact date of Brenner's death is unknown.) While the deserving W. H. Pickering has been stricken from the map of the Moon, Brenner remains ensconced there to this day.

Meanwhile, Fauth collaborated with Adolf Mang on two popular works on general astronomy, *Wegweiser am Himmel* (Signposts in the Heavens) and *Einfache Himmelskunde* (Simple Knowledge of the Heavens), which enjoyed modest commercial success and led Wilhelm Foerster, director of the Berlin Observatory and a founder of the "Urania Society" for the popularization of astronomy, to recommend him for a position as observing assistant at the newly established national observatory in Mexico City. After much soul-searching Fauth declined, reluctant to abandon his homeland and his teaching career.

It was a fateful decision. Characterizing Fauth as a "lone wolf," Joseph Ashbrook claimed that the key to understanding the evolution of his ideas was "the isolation in which Fauth worked, leading to an extreme independence of outlook... He remained an amateur astronomer all his life and never took part in the collective life at a professional observatory."[15] It is hard to imagine him doing so.

Foerster did succeed in encouraging Fauth to write a book devoted entirely to selenography, *Was Wir vom Monde Wissen* ("What We Know About the Moon"). It was translated into English as *The Moon in Modern Astronomy* in 1907, and in the English-speaking world it remains Fauth's best-known book. The tactlessness of many passages must have done little to endear its author to British readers, however. For example, Fauth denounced such revered selenographers as Edmund Neison and Thomas Gwyn Elger as "the most superficial of previous observers," while boasting that his own maps gave "fresh material every time, even in reduced forms."[16] This combination of vitriol and bombast is a hallmark of his writings.

While granting the unfailing accuracy of mutual position, relative size, and shape of the coarser lunar features as captured in the best photographs of the era, Fauth defended the enduring value of visual observations of the Moon:

> It is a very different thing to study pictures and to observe the real Moon in all its brilliancy and colour. On the Moon we have light in every possible tone; there is a complete scale from the glowing white that dazzles the eye, to the coldest and deepest black, all over the Moon, except when examined at full. But even the finest pictures can only give a clear reproduction of the *middle* tones. What the eye takes in at a glance can only be gradually approached by the photograph... Thus selenologies based on photographs are open to question. If it is ever necessary to go into detail, this is certainly the case as regards the Moon... But the eye can, under moderately good conditions, perceive things and judge their size, posture, and shape, when the photographs will give nothing but a hazy spot. In other words, ocular vision is clearer, truer, and finer, and is able to penetrate precisely into those regions which it is indispensable to explore if we are to have a sound theory of the Moon.[17]

Fauth denounced as "degenerate selenology" the persistent preoccupation with possible changes that had so long motivated many British lunar observers. This was "the evil star which has ruled lunar observation in a transitional period." He took pains to debunk the alleged changes on a case-by-case basis, and firmly stated his own conclusion: "As a student of the Moon for the last 20 years, and as probably one of the few living investigators who have kept in practical touch with the results of selenography, he is bound to express his conviction that no eye has ever seen a physical change in the plastic features of the Moon's surface."[18] To Fauth, "the only useful result" of the Linné furor "was that at last a large number of astronomers... were induced to turn their large telescopes on the object, and so a number of finer features were detected."[19] Equally baseless, in his view, were Klein's report of the change in "Hyginus N," Krieger's later report of a change in nearby "Hyginus N1," and the claims by Brenner that eight to ten "newly-formed" minute pits had appeared in the same vicinity. All of these cases only served to

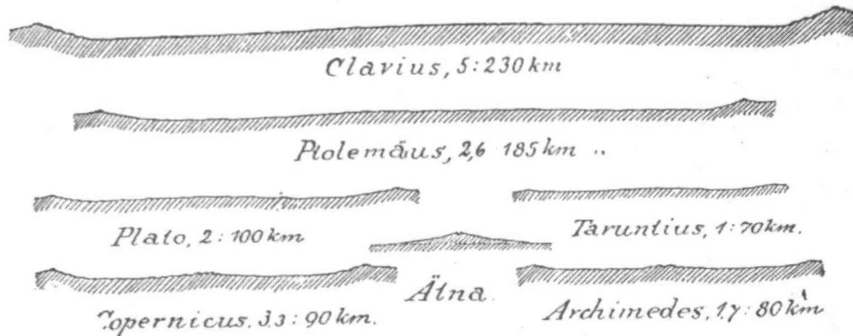

Figure 16.13 Cross-sections of various lunar craters as drawn by Fauth from his careful measurements, showing their characteristically shallow depth-to-diameter ratios. He included a cross-section of Sicily's Mt. Aetna for comparison. From *The Moon in Modern Astronomy* (1907).

demonstrate that "with practice and perserverance we may discover many things in the middle of 'known localities,' yet they will not be 'new formations.'" Fauth noted that he was able to find hints of Hyginus N and Hyginus N1 in old drawings by Gruithuisen, Mädler, and Lohrmann. So much for alleged changes.

Fauth was assuredly swimming against the tide. Indeed, the search for lunar changes during the very years that Fauth was penning these words led to the development of two novel instruments for detecting small differences between successive photographs of the same scene. Invented by an engineer with the Carl Zeiss firm named Pulfrich, the "stereo comparator" simultaneously views two exposures with binocular vision. Any feature which differs on the photographs will appear to stand out of the plane of the resulting composite image and is thus immediately brought to the attention of the observer. In the case of the more refined "blink comparator," introduced in 1904, the two exposures are observed alternately in very rapid succession rather than simultaneously. Objects with identical positions in both exposures remain stationary, but those that have altered their position appear to jump back and forth. Objects that have changed in brightness appear to pulsate. Pulfrich's inventions made it possible to quickly recognize variable stars, novae, comets, and asteroids on photographs of star fields. Indeed, Clyde Tombaugh's discovery of Pluto in 1930 would have been all but impossible without the aid of a Zeiss blink comparator. But when Pulfrich exhibited his stereocomparator at an international astronomical convention at Göttingen in 1902, he touted it first and foremost as the perfect means by which to critically compare lunar photographs in order that "the often-debated question whether changes are still occurring on the Moon might be answered relatively easily and definitely."[20]

Fauth mounted a bold frontal assault on those elements of Pickering's "new selenography" that represented the culmination of the uncritical acceptance of claims of active volcanism and ongoing changes on the Moon. He deplored Pickering's having chosen to describe as "canals" the linear features he detected in Eratosthenes under a high Sun: "These ideas have introduced error enough in the

Figure 16.14 Fauth's drawing demonstrating how even a gently-sloped lunar hill can cast a long spire of shadow under conditions of glancing illumination, creating the illusion that lunar peaks are jagged crags. From *Der Mond und Hörbiger's Welteislehre* (1925).

case of Mars. We may say, in fact, that the object depicted by Pickering does not exist in that form."[21] The alleged tenuous lunar atmosphere—the very cornerstone of Pickering's "new selenography"—was to Fauth "equivalent to what is called a vacuum under the glass bell of an air-pump… It seems as if some experts are straining the evidence to retain an atmosphere at all costs on the Moon, because they take their stand on the plutonic theory of its origin, and this involves the formation of a gaseous mantle about the cooled globe."[22]

The depth of understanding of the nature of lunar topography demonstrated by Fauth was certainly far superior to that possessed by the vast majority of his contemporaries. He had followed up on Julius Schmidt's pioneering investigations by methodically measuring the depth-to-diameter ratios of hundreds of lunar craters and the slopes of their exterior and interior walls (Figure 16.13). These careful studies—along with his graphic demonstration of how the long shadows cast under conditions of low solar illumination could give exaggerated impressions of the ruggedness of the surface relief (Figure 16.14)—ought to have demolished forever the needle-like pinnacles in plaster-of-Paris of Nasmyth and Carpenter, which Fauth noted "could only deceive the inexpert." The various plutonic theories of the origin of lunar craters by some form of volcanism Fauth regarded as utter folly:

> The dominant forms look like the mouths of craters, or at least like deep cauldrons with very prominent walls. The first observers embodied this impression in their phrases, and so we still convey wrong ideas in the names we give these structures. The circular forms on the Moon are not mouths, or cauldrons, or even depressions of a dish-like character. When we do find considerable depressions, it is among the smallest structures that our pictures reproduce. Even the hollow of a flat dessert-dish would convey an exaggerated idea of the large and medium-sized circular structures on our satellite… As a matter of fact the "mouth" is in many cases so incredibly shallow that the eye of an observer on the crest would hardly be able to see the crest on the opposite side, because the depression is so slight that the curvature of the Moon's surface covers the opposite wall.[23]

Like Edmund Neison, Fauth objected to the very use of the term "crater" to denote the most distinctive of lunar features. Instead he urged the adoption of such neutral terms as "ring-mountain,"[24] in keeping with a maxim of the Scottish philosopher Dugald Stewart which Mädler had also adopted as his motto: "Phenomena should always be described by names involving no theory as to their causes." In fact, Fauth repeatedly took pains to emphasize the differences in morphology between terrestrial and lunar crater-forms:

> In the first place the plan of its construction is quite the reverse to that of a volcano. Its "crater" does not lie on the summit of a cone; it is merely a wall, generally composed of a ring of elevations, within which there is a more or less level surface, often strewn with mountains, hills, bosses, cavities, small craters, ridges and bridges, and so deeply excavated that it lies below the general level of the Moon. Lunar craters, therefore, are cavities or depressions, not elevations. In order to reach a terrestrial crater one has to ascend mountains thousands of yards high, then to descend only a moderate depth; to get into the depths of a lunar crater one would first have to ascend a broad upland of moderate height (the wall), and one would look down from a ridge on a succession of terraces, leading down with fair steepness, steeper as a rule than the descent from the higher plains of Mexico to the coast… In the terrestrial case we have a slight depression at a great elevation; on the Moon a moderate elevation and then a deep cavity.[25]

All of this is eminently reasonable. And yet these opinions were accompanied with patronizing remarks like the following: "It is very important to bear well in mind all that was discovered by German observers, so that further investigation may build on this, rather than losing itself on a false path. The first epochal works on the Moon came from German writers, and the greatest map of the Moon was constructed in Germany."[26]

While such overt displays of nationalism were not unusual in the years preceding the First World War—and indeed German and British astronomers had been engaged in a kind of "Moon Race" for much of the nineteenth century—they were still comparatively rare in scientific discourse.

Fauth's abrasive manner undoubtedly detracted from the persuasiveness of his arguments, but far more damage was done by one highly idiosyncratic notion that he advanced. Although he usually parted company with Pickering, he did agree with the Harvard astronomer that most lunar landscapes "bleach very visibly, forming and depositing a light hoar-frost during the 14-day lunar 'day,' and are dull when they pass away into the lunar night."[27] Pickering, of course, had held that the last vestiges of lunar volcanism were the source of the water vapor that condensed as hoarfrost and even fell as snow in certain localities. But Fauth's views were even more extreme:

> The Moon is covered with a thick layer of ice… Underneath this there is, perhaps, an ocean… Its nucleus, the real body of the Moon, which we look upon as covered with the ocean we have suggested, must lie in it like the yolk in the white of an egg. If we assume for this globular nucleus—which we take, on terrestrial analogy to be metallic-earthy—an average density of 4.5 (the Earth's is 5.5), we need, as a very simple

> calculation will show, an ocean 115 miles deep... in order to give the well-known average [specific] gravity of 3.5.[28]

Fauth was not the first to advance such an idea. An icy Moon had been proposed by the Norwegian Ericsson in 1885.[29] A similar view was also championed by the amateur astronomer S. E. Peal, a tea-planter in Ceylon, who suggested that lunar craters were the sites of long-frozen pools created by former internal sources of heat, their encircling ramparts the result of rising vapors that precipitated as vast deposits of snow.[30] Gilbert had passed cursorily over such aberrant views, but Fauth received them with the tenacity of a fatal attraction.

Fauth was familiar with the work of meteorite-impact theorist Anton Meydenbauer, who had produced structures resembling those in the Moon by letting small quantities of dust fall onto a layer of powder.[31] But Fauth rejected the impact theory in the form offered by Meydenbauer on the spurious grounds that the almost universally circular form of lunar formations required that the impacts must have been nearly "vertical, or in the direction of the Moon's radius."[32] Instead, he offered a modified form of the theory. In his view, the key to a viable impact theory was not the trajectory of the impacting bodies but the properties of the target, based on the premise that "the sudden arrest of a cosmic movement at the rate of ten or twenty miles a second would cause an enormous generation of heat, so that a meteor would neither dig up the material of the Moon, nor deposit its own material in a circular mound. The visible product of any such meteoric impact would rather be a spot molten with the heat and traces of the explosive expansion of the vaporized mass of the meteor—an indication of an explosion, but not a circular mountain."[33]

But if the Moon were awash in an ocean covered in turn with a rind of ice, cosmic bombardment would repeatedly penetrate the icy shell, "with massive effects on the intensely cold surface, and the fact of the repetition of the catastrophe enables us to understand the overflows that we call 'seas' and the gradual construction of the remarkably flat walled plains with hollow interior, which would be washed out, so to say, by the ebb and flow of warmer water from below."[34]

Under airless lunar conditions the water vapor released by these events would be converted into clouds of ice-dust. Fauth imagined that this material would often seep through breaches in the surrounding walls and gradually sink outward under the influence of the Moon's feeble gravity to produce a halo of bright radial streaks—the mysterious lunar rays. He further claimed that his model handily explained "why the Moon alone is so pock-marked, while the Earth, which is far more active in attracting foreign bodies, shows no trace of impacts."[35]

To Fauth the central peaks of the large ring formations and isolated mountains like Piton and Pico represented islands of ice that once floated in the midst of liquid surroundings. As in icebergs, the greater portion of their substance lay below the surface. The rilles that appear around the shores of the maria and in areas of subsidence like the interior of the crater Gassendi were cracks resulting from the bursting of the Moon's shell that "help us to appreciate the brittleness of the solid matter of the Moon," while the sinewy wrinkle ridges were "nothing more than earlier ruptures, filled up to overflow with a fluid that issued from below and solidified."[36] This was Fauth's view, then and for the rest of his life. He remained, so to speak, frozen in his vision of an ice-covered Moon.

With no small measure of irony Fauth cautioned: "If anywhere, the Moon is the place for a man to discover whatever he wishes." This was particularly true in the case of the intriguing "variable spots" like those in Alphonsus and Eratosthenes that appear to darken under a high Sun—the very phenomenon that Pickering attributed to the growth of vegetation. But Fauth had a radically different explanation:

> Ice naturally assumes a crystalline and transparent form, as can be seen in any pond. Why these interesting localities [in Alphonsus and Eratosthenes] do not engender any hoar-frost under the 14-day ardour of the Sun, as the other parts of the Moon's shell consisting of amorphous white ice do, may be explained by the fact that smooth surfaces reflect light almost as well as a mirror, and change very little under the action of heat, and so do not form hoar[frost].[37]

Of course, Fauth's explanation overlooked the fact that these features failed to give a brilliant specular reflection under suitable angles of illumination. But this objection was the least of the many difficulties with his conception of an ice-encapsualted Moon.

In fact, Fauth fell victim to a well-known illusion. Seen enthroned amidst a dark sky, the Moon appears dazzlingly bright, but by day it is no brighter than other sunlit rocks and mountains. Its albedo—the efficiency of its surface in reflecting the sunlight that falls upon it—is far less than it would be if it were covered with ice. Half a century earlier Johann Karl Friedrich Zöllner (1834–1882) had employed a polarizing photometer to determine that the Moon reflects less than twelve percent of the sunlight that strikes it.[38] By positing the presence of a thin veneer of meteoric dust, however, Fauth tried to mitigate the evidence of Zöllner's photometer.

Yet another argument against the ice-theory derived from studies of alpine glaciers, which had shown that ice behaves like an extremely viscous fluid rather than like a rigid solid. Like pitch, it deforms and flows when subjected to strain, a process then referred to as regelation. Given these mechanical properties, any lunar mountains and crater walls made of ice should long ago have flattened out under their own weight, a process that geologists now describe as "gravitationally-driven viscous relaxation."[39] But according to Fauth, lunar temperatures were so low that ice would exhibit rigidity sufficient to maintain topographic relief, as the Voyager spacecraft have indeed found on the remote frigid satellites of Saturn, Uranus, and Neptune. He based this belief on the curious misconception that in the absence of an atmosphere, sunlight loses its ability to heat objects exposed to it.[40] This was in spite of the fact that early measures with a thermopile by the fourth Earl of Rosse in 1869–70, and more accurate ones by Frank Washington Very with a thermocouple at Pittsburgh's Allegheny Observatory in 1898, had succeeded in detecting radiant heat from the Moon and suggested that midday lunar surface temperatures might be hotter than the boiling point of water. Of course, as soon as the Sun sets

Figure 16.15 Fauth poses proudly beside the dome of his beloved 15.1-inch Schupmann-Medial apochromatic refractor, an instrument of legendary performance which disappeared under mysterious circumstances in the closing days of the Second World War. Its whereabouts remain unknown to this day. Courtesy José Olivarez.

over a region on the Moon, the temperature plummets rapidly to well below zero.[41]

Neither Fauth's book nor its translation won many converts to his theory of an icy Moon.[42] But the author did derive one unexpected and enduring benefit from their publication. In the preface he had complained that he had exhausted the capabilities of his 6.4-inch refractor: "I have, after seventeen years of use, grown to love it, as one does grow to love such instruments, for the endless intellectual enjoyment it has afforded me and the work it has done for Science; yet I have regretted that I was unable at times to apply a more powerful instrument... I have reached the limit of my faculties, and my apparatus can hardly show me anything new in the regions I have studied."[43]

Fauth's lament struck a sympathetic chord in Ellen Waldthauser, a wealthy reader in the town of Königswinter, near Bonn. In a display of extraordinary generosity rivalling that of the patron who had bestowed a chateau and observatory to the French astronomer Camille Flammarion, she donated a princely sum that made it possible for Fauth to acquire his last and greatest telescope, a 15.1-inch (385-mm) Schupmann-Medial refractor fabricated by the Munich firm of Reinfelder und Hertel in 1911. Invented just a dozen years earlier by Ludwig Schupmann (1831–1920), a professor of engineering at the Technische Hochschule at Aachen, the novel optical configuration employed a singlet objective lens of crown glass combined with a compound element known as a Mangin mirror located near the focus. The Mangin mirror is the heart of Schupmann's ingenious invention, eliminating the residual chromatic aberration that proves so bothersome in conventional achromats of this aperture. Although the design provides a highly corrected field of view of rather limited dimensions, its axial images are of superb quality, almost as free of color error as a reflector, and also free of the harmful diffraction effects of a central obstruction caused by the presence of a reflector's secondary mirror.[44]

Thus Fauth came to possess one of the finest telescopes for lunar and planetary work ever wielded by an amateur astronomer. He housed the instrument in an imposing dome constructed from sheet zinc, sited in a grove of young pine trees 600 feet to the south of his stone tower on the Kirchberg (Figure 16.15). Here he would play host to thousands of visitors during the next twelve years. "In the truest sense, all of his observatories were public observatories," recalled his son Hermann.[45] Hundreds of letters giving practical advice and sound counsel to fellow observers bear witness to a kinder, gentler Fauth. To this day he is often cited as a respected authority in German observing handbooks.

Despite the acquisition of his powerful telescope, Fauth's energies were increasingly diverted into an ill-starred collaboration with a fellow amateur astronomer living in Vienna, Hanns Hörbiger (1860–1931) (Figure 16.16), with whom he had been corresponding ever since 1894. A former blacksmith's apprentice who had taken up engineering and become a successful designer of valves, pumps, and mining equipment, Hörbiger, like Fauth, had long harbored notions of an icy Moon—the first of many astronomical theories that came to him in flashes of intuition, visions, and vivid dreams, as if they were the products of mystical illumination. Hörbiger's belief in his theories was that of the delusional psychotic, and he embarked on a flurry of manic activity, bombarding observatories throughout central Europe with letters and telegrams, often followed up with what must surely have been unwelcome personal visits. His efforts were so all-consuming they had to be interrupted by a rest cure taken on the advice of a physician.

Generally shunned, he found a receptive audience in Landstuhl. Fauth was quickly converted, and became Hörbiger's staunchest disciple. It may be that Fauth's painstaking charting of the Moon—a body he regarded as an unchanging and frozen waste—was made to seem less wearisome by his conviction that "every theory of the formation of cosmic bodies" could be tested there,[46] and by the promise that "the solution of the Moon's hieroglyphics is sure to lead to general cosmic as well as specifically lunar results."[47]

For the next decade Fauth's work on his lunar map ground to a virtual standstill as he and Hörbiger labored to produce a *magnum opus*. The strange product of their collaboration was *Hörbigers Glazial-Kosmogonie* ("Hörbiger's Glacial Cosmogony"), written mostly by Fauth but containing lengthly passages contributed by Hörbiger. It is a turgid, 790-page tome printed in double columns, replete with 212 illustrations.[48] This length was rather modest, however, when one considers that the authors were attempting to describe nothing less than a "gigantic, monolithic, and self-consistent view of the world" that would explain the principal mysteries of astronomy, meteorology, geology, and paleontology—an ambition worthy of an Alexander von Humboldt. The imposing subtitle read: "A New History of the Development of the Universe and of the Solar System, Based on the Realization of a Constant Struggle of Cosmic Neptunism Against an Equally Universal Plutonism, Prepared with Consideration of the Latest Results of All the Ex-

Figure 16.16 An Austrian postage stamp honoring the eccentric Hanns Hörbiger's achievements as an inventor and engineer without calling attention to his crank cosmological theorizing. Courtesy Martin Stangl.

act Sciences and Supported by Personal Knowledge and Experiences." Published in 1913 on the eve of the First World War, Fauth called the book "my second life's work."

Proof of Winston Churchill's quip that "there is no such thing as unutterable nonsense," Glacial Cosmogony incorporated the Hegelian dialectic in the form of a perpetual struggle of ice versus fire. According to Hörbiger, space is not the near-perfect vacuum generally imagined, but instead is filled with gaseous hydrogen, tenuous yet sufficiently dense as to render orbits unstable by acting as a resisting medium.[49] As a result, planets slowly spiral into the Sun and satellites wind into their parent planets.

As conceived by Hörbiger, the original Solar System contained at least eighteen planets. Six which circled inside the then much larger orbit of Mercury had already plummeted into the Sun and been consumed, replenishing the Sun's energy. Five additional planets once located between the Earth and Mars had been successively captured by the Earth as their orbits decayed. These earlier moons were torn asunder by tidal effects, and the resulting debris showered onto the Earth to cause the Ice Ages and form the major geological strata. The break-up of the last moon was dimly remembered in the Norse myth of Ragnarok and in ancient traditions of titanic battles waged between gods in the sky. The Greek philosophers Heraclitus, Democritus, and Anaxagoras were all said to have taught that in the very remote past there was no Moon in the sky, but Hörbiger's notion of a succession of moons was likely derived from the "Secret Doctrine" of Madame Helena Petrovna Blavatsky, the occultist whose Theosophical Society had an active chapter in Vienna. According to Hörbiger, the acquisition of our present satellite only 12,000 years ago brought about yet another series of terrible cataclysms, memories of which were still preserved in the flood legends of many cultures and in Plato's tale of the doomed continent of Atlantis.

The capture of one planet by another is an exceedingly improbable event. Without assuming very special circumstances, a planetary close encounter leads, in general, to the perturbation of the orbit of one body by the other but not to a permanent association. Whenever he was confronted with the fact that his theories violated well-established principles of celestial mechanics, Hörbiger refused to dignify the argument. Instead he would retort: "Calculations can only lead you astray."

According to Hörbiger, much of the gaseous hydrogen pervading space had chemically combined with oxygen to form ice, the "cosmic form" of water. The sheer scope of his fantasies (like those of the later cosmic-mystic Immanuel Velikovsky) all but guaranteed that they would contain occasional kernels of truth, and so he must be credited with intuiting that the rings of Saturn and the nuclei of comets are composed of ice. But he also claimed that the Milky Way is not composed of stars but consists of a ring of ice that surrounds the Solar System. Occasionally blocks of this ice spiral into the Sun, causing sunspots to appear at the impact sites.[50] Jets of water vapor propelled from the fiery maws of sunspots quickly froze into clouds of finely-divided ice particles, which coated the surfaces of Mercury and Venus. The Earth was too distant from the Sun for it to have intercepted much of this ejecta, but when it did so, cirrus clouds appeared. Meteors were not simply motes of cosmic dust burning up in the upper atmosphere, but sunlight glinting off passing fragments of cosmic ice. Damaging hailstorms, tornadoes, and even typhoons occured whenever the Earth encounters these *Eislings*.[51]

And so on it went for hundreds of pages, one extravagant claim piled upon another. Hörbiger and Fauth attributed the swift and decisive rejection of their theories by "reactionary" astronomers to simple jealousy. Wanton alienation of the astronomical community ensued, with irreparable damage to Fauth's reputation. To many his name became anathema. Even those who admired his talents as an observer were forced to regard him as a veritable *idiot savant*.

The world's attentions were soon focused on the war which would kill or cripple an entire generation of Fauth's former students. Beside his observatory atop the Kirchberg, he often watched the muzzle flashes of heavy artillery on the Western Front 80 miles to the west, flickering on the horizon like the lightning of a distant thunderstorm. In 1916, the year of the mass slaughter at Verdun, Fauth published *25 Jahre Planetenforschung* ("25 Years of Planetary Research"), a work replete with interpretations drawn from Glacial Cosmogony. In a patriotic gesture, he donated all proceeds from the sale of the book to the war effort.

Germany's defeat in the autumn of 1918 had a devastating effect on Fauth, who recorded in his diary: "The world is turned on its head… If one didn't have a bit of science one could go crazy!"

In accordance with the terms of the Versailles Treaty, French troops occupied the Rhineland. Fauth took part in the "passive resistance" to the French presence. Harassed by the authorities, he left in 1923 to assume a teaching position in Munich. Seven years would pass before he re-established his observatory at Grünwald, a site nine miles south of Munich, though in the interim he regularly used the 11.8-inch (300mm) Zeiss refractor of the Deutsches Museum in Munich. He also made extensive use of transportable 7.9-inch (200mm) and 10.2-inch (260mm) Newtonian reflectors (Figure 16.17) made with mirrors purchased from

Figure 16.17 Fauth beside his transportable 10.2-inch Newtonian reflector atop the Rauhe Alb in Swabia during the early 1930s. The collapsible yoke mounting, made from wooden planks and spars, is an example of the design known as the "parallactic ladder." Renowned for stability combined with economy and ease of construction, it was once very popular with British amateurs as well. Courtesy José Olivarez.

Bernhard Schmidt (1879–1935), an obscure Estonian optician who would be eventually become famous for his invention of the Schmidt camera in 1931.[52]

In 1931, Hörbiger died, embittered by the failure of the scientific community to embrace Glacial Cosmogony. In his later years he had taken on the appearance of an Old Testament prophet. Almost as if a spell had been broken, a year after Hörbiger's death Fauth issued a 16-section regional lunar atlas, the labor of an extended period of convalescence from a severe illness that had interrupted his observing. These magnificent charts were depictions on a huge scale of 1:200,000 of Copernicus, Eratosthenes, Ptolemaeus and other notable features, carefully corrected for foreshortening, and some rendered in carefully estimated contour lines rather than the hachures of his earlier work (Figure 16.18). He also announced that pencil drafts of the 22 sheets of his long-awaited *Grosse Mondkarte* ("Large Moon Map") were all but complete. Its scale of 1:1,000,000 would correspond to a diameter of 3.5 meters (11.5 feet) and surpass any previous achievement in lunar cartography not only in uniform richness of detail but in positional accuracy as well, since Fauth incorporated almost 5,000 reference points, some based on measurements made with a visual micrometer by Julius Heinrich Franz (1847–1913) at the Königsberg and Breslau observatories,[53] others derived by Samuel Arthur Saunder (1852–1912) from photographic negatives obtained at the Paris and Yerkes observatories.[54]

When Hitler came to power in 1933, Hörbiger at last received the posthumous recognition he had never known in his lifetime. Hitler and key members of the Nazi inner circle made no secret of the fact that they greatly admired his theories.[55] Hitler intended to make Linz, Austria, the great central city of a future

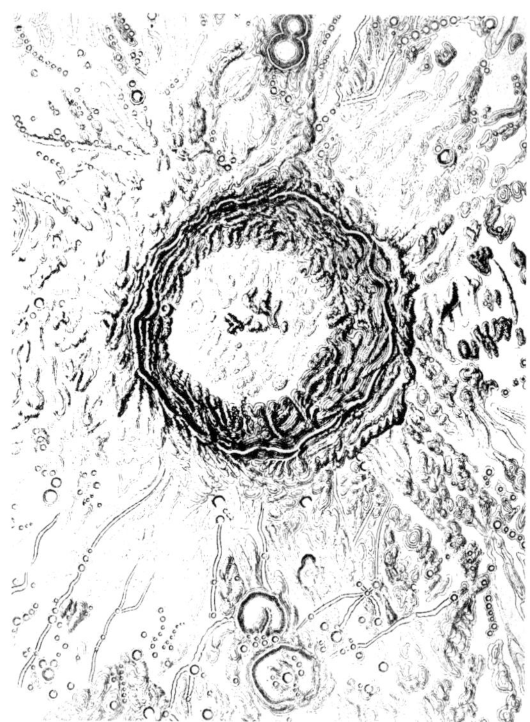

Figure 16.18 Fauth's 1932 chart of the crater Copernicus. The scale of the original is 1:200,000, with each contour line representing a 200-meter increment of elevation. Courtesy José Olivarez.

Greater German Reich, and his plans for the metropolis included a magnificent observatory dedicated to the memory of history's three greatest astronomers, the Germans Copernicus, Kepler, and Hörbiger.[56]

The Graeco-Roman appelation *Glazial-kosmogonie* was soon abandoned in favor of the Teutonic *Welteislehre* ("Cosmic Ice Doctrine"), customarily abbreviated as "WEL" and disdainfully referred to as "The Circus" by members of the astronomical community. The WEL began to assume the proportions of a vast popular movement. Astronomical meetings were disrupted with shouts of "Out with orthodoxy! Give us Hörbiger!"

It was against this backdrop of arrant irrationality that the most rational and valuable of Fauth's works, *Unser Mond* ("Our Moon"), appeared in 1936. Virtually unknown to an English-speaking readership, it has been described by Joseph Ashbrook as "the best of all observing guidebooks to the Moon's surface."[57] Subtitled *Neues Handbuch für Forscher nach Erfahrungen aus 52 Jahre Beobachtung* ("New Handbook for Researchers based on an Experience of 52 Years' Observation"), it contains topographical descriptions of every major lunar formation, complete with summaries of their observational histories. "The description is by far the most detailed which has ever been published… In that respect the new book stands far above any other," T. L. MacDonald wrote in a review for the *Journal of the British Astronomical Association*.[58] It was meant to serve as the companion text to the still unfinished 1:1,000,000 map. Instead it appeared in conjunction with a map of one-fourth that scale in sixteen sections (the *Obersichtkarte des Mondes* or "Overview Map of the Moon") which was intended to serve as a guide to nomenclature. Notable among Fauth's ill-fated additions was "Hörbiger" for the incomplete remains of an enormous ruined crater in the south-central highlands then known as "Hell Plain" (today designated "Deslandres") and "Schupmann" after the inventor of his beloved telescope for the crater that is still known as "Hell B."

While the Glacial Cosmogony was virtually banished to one of the appendices, age had not completely mellowed Fauth. Most of the book is free of the outright venom of his earlier work, but it is sprinkled liberally with remarks like the following, accompanying his description of the crater Gassendi:

> It truly causes grief to ascertain how unfortunately unmethodical and aimless so much work has been... Mädler and Neison managed contours and rilles without fine features; Schmidt once again lapsed into crudeness; Brenner would even refine and register tendencies of direction and relation more than is permissible; Krieger was deceived in form, location, and distinguishing features; Klein missed the topography, let alone the pattern of the rilles. Is it any wonder that when I carefully checked my own chart, I had to depend on myself and myself alone?[59]

The same year the WEL fell under the "protection" of Heinrich Himmler, the notorious Reichsfuehrer of the Nazi SS, the dreaded secret police. Himmler had long been fascinated with a host of crackpot subjects, notably homeopathy, medieval herbalism, spiritualism, astrology, reincarnation (he fancied he had been the tenth-century Saxon king Henry the Fowler in a previous life), and notions of "racial hygiene" that would culminate in the appalling genocide committed under his direction. Himmler served as curator of the *Deutsches Ahnenerbe* ("German Ancestral Heritage"), an SS institute devoted to financing and publishing "Germanic" research, which consisted largely of mythology-inspired archaeology and attempts at buttressing the doctrine of Nordic racial superiority.[60] Meeting at Bad Pyrmont, Fauth and other leading WEL luminaries signed a document known as the "Pyrmont Protocol," which declared:

> The undersigned are convinced that the *Welteislehre* of Hanns Hörbiger is in its basic form the intellectual gift of genius, and extremely valuable for all mankind both from the practical point of view and from that of *Weltanschauung* [ideology]; and for us Germans as a true Aryan treasure of the intellect.[61]

The rest of Fauth's career can be briefly summarized. In 1937 he issued a large collection of drawings of formations located near the lunar limb, observed under conditions of especially favorable libration. Progress on the *Grosse Mondkarte*, however, remained painfully slow, since every night at the telescope revealed new features that he felt compelled to add. He wrote: "I am constantly finding peculiarities that add fresh interest to the work. In a way, all lunar maps are premature, for the detail is truly inexhaustible."[62] When Fauth died in 1941, he was satisfied with the state of only 5 of the 22 sheets of the 1:1,000,000 map. He was in the midst of preparations to move his observatory yet again—from Grünwald to a more favorable site at Rauhe Alb in Swabia—and had just begun to commit his thoughts to paper for a final work, to be entitled *Selenographie, Ein Weg zur Aufhellung von Welträtseln: Mein Bekenntnis und Vermächtnis an Künstige Mondbeobachter* ("Selenography, A Path to Shed Light on the Riddles of the Universe: My Testament and Bequest to Future Lunar Observers").

Fauth did not live to see Germany's defeat in the Second World War—an outcome abetted by meteorological forecasts based on WEL principles by a certain Dr. Hans Robert Scultetus, who predicted a mild winter following the German invasion of the Soviet Union in the summer of 1941. Issued totally inadequate uniforms, the soldiers of the Wehrmacht found themselves facing brutal cold and howling blizzards at the gates of Moscow, their mechanized equipment paralyzed as lubricants and fuel lines froze. The hinge of the fate of nations may have turned on an eccentric astronomical theory.

Fauth's 1:1,000,000 map was completed by his son Hermann, and finally printed in 1964 (Figure 16.19).[63] However, the son did not draw with the skill and assurance of the father, so the final and long-awaited result was a disappointment.[64] It was also an anachronism. By then, the U.S. Air Force Aeronautical Chart and Information Center had undertaken the preparation of Lunar Astronautical Charts on the same 1:1,000,000 scale as the *Grosse Mondkarte.* These beautiful airbrushed maps—the product of inserting minute details glimpsed visually through the Lowell Observatory's 24-inch (610-mm) refractor onto outlines of coarser features derived from the finest photographs obtained at several observatories—represented eight years of work by a 22-member staff that included a dozen professional illustrators and cartographers.[65]

When one considers that, instead of being the work of a team of cartographers, Fauth's map was the result of solitary effort, his achievement seems all the more remarkable. And yet his diligence was, as Richard Proctor had forewarned, a "barren labour." Deprived of the motive of furnishing a basis to detect changes, lunar mapping had been reduced to little more than uninspired drudgery, a mania to record ever more minute features near the limit of telescopic vision. Fauth sustained his obsessive enterprise with the aid of a bizarre illusion. The advance his work represented over Mädler and Schmidt came at a terrible personal price, and his crank theorizing robbed his often sober and incisive arguments against lunar volcanism of the authority they might have commanded. In the end his refutations of the claims of ongoing small-scale eruptive changes were largely without influence. Outside of Germany his legacy was slight, and most selenographers continued steadfastly down the path cleared by Webb, Phillips, Birt, and Pickering, hot on the heels of a chimera.

References

1. Rolf Riekher, *Fernrohre und ihre Meister* (Berlin: Verlag Technik, 1990), p. 213.
2. Hermann Fauth, *Erläuterungen zum Fauth'schen Mondatlas* (Bremen: Olbers Gesellschaft, 1964), p. 5.
3. *Philipp Fauth—Leben und Werk,* autobiography collected by Hermann Fauth and edited by Freddy Litten, (Munich: Institüt für Geschichte der Naturwissenschaften, 1993), p. 36.

4. Joseph Ashbrook, "J. N. Krieger: The Moon half-won" in *The Astronomical Scrapbook,* p. 264. Despite Krieger's exceptional results, his arduous technique of depicting lunar formations inspired only a few imitators, notably Strathmore R. B. Cooke (1907–1985), a New Zealand-born metallurgist who later became a professor of geology at the University of Minnesota. For many years his work appeared in *The English Mechanic and World of Science* and in the memoirs of the Lunar Section of the British Astronomical Association. See "The Walled Plain Schickard: And a Useful Method of Making Lunar Drawings" *Memoirs of the British Astronomical Association,* 36 (1950) 22–24.
5. Krieger's work was collected and edited by his friend Rudolf König (1865–1927), an Austrian businessman who was a mathematician and amateur astronomer of rare ability. König published the two lavish volumes of *Joh. Nep. Kriegers Mond-Atlas* (Vienna: Carl Überreutersche Buchdruckerei, 1912), but only 18 of the 58 plates had been completed by Krieger, the remainder being little more than rough outlines. König intended to produce a lunar atlas of his own, and over the years carefully measured thousands of positions from photographic negatives. This project, like Krieger's, was forestalled by his untimely death.
6. Philipp Fauth, *The Moon in Modern Astronomy* (London: A. Owen & Co., 1907), pp. 19–20.
7. Fauth, *Beobachtungen der Planeten Jupiter und Mars aus den Oppositionen von 1896/1897 auf der Privatsternwarte zu Landstuhl* (Kaiserlautern: Buchdruckerei Heinrich Köhl, 1898).
8. J. B. Sidgwick, *Observational Astronomy for Amateurs* (London: Faber and Faber, 1971), p. 102.
9. Both, *A History of Lunar Studies*, p. 28.
10. Hermann Fauth, *Erläuterungen*, pp. 5–6.
11. Martin Stangl, "The Forgotten Legacy of Leo Brenner," *Sky & Telescope,* 90, 2 (1995), 100–102.
12. J. Ashbrook, "The Curious Career of Leo Brenner," *Sky & Telescope*, 56, 6 (1978), 515–516.
13. Philipp Fauth, "Linné und Mondveränderungen," *Astronomische Rundschau*, 3 (1901), 172–176.
14. Fauth, *The Moon*, p. 149.
15. J. Ashbrook, "Philipp Fauth and his Moon Atlas" in *The Astronomical Scrapbook*, p. 268.
16. Fauth, *The Moon in Modern Astronomy*, pp. 47–48.
17. Ibid., pp. 45–48.
18. Ibid., p. 156.
19. Ibid., pp. 145–146.
20. See J. Ashbrook, "C. Pulfrich and a Short Footnote to Lunar History," *Sky & Telescope*, 35, 4 (1968), 215.
21. Fauth, *The Moon in Modern Astronomy*, p. 15.
22. Ibid, p. 137.
23. Ibid, pp. 65–66.
24. Ibid, pp. 71–72.
25. Ibid, p. 99.
26. Fauth, *The Moon in Modern Astronomy*, p. 158.
27. Ibid. p. 140.
28. Ibid, pp. 141–142.
29. John Ericsson, "The Lunar Surface and its Temperature," *Nature*, 34 (1886), 248–251.
30. S. E. Peal, "Lunar Glaciation," *Nature*, 35 (1886), 100–101.
31. A. Meydenbauer, "Uber die Bildung der Mondoberfläche," *Sirius*, 10 (1877), 180 and "Die Gebilde der Mondoberfläche," *Sirius*, 15 (1882), 59–64.
32. Fauth, *The Moon in Modern Astronomy*, p. 37.
33. Ibid.
34. Ibid, pp. 141–142.
35. Ibid, p. 37.
36. Ibid, p. 131.
37. Ibid, p. 141.

38. J. K. F. Zöllner, *Photometrische Untersuchungen* (Leipzig,1865). Modern values of the lunar albedo are even lower—just seven percent. Most of the samples of lunar soil brought back by the Apollo astronauts look like crushed coal or powdered graphite.
39. Ronald Greeley, *Planetary Landscapes* (Boston: Allen & Unwin, 1987), pp. 243–247.
40. Such muddled reasoning about the temperatures on other worlds had been commonplace early in the nineteenth century. Notable in this regard are remarks by H. Flaugergues, a French amateur astronomer remembered for his discovery of the Great Comet of 1811. Noting that the seasonal melting of the polar caps of Mars is "more prompt and much more complete than with our own terrestrial ice-caps," he concluded that "the heat on Mars is greater than on Earth, though because of the planet's greater distance from the Sun it ought to be less in the ratio of 43 to 100. This is an extra reason to add to those which have made the most skillful physicists believe that the rays of the Sun do not in themselves cause heat but are only the indirect cause of heat." See H. Flaugergues, "Les taches de la planete Mars," *Journal de Physique*, 69 (1809).
41. Rosse's earliest experiments, which involved placing a thermopile at the focus of his 3-foot aperture reflector at Birr Castle, showed "that the heat radiated from the lunar surface could not only be detected with certainty, but also measured with more or less accuracy." By interposing before the thermopile a sheet of glass opaque to long-wavelength infrared radiation but transparent to visible wavelengths, he further demonstrated that this heat was largely absorbed by the lunar surface. Lord Rosse, "Researches on Lunar Heat," a paper read before the British Association for the Advancement of Science at Capetown on August 18, 1905.
42. A notable exception was F. J. Sellers (1875–1960), a future president of the British Astronomical Association who specialized in solar work. See F. J. Sellers, "A Plea for the Glaciation Theory of the Lunar Surface," *Journal of the British Astronomical Association*, 36 (1925), 51–55.
43. Fauth, *The Moon in Modern Astronomy*, p. 20.
44. Christian Wolter and Rainer Merz, "The Neglected Schupmann Refractor", *Sky & Telescope*, 71, 3 (1983), 273–278.
45. Hermann Fauth, *Erläuterungen*, p. 7.
46. Fauth, *The Moon in Modern Astronomy*, p. 24.
47. Ibid, p. 160.
48. Philipp Fauth, *Hörbiger's Glazial-Kosmogonie* (Leipzig: R. Voigtländer Verlag, 1913).
49. This notion was a revival of Johann Franz Encke's old idea of a resisting medium pervading interplanetary space that reveals its presence by subtly altering cometary orbits.
50. This tenet of Glacial Cosmogony no doubt accounts for the fact that Fauth made literally thousands of drawings of sunspots.
51. In 1986 University of Iowa physicist Louis A. Frank claimed that transient dark spots that appeared in ultraviolet images of the Earth's atmosphere taken by the Dynamics Explorer 1 satellite suggested to him that no fewer than twenty 100-ton icy "mini-comets" slam into our planet every minute. He soon issued a book devoted to the subject, *The Big Splash* (New York: Birch Lane Press, 1990).The subtitle as well as the subject matter are certainly in the Hörbiger tradition: "A Scientific Discovery that Revolutionizes the Way We View the Origin of Life, the Water We Drink, the Death of the Dinosaurs, the Creation of the Oceans, the Nature of the Cosmos, and the Very Future of the Earth Itself."
52. A. Wachmann, "From the Life of Bernhard Schmidt," *Sky & Telescope,* 7, 5 (1955), 5.
53. Julius Heinrich Franz worked as an observing assistant at the Neuchâtel and Königsberg observatories, then served as director of the Breslau Observatory. His principal work was the determination of precise longitudes, latitudes, and libration constants of the Moon. His positions of limb formations from heliometer measures and photographs were employed in the International Astronomical Union's *Atlas of Named Lunar Formations*. Franz authored *Die Randlandschaften des Mondes* (The Limb Formations of the Moon) and a popular booklet, *Der Mond* (The Moon).
54. Samuel Arthur Saunder, for thirty years a mathematics master at Wellington College, cataloged 2,885 lunar positions with an accuracy far greater than any achieved before his time. His suggestions eventually led to a standard authoritative system of lunar nomenclature.

55. See Brigitte Nagel, "Die Welteislehre: Hörbigers Glazial-Kosmogonie in nationalsozialistischen Staat", *Sterne und Weltraum*, 26 (1987), pp. 9–13. Notable among the adherents of Glacial Cosmogony was the addled Rudolf Hess, the number three man of the regime (behind Hitler and Reichsmarschall Hermann Göring) until he made an unauthorized flight to England on an abortive peace mission in May of 1941. Sentenced to life imprisonment by the Nuremberg Tribunal in 1946, Hess was permitted one book per month during his four decades of solitary confinement. He often requested astronomical titles, and a map of the Moon adorned the wall of his tiny cell in Spandau Prison.
56. In National Socialist Germany, Copernicus was regarded as a good German, even though Poland had (and has) an equally valid claim on him.
57. Ashbrook, "Philipp Fauth and his Moon Atlas," p. 266.
58. T. L. MacDonald, "Unser Mond, by Philipp Fauth", *Journal of the British Astronomical Association,* 47 (1936), 34–37.
59. Philipp Fauth, *Unser Mond: Neues Handbuch für Forscher* (Breslau: Verlag Dr. Hermann Eschenhagen, 1936), p. 462.
60. James Webb, *The Occult Establishment* (LaSalle, Illinois: Open Court, 1976), pp. 326–329.
61. Ibid.
62. Hermann Fauth, "Philipp Fauth and the Moon," *Sky & Telescope*, 19, 1 (1959), pp. 21–24.
63. Ashbrook, "Phillip Fauth and his Moon Atlas," in *The Astronomical Scrapbook*, p. 270. Hermann Fauth's completion of the map was, unfortunately, somewhat careless and haphazard. For example, the Hadley Rille is omitted completely, while the nearby Bradley and Fresnel rilles are included. Such a lapse would have been very uncharacteristic of his fastidious father.
64. The eminent British selenographer Harold Hill writes: "One does feel a certain justification for being as mercilessly critical as Fauth was in his often vitriolic attacks on the insufficiency of other observers... One has only to apply oneself in a close study of the 1:1,000,000 atlas to realize that there are many errors both in detail and position and, it has to be said, in certain areas of the Moon tell-tale signs of "filling in" with fictitious rings to give a semblance of uniformity quite inadmissible." Harold Hill to Thomas Dobbins, personal communication.
65. Wilhelms, *To a Rocky Moon*, p. 52.

Chapter 17: Afterglows

Among professional astronomers, the Moon had become largely passé by the early years of the twentieth century. The speculative excesses of Pickering's "new selenography" had cast a pall of suspicion around the Moon as a subject for sober scientific investigation. Other concerns were taking precedence, and after 1910 there was a headlong rush into stellar astronomy and the all-absorbing problems of astrophysics. In terms of the "big picture," the Moon was but a trivial detail. It was utterly insignificant in the grand scheme of things. The Yerkes Observatory director Edwin B. Frost (1866–1935), both an active popularizer of astronomy and editor of the prestigious *Astrophysical Journal*, fulminated that lunar studies were an example of "palaeontological astronomy" that should be relegated exclusively to amateurs.

A few professionals still occupied themselves with the Moon, like the colorful eccentric Thomas Jefferson Jackson See (1866–1962) of the U.S. Naval Observatory (Figure 17.1), who advanced a version of the impact theory for the origin of lunar craters in 1910.[1] There was also the Carnegie Institution's "Committee on Study of the Surface Features of the Moon," established in 1925. Chaired by Frederick E. Wright (1877–1953) of the Geophysical Laboratory of the Carnegie Institution, its other members included the Mt. Wilson Observatory astronomers Walter S. Adams (1876–1956), Francis G. Pease (1881–1938), and Edison Pettit (1890–1980), as well as the renowned Princeton astrophysicist Henry Norris Russell (1877–1957). The Committee's approach was very cautious and deliberate, but their ambitious program of photometric, polarimetric, and thermal investigations, as well as extensive measurements of elevations and slopes combined with statistical analyses of crater distribution, was disrupted by the gathering storm clouds of World War II. It disbanded in 1940.

During the interwar years, the Soviet astronomers Vsevolod Sharonov (1901–1964) and Nikolai Barabashov (1894–1971) energetically pursued independent investigations of the colorimeric and polarimetric properties of the lunar surface. Despite the modest instruments at their disposal, they managed to make contributions of enduring value.

Geologists, who might have been expected to assume the initiative, had problems and preoccupations of their own. By the early twentieth-century geology was, as historian Mott Greene has pointed out, "a science deep in controversy."[2] The theory of secular cooling and contraction of the Earth, around which a con-

Figure 17.1 A native of Missouri, Thomas Jefferson Jackson See earned his doctorate at the University of Berlin, where he studied under some of Europe's leading astronomers, physicists, and mathematicians. He was outspoken in his conviction that his abilities far surpassed those of his colleagues, a trait that did not serve him well. After brief but stormy tenures at Yerkes and Lowell observatories, See joined the staff of the U.S. Naval Observatory. He was soon exiled to the timekeeping station at the naval shipyard at Mare Island, California, where he remained in relative obscurity until his retirement in 1930.

Figure 17.2 New Zealand's Algernon Charles Gifford, whose insights into the mechanism of cosmic impacts were able to convincingly account for the fact that virtually all lunar craters have a circular form. Courtesy George Jones.

Figure 17.3 A. W. Bickerton, an eccentric but influential professor at Christ College, Christchurch, championed the notion that novae, variable stars, and planetary systems were all the products of "partial collisions" between various bodies in space. He not only inspired his younger colleague Gifford but also New Zealand's sole Nobel laureate, Ernest Rutherford, who studied colliding atoms rather than stars or planets—studies that led to the discovery of the atomic nucleus. Among his other projects, Bickerton is remembered for establishing a sanatarium in which the therapy consisted of lectures in the natural sciences, with an emphasis on astronomy. He hoped that if the mentally ill could be suitably inspired, their maladies would fade into insignificance. Predictably, the facility closed after less than a year. Sheehan's Collection.

sensus had formed for European geologists during much of the nineteenth century, was everywhere under attack by 1910. Geologists had realized, for instance, that Ice Ages were not only a recent, (i.e., late Neocene) phenomenon, but had also occurred during the Late Paleozoic, Ordovician, and even the Precambrian periods. The cooling of the Earth's climate was obviously a more complicated—and uneven—affair than expected from the thermodynamic intepretation of planetary evolution, at least in its most straightforward form.

The collapse of the theory of secular cooling, Greene adds, "was followed by a period of theoretical pluralism, confusion, and even despair,"[3] and geology became "as fragmented, from a theoretical point of view, as at any time since its eighteenth century beginnings."[4] As the sense of a working consensus fell apart, geology lost prestige among the scientific community. Even after Ernest Rutherford's discovery of radioactivity demolished Lord Kelvin's calculations of the age of the Earth based on heat loss, geologists continued to accept physical arguments for the Earth's rigidity—with the result that, according to historian Stephen Brush, "when Alfred Wegener proposed his hypothesis of continental drift in 1912, he could make little headway against the prevalent belief in a rigid Earth."[5] Continental drift would not be taken seriously until the discovery of seafloor spreading in the 1960s. Without a coherent and universally accepted paradigm for explaining the Earth's features, however, most scientists lacked the confidence to apply their methods to the distant Moon.

A notable exception was Rutherford's compatriot, the New Zealand mathematician and astronomer Algernon Charles Gifford (1861–1948) (Figure 17.2), who had the vital insight that cosmic collisions were far more violent than had been thought. Giffford was an interesting man. He was born at sea while his parents were in transit from Labrador to New Zealand, and was educated at Oamaru, on the South Island, until the age of 15, when he went to England. There he graduated from St. John's College, Cambridge, winning the prestigious Herschel Prize for mathematical astronomy. After returning to New Zealand in 1883, he taught mathematics and astronomy at Waitaki High School and Christ College,

Figure 17.4 "No-Man's Land" between the opposing lines of trenches during World War I: an aerial photograph of the Meuse-Argonne battlefield in 1918, showing the numerous depressions excavated by exploding artillery shells. These horribly scarred landscapes seemed starkly lunar to some of the men who fought there, notably the poet Wilfred Owen, who described the battlefield as "cratered like the moon... All pitted with great pocks." They also gave increasing credibility to the impact theory of the origin of lunar craters, which experienced marked surges after each of the world wars, in the work of Gifford after World War I and of Baldwin after World War II. U.S. Army Signal Corps photograph.

Christchurch, where he fell under the spell of Professor Alexander William Bickerton (1842–1927) (Figure 17.3), a charismatic teacher who in 1877 had suggested that novae, like the one which had appeared in the constellation Cygnus a year earlier, were caused by grazing collisions of dark stars. Bickerton later extended this "insight" into a wide-ranging theory of cosmic encounters involving every class of body and system in the universe. Not only did the impacts of stars produce novae and variable stars, he supposed, but they occasionally produced planetary systems as well. Even after Bickerton's downfall—he was forced to renounce his theories or resign his chair at Christ College in 1902—Gifford remained loyal to him and applied his formidable mathematical skills in an attempt to prove Bickerton's idea of "partial impact" as a cause of planetary formation. In pursuit of this eccentric theory he was thus led to the study of lunar craters and the mechanics of the celestial collisions that had formed them.

"The fact which has not been taken into account hitherto in considering the meteoric hypothesis," Gifford explained in 1924, "is that a meteor, on striking the Moon, is converted, in a very small fraction of a second, into an explosive compared with which dynamite and T.N.T. are mild and harmless."[6] Instead of resembling bullet-holes as had been so long assumed, lunar craters were like the cavities formed by exploding artillery shells during World War I. Such cavities dotted the splattered landscapes of No-Man's Land between the opposing lines of trenches (Figure 17.4), and were hauntingly described by the British soldier-poet Wilfred Owen:

> He saw the earth face grey and sunk with death
> and cratered like the moon with hollow woe
> All pitted with great pocks.[7]

The incredibly high pressures produced by an explosion would be equal in all directions. Ahead of it, the pressure wave would encounter the resistance of the lunar crust and produce compression of rocks; laterally, it would shock rocks and eject debris in an outward direction. The end result, Gifford realized, would be—at all but the most oblique angles of incidence of the impacting meteorites—the excavation of a roughly circular cavity. The awkward necessity of assuming that the impacting bodies had somehow been gathered in the equatorial plane of the primordial Moon was thus done away with.

With the exception of the lone wolf Gifford and a few other professional scientists, the Moon, for several decades, was virtually consigned to the lonely efforts of amateur astronomers. Certainly the level of observational activity during the interwar years was far less than in the closing years of the nineteenth century. Walter Goodacre (1856–1938) (Figure 17.5), Elger's successor as director of the Lunar Section of the British Astronomical Association and the creator of yet another large Moon map, could complain during the 1920s that few serious students of the Moon remained.

In part this could be attributed to the perceived aridity of the subject matter. "One of the chief sources of pleasure to the lunar observer," Goodacre wrote, "is

Figure 17.5 Walter Goodacre succeeded Thomas Gwyn Elger as Director of the British Astronomical Association's Lunar Section in 1897, a position he held for the next 40 years. His two privately published Moon maps were based on careful measurements of photographs. Although they were artistically inferior to the work of Mädler and Schmidt, they were far superior in positional accuracy. Courtesy John Koester.

to discover and record, at some time or other, details not on any of the maps. It also follows that in the future when a map is produced which shows all the detail visible in our telescopes, then the task of selenography will be completed."[8] One gets the sense of diminishing returns as selenography became an asymptotic exercise in exhaustion. Finding themselves in the unenviable position of "late-comers,"[9] faced with a claustrophobic sense of the closing in of the frontier where each discovery depleted the range of future possibilities, selenographers occupied themselves with observing smaller and smaller features.[10] Like giraffes, they lived off higher and higher branches of the trees.

What plan, what organizing principle, apart from that of mere taxonomy, was to be applied to this growing welter of detail? And what attention was given, apart from the still generally accepted notion that the confused and complicated Moon was a vast volcanic landscape, to the origin and meaning of these features? The answer was—very little.

There is an analogy, fanciful perhaps, between the making out of a geological landscape—whether it be the Grand Canyon or the Earthward hemisphere of the Moon—and a work of music which consists of a series of melodic fragments, "repeated, modified, and intertwined to form a long melodic landscape that is more difficult to grasp than a simple tune."[11] It takes time for the brain to perceive the extended relations between these melodic elements. As with an uncomprehending critic who fumed at the first performance of Debussy's *La Mer*, "Does anybody for a moment doubt that Debussy could not write such chaotic, meaningless, cacophonous, ungrammatical stuff, if he could invent a melody?"[12] the Moon's surface appeared chaotic, meaningless, cacophonous, ungrammatical—almost without a melody.

All the same, it was not entirely confused and without pattern. The most obvious element of organization consisted of the circular forms of craters, bays, and seas. The other was the existence of what Webb had referred to as a "tendency to parallel direction." Small craters were "entangled" in the general pressures that had distorted the lunar surface and "squeezed into oval form" so as to produce an effect "like that of an oblique strain upon the pattern of loosely woven fabric."[13] Beer and Mädler had also called attention to linear elements on the Moon, especially prominent in the mountain ridges between Mare Tranquillitatis and Mare Vaporum[14] where Gruithuisen had described his "city." That latter had been so striking in appearance precisely because of the remarkable parallelism of the supposed "ramparts." Admittedly, the early observers had usually discussed individual formations, not wider patterns. [15] Gilbert had been the first to discern a larger system in the "parallelism of direction" of his "grooves and furrows," and to trace the trend lines of the oval forms of the "sculpture" back to an origin in the middle of Mare Imbrium.

Later investigators, unaware of Gilbert's seminal paper, independently rediscovered his system of lineaments, notably the well-known English amateur W. H. Steavenson (1894–1975) and especially the American mining geologist Josiah Edward Spurr (1871–1950), who in a series of privately published monographs called attention to what he called the "lunar grid," a network of lineaments which he regarded as tectonic features.[16] The grid consisted of wrinkle ridges and faults running predominantly north-south and east-west, presumably due to stresses from tidal deformations of the Moon's figure. However they had been formed, their importance to proponents of the volcanic theory was obvious, since if the major features of the Moon's surface could indeed be shown to conform to a global tectonic pattern, "they must be endogenic, and all theories of extra-lunar origin for them may be rejected."[17]

During the 1940s, the lineaments around Mare Imbrium were independently noticed by another American, Ralph Belknap Baldwin (b. 1914). Long after earning his doctorate in astrophysics at the University of Michigan, Baldwin became interested in the Moon while waiting to give a popular lecture at the Adler Planetarium in Chicago. Passing the time by studying photographic transparencies of the Moon on display in the planetarium's gallery, he became intrigued by the groove-like troughs and ridges around the periphery of Mare Imbrium. At the time he had not read Gilbert's 1893 paper, nor would he do so until 1948. (He would later remark that it was "one of the great tragedies of lunar study that Gilbert's pioneering paper was published in the obscure *Bulletin of the Philosophical Society of Washington*."[18]) The widespread neglect of the Moon by professional astronomers is underscored by the fact that Baldwin's first paper on the Moon, published in 1942, was rejected by the professional journals. In the end he had to settle for the forum favored by the late W. H. Pickering—*Popular Astronomy*.

Above all else, Baldwin was fascinated by the origin of the Moon's features. In an attempt to decide between the competing volcanic and impact theories, he looked for formations which could be accounted for only in terms of one theory or the other. There were about a dozen craters—some in Mare Vaporum, others reaching eastward from the Apennines to Mare Tranquillitatis—which were noticeably elongated in a southeast-northwest direction. There was also a series of

Figure 17.6 Walter Haas, founder and Director Emeritus of the Association of Lunar and Planetary Observers, during a 1999 visit to the observatory of co-author T. A. Dobbins. Haas spent the summer of 1935 as a guest of W. H. Pickering, travelling to and from Jamaica as a steerage passenger aboard a banana boat. He recalls the crew's attitude as "every banana a guest, every passenger a pest." The youthful Haas acquired Pickering's infectious enthusiasm for lunar changes, but his views have moderated in his recent years. Photograph by Karen Dobbins.

long sharp grooves trending north-south in the region of the craters Hipparchus, Ptolemaeus, Alphonsus, and Albategnius, and a group of ridges ranging from northwest of Ptolemaeus to beyond Copernicus. The axes of all these features intersected in or, as Baldwin put it, "pointed accusingly to," a common origin in the middle of Mare Imbrium. "Their nature, location, and orientation mark them as a unique type and are overpowering evidence of a common origin. Formations like these certainly are not volcanic in nature. They could only have been formed by the almost tangential impacts of tremendous masses of material."[19] He concluded:

> The existence of so many features which are oriented toward this lunar sea is evidence that some terrific explosion occurred at that point of the Moon's surface... The violence of this explosion must have been such that we may associate it with the actual formation of Mare Imbrium. In this view, Mare Imbrium is nothing more nor less than the largest crater on the Moon.[20]

Effects of the blast could be discerned in the southern and southwestern walls of Mare Serenitatis, which were "extensively mutilated" by material ejected from Mare Imbrium. This proved that Mare Serenitatis had formed before Mare Imbrium; it was presumably another large fossil crater of the same type, and so were Mare Crisium and the circular bay Sinus Iridum. The colliding asteroid that had produced Mare Imbrium was, Baldwin estimated, probably on the order of seventy-five miles in diameter. Upon impact it must have liquefied an enormous amount of rock and smashed through the crust to the still-molten interior. Lava subsequently welled up and out onto the surface "where it practically filled up

Mare Imbrium and then burst its bonds," then plunged through the gap between the Apennine and Caucasus mountains and into Mare Serenitatis and Mare Tranquillitatis. In the western part of Mare Imbrium, the flood spread over Oceanus Procellarum and Mare Nubium. Craters superimposed on the lava flows—Copernicus, Archimedes, Autolycus, and Aristillus—were obviously formed after the lava flows themselves. In Baldwin's view, "vulcanism is thus a follow-up mechanism in the formation of the major lunar features." However, "the very great majority of lunar craters... were produced by a violent explosion attendant upon meteoritic action."[21]

After World War II, Baldwin worked out a greatly expanded version of his theory, published in 1949 as *The Face of the Moon*—one of the most prescient books in the history of science. And yet it cannot be pretended that Baldwin's book came close to settling the question of the origin of lunar craters. Well into the 1950s, and beyond, the debate between meteorite-impact and endogenic origin continued to rage, the partisans of each view holding forth with almost religious fervor.

William Henry Pickering's ghost, like Julius Caesar's, remained mighty yet. Though Pickering had become *persona non grata* at Harvard—after his death his observations and manuscript materials were discarded wholesale by his brother's successor as director, Harlow Shapley—he was fondly remembered by many amateurs who shared his love of visual lunar observing, his fascination with the question of lunar change, and—lamentably—his implausible interpretations of his observations.

Pickering had passed the torch, in a sense, to Walter Haas (b. 1918) (Figure 17.6). A native of Ohio, Haas was trained as a mathematician and as a young man spent five formative months with Pickering in Jamaica. After the Second World War, Haas founded the Association of Lunar and Planetary Observers (ALPO), an American organization founded on the model of the British Astronomical Association. Unlike Pickering, Haas was reserved in his personal manner, to all appearances restrained, judicious, and cautious. But his association with Pickering had profoundly influenced his views on lunar conditions. In 1942—the year Baldwin published his first paper in *Popular Astronomy*—Haas published "Does Anything Ever Happen on the Moon?"[22] He contrasted the amateur view with the prevailing consensus among professionals, which he typified with a reference to Simon Newcomb's description of the Moon as "a world which has no weather, and on which nothing ever happens." To the contrary, Haas argued, the observational record, beginning with the supposed change of Linné (which he accepted as highly probable), seemed at odds with the bleak professional assessment. "The great majority of those astronomers who have made special observational studies of the Moon, so-called selenologists, have been of the opinion the changes do occur on the Moon or at least that there exist lunar phenomena not explicable by known physical laws," he wrote.[23]

Haas presented a sympathetic review of Pickering's evidences for periodic changes—clouds, snow-fields, and vegetation. He maintained that the fluctuations in the visibility of detail in Plato, whose floor he had carefully mapped between 1935 and 1940, were probably due to local obscurations by "otherwise unrevealed lunar vapors."[24] He concluded by recommending the study of lunar changes to his colleagues with the encouragement: "He who goes into this subject may decide that the Moon is much less dead than he has been told."[25]

None of these studies broke new ground; they were in the long-established amateur tradition of lunar observation. The investigation into what the British lunar observer Rob Moseley has called the "dodgy reputation of Plato's floor"[26] dated back to Birt and Arthur Stanley Williams, if not indeed to Gruithuisen, who had wondered whether variability in the visibility of the craterlets that he discovered might not be due to the occasional presence of fog. Seldom stressed was the fact that the appearance of a crater like Plato changes very dramatically in the span of a few hours under sunrise or sunset conditions. The Sun travels through half a degree every hour from a lunar vantage point, so shadows can often be seen to advance or retreat within an interval of only a few minutes. At Plato's latitude of 50° North, the terminator moves across the lunar surface at a rate of about 10 kilometers per hour, and the crater's appearance is also markedly affected by libration. A nearly exact duplication of the same observing circumstances occurs—not once every lunation—but only once every complete saros cycle of 18 years.[27]

Haas's counterpart in Britain was a Welsh civil servant with a doctorate in engineering, Hugh Percy Wilkins (1897–1960) (Figure 17.7), who served as Director of the BAA Lunar Section from 1946 to 1954. A remarkably maladroit draughtsman (by comparison, Pickering was a veritable Rembrandt), Wilkins unhappily chose for his life's work a task for which he was singularly ill-suited, the compilation of the largest lunar map yet made, on a scale of 300 inches (7.62 meters) to the Moon's diameter (Figure 17.8). It was, according to lunar geologist and cartographer Don Wilhelms, based on "detailed but very unrealistic line drawings laboriously prepared over decades."[28] Wilhelms was not alone in harsh appraisal; Ernst Both's verdict was that "neither positional nor artistic quality was at all commensurate with the quantity of detail represented."[29]

In 1953 Wilkins and a younger colleague, Patrick Moore (b. 1923) (Figure 17.9), who went on to fame and fortune as an enduring BBC television personality and one of the most prolific astronomy popularizers of the twentieth century, observed the Moon with the great 33-inch (840-mm) Meudon refractor (Figure 17.10), Europe's largest, located in a suburb of Paris. With Wilkins and Moore making the channel crossing from Britain to France on every available weekend, "their reports were read as eagerly as dispatches from the front in wartime."[30] Much of the time they spent closely examining the central peaks of craters in the hope of finding minute summit craterlets. Baldwin had calculated that only about 15 examples of such features, formed by random impacts, ought to exist on the visible surface of the Moon.[31] But Wilkins and Moore found no fewer than 52, and they cited this number to buttress their belief that the central peaks were volcanic cones.[32] Parenthetically, it should be noted that the lion's share of these summit craterlets have proven to be illusory, "merely the effects of shadows cast by parts

Figure 17.7 Hugh Percival Wilkins (right) appears in this photograph taken in front of his telescope with Patrick Moore and Buttons the Cat. Courtersy John D. Koester.

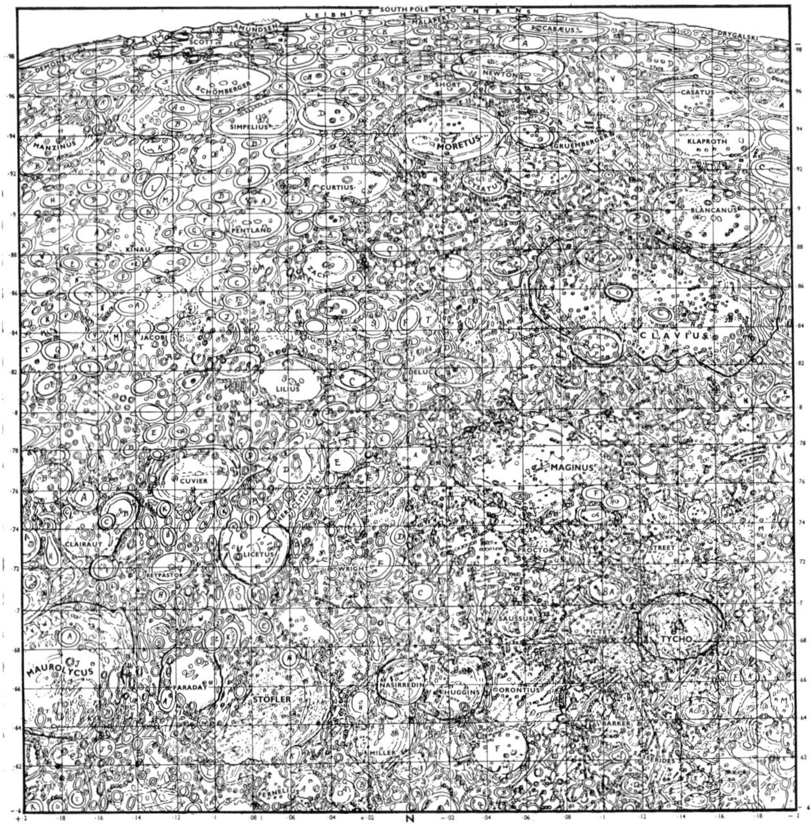

Figure 17.8 Section 23 of Wilkins' 300-inch Moon map, depicting the rugged southern highlands near the prominent craters Tycho and Clavius. Even a cursory inspection confirms the verdict of the historian of selenography Ewen Whitaker that "the addition of fictitious fine detail has so cluttered the map that it is virtually uninterpretable."

Figure 17.9 Co-author William Sheehan with Patrick Arthur Caldwell-Moore (better known as Patrick Moore) and the 15-inch Newtonian reflector at his private observatory at Selsey in Sussex. This fine telescope has served as Moore's principal instrument since 1970. One of the twentieth century's leading observers of the Moon, Moore served as a navigator in the Royal Air Force's Bomber Command during the Second World War. In the early 1950s he collaborated with H. P. Wilkins, frequently crossing the English Channel to make lunar observations with the great refractor of the Meudon Observatory on the outskirts of Paris. With Wilkins he co-authored the book which was long regarded as the lunar observer's bible, *The Moon* (1955), which included a reduced-scale version of Wilkins's map. Later Moore mapped the Moon using the 24-inch refractor of the Lowell Observatory as a member of the U.S. Air Force mapping team. He is best known as a prolific astronomy popularizer and host of the long-running BBC television series *The Sky at Night*. Photograph by Deborah Sheehan, 1993.

of the peaks, which, in a large crater, cluster around a depression as do the points of a molar tooth."[33]

Wilkins' popular book *Our Moon*, published in 1954, is rife with the kind of arguments that Schroeter had employed a century and a half earlier. After recounting that on April 18, 1953, Moore had discerned a small pit never before recorded on the central peak of Theaetetus, he remarked: "Now the newly-found peak is not so small as to justify our saying that it requires a large telescope to be seen. It should have been recorded by others but was not."[34] There were, not surprisingly, a host of similar instances, leading Wilkins to consider it likely some of them might have been newly formed. He mentioned nothing of the far more plausible explanation that these discoveries might be accounted for simply by the closer attention later observers paid to such features.[35]

Wilkins also puzzled over the non-mystery of why Nasmyth, whose "instrument was certainly adequate to reveal the more minute features," had found no domes on the Moon other than the one to the north of Birt, near the showcase Straight Wall.[36] Wilkins and Moore managed to locate over a hundred domes. This

Figure 17.10 Europe's largest refractor, the 33-inch "Grand Lunette" at the Meudon Observatory, which is perched on a hill overlooking Paris that commands an excellent view of the Eiffel Tower. This instrument was frequently used by the British amateurs H. P. Wilkins and Patrick Moore to observe the Moon during the 1950s. Its unusual rectangular tube houses two objective lenses, one corrected for visual work, the other optimized for use with the early blue-sensitive photographic emulsions. This photograph shows the telescope as it appeared in 1996. Despite its excellent state of preservation, it is no longer in service because the dome above it can't be rotated. Courtesy David Graham.

Figure 17.11 The accomplished British amateur Richard M. Baum beside his 4.5-inch Cooke refractor. A native of Chester, Baum is renowned for his artistic renderings of the Moon and planets. During the 1940s he was a TLP enthusiast, but like Walter Haas he too has grown increasingly skeptical of the reality of these phenomena. 1996 photograph by Julian Baum.

led to the inevitable question: Were some of them new? As a case in point, Wilkins recalled that the skilled observer F. H. Thornton had found "a low dome with a summit pit" near the crater Picard on Mare Crisium where the Reverend T. E. Espin had noted only a white patch marking a shallow depression. Wilkins wondered: "Is it possible that it has only become a dome recently? Has the ground here swollen up owing to the pressure of gas underneath?"[37] Of course, it would have been quite remarkable if, after aeons of existence, the Moon had somehow begun to sprout domes just as observers began to train powerful telescopes on it.

The dusky radial banding in craters like Aristarchus, Birt, and Bullialdus posed yet another mystery. John Phillips had recorded the bands in Aristarchus, but the others went unnoticed by even the most diligent nineteenth-century observers. By 1955 no fewer than 188 examples of banded craters had been reported.[38] "All this is mysterious enough," wrote Wilkins, "almost sufficient for us to sympathize with the idea often expressed by Schroeter, that some of these appearances are caused by the 'Industrial Activities of the Selenites'! We cannot subscribe to this idea because without air to breathe it is exceedingly difficult to contemplate the existence of selenites let alone to speculate as to their possible activities, industrial or otherwise. It is equally difficult to explain these things on natural grounds."[39]

There were also bright points of light, obscurations, and colored glows—a hodge-podge of the kind of curiosities that Birt had catalogued. Thus an American observer, David Barcroft (1897–1974), had seen the crater Timocharis "filled with vapor and very indistinct near full Moon," while the Spanish astronomer José Comas Sola (1868–1937) once saw the crater Reiner "as a white patch when it should have been sharply defined."[40] There were strange blue or violet glows, particularly around brilliant craters like Aristarchus and Proclus, and reddish glows were also not infrequent. One of the events which Wilkins referred to as "mysterious happenings" on the Moon in 1948 was an observation by a youthful enthusiast, Richard M. Baum (b. 1930) (Figure 17.11). Studying Philolaus, a crater near the Moon's north pole, on May 20 Baum noted a reddish glow to the northeast, which he watched for fifteen minutes before it faded from sight. Three years later he observed another red glow west of Lichtenberg.[41]

In evaluating these observations, Wilkins hardly found a case that failed to satisfy his credulity. In the end, however, it would be his very credulity that would lead to his downfall.

In July 1953, John J. O'Neill, science editor of the *New York Herald Tribune*, was casually surveying the Moon through a 4-inch refractor when he noticed a slowly retreating, fan-shaped patch of light that diverged from a small bay on the shore of Mare Crisium. He interpreted this feature as the rays of the setting Sun streaming beneath "a gigantic natural bridge having the amazing span of about twelve miles from pediment to pediment." O'Neill's report created a brief sensation. The existence of the arch seemed to be confirmed by Wilkins in observations under similar conditions of light on August 26; he used his 15-inch Newtonian reflector at Bexley Heath, near London (Figure 17.12). However, a number of other observers, including Paul Roques, who photographed the region with the 12-inch Zeiss refractor at the Griffith Observatory in Los Angeles (Figure 17.13), were skeptical. According to the interpretation Roques gleaned from his photographs, the Promontorium Lavinium and the Promontorium Olivium, the mountainous formations on either side of the supposed arch, diminish in elevation as they converge. O'Neill's "fan of light" could be explained as sunlight shining through the pass and falling on gently rising terrain to the east. Before long, a number of prominent British amateurs had verified Roques' interpretation.

There matters might have rested were it not for an unfortunate turn of events. Wilkins gave a wide-ranging radio interview on the British Broadcasting Corporation—the topics discussed included flying saucers—in which he happened to mention the bridge. In the edited version of the interview that was broadcast, it seemed that he not only supported the idea that it was real, he was even suggesting that it might be artificial! (He would later complain, "My experience has been that it does not matter how carefully you explain things to most reporters, they always get things wrong!") He was immediately and roundly attacked by fellow members of the BAA, who were horrified that one of their most prominent figures seemed to be spouting such nonsense. However, Wilkins—then and later—did not withdraw his claim that something extraordinary was present at Promontorium Olivium. During a tour of the United States in the summer of 1954, he once again

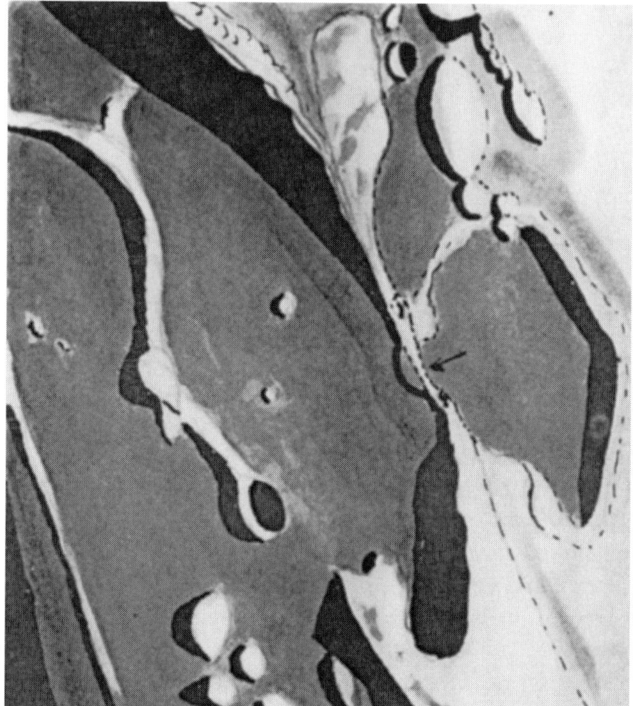

Figure 17.12 O'Neill's "Bridge," as drawn by H. P. Wilkins with his 15-inch Newtonian reflector on August 27, 1953, one lunation after its discovery by *New York Herald Tribune* science editor John J. O'Neill. Courtesy of Richard Baum and the British Astronomical Association.

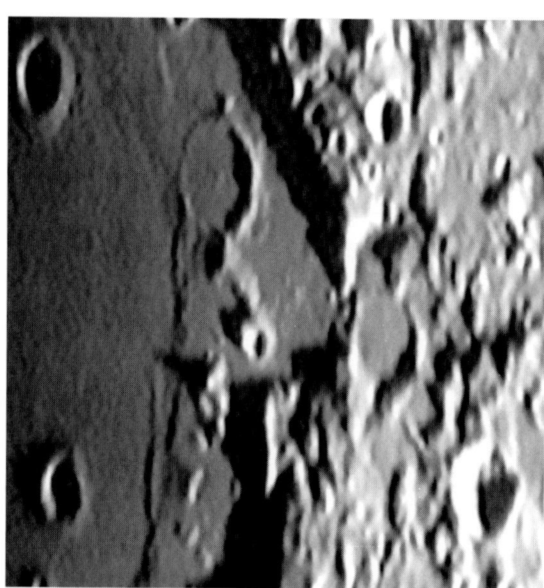

Figure 17.13 This CCD image of the O'Neill's Bridge region by T. A. Dobbins captures the blunt, beak-like tip of Promontorium Lavinium and the long, narrow extension of Promontorium Olivium that extends to the west of Promontorium Lavinium, separated by a narrow gap. With sufficient resolution, all hints of the vast arch alleged by O'Neill disappear.

"confirmed" the existence of the bridge in observations with the 60-inch reflector on Mt. Wilson (though he did revise its dimensions downward to a more manageable span of 1½ miles, causing W. H. Steavenson, a Cambridge University lunar specialist, to remark, "If the bridge really is decreasing in length in this way, I fear it may now be too late for us to catch it!" He added, referring to the large instrument wielded by Wilkins on this occasion, "The aperture of the telescope is not the only thing to be taken into account. There is also the man at the small end."). A minor scandal ensued, culminating in an acrimonious debate at the November 1954 meeting of the BAA in London, from which Wilkins emerged with his credibility so irreparably damaged that he would soon resign.[42] He died in 1960.

Although Wilkins had been discredited, many of his cherished ideas not only endured but flourished. For that matter, with the Earthward hemisphere of the Moon mapped in exhaustive detail, what "useful work" remained for lunar observers equipped with modest telescopes who still aspired to make a contribution, however modest, to lunar science? In the search for some useful project in which to invest their energies, an increasing number of amateurs and a handful of professionals began to concern themselves with what came to be known as "transient lunar phenomena" or "TLP"—the luminous spots, colored glows, mists, and obscurations of topographic features which, if taken at face value (as they all too often were), seemed to suggest that the Moon might be still geologically—and perhaps even meteorologically—active.

One of the few professional astronomers to specialize in lunar work during these years was Dinsmore Alter (1888–1968), who left a professorship at the University of Kansas in 1935 to assume the directorship of the Griffith Observatory in Los Angeles. Administered by the city's Department of Recreation and Parks, this institution was devoted to the popularization of astronomy rather than to research. A capable teacher and lecturer, Alter was a powerful stimulus to the growth of amateur astronomy in southern California. He wrote many articles and several popular books on the Moon, all liberally sprinkled with remarks like the following:

> There is no reason why there should not be a small leakage of gas from the rocks of the floor… of [Plato] and of other craters. Such an atmosphere well might turn to a "fog" as the moon cools toward sunset. With the Sun still lower it might solidify, leaving the floor clear again. After sunrise the reverse process would take place.[43]

The Pickeringesque overtones are hardly a coincidence, for Alter was an admirer of the fallen-from-grace Harvard astronomer. He confided to one correspondent: "W. H. Pickering was one of the most skillful observers who ever spent much time in an examination of the Moon."[44]

The Griffith Observatory's largest telescope was a 12-inch (305-mm) refractor, but Alter was often permitted to use the 60-inch (1.52-meter) reflector at nearby Mount Wilson Observatory (the same instrument, by the way, that Wilkins had used in 1954 to "confirm" the existence of the supposed lunar bridge). With this powerful instrument Alter took hundreds of pairs of photographs of the Moon in

Figure 17.14 The Russian astronomer Nikolai Alexandrovich Kozyrev, whose 1950s observations of TLPs in the crater Alphonsus continue to be cited as "Exhibit A" by those who still believe in ongoing volcanic activity on the Moon. Courtesy *Sky & Telescope.*

violet and infrared light in the hope that localized lunar outgassing would become evident by Rayleigh scattering, the same phenomenon responsible for the blueness of the daytime sky on Earth. In 1871 the British physicist John William Strutt, Lord Rayleigh (1842–1919) had demonstrated that the scattering of a beam of light transmitted through a gas is inversely proportional to the fourth power of its wavelength. Hence, violet light with a wavelength of 400 nm is scattered 16 times more than infrared light with a wavelength of 800 nm. Alter hoped that even a very tenuous lunar cloud would be revealed as a localized blurring on violet plates that would not appear on the infrared plates. This technique is far from foolproof, however, for in violet light *all* lunar features are invariably rendered more diffuse by scattering in the Earth's atmosphere. At best, one might record a more pronounced lack of clarity in a particular location that would signify the presence of gas or dust.

On the morning of October 26, 1956, Alter found the seeing "unusually good." He recorded in the observing log that the "fluctuations of the image, as observed visually with a telescopic power of about 700... were characterized by very small, extremely rapid vibrations."[45] He took advantage of the opportunity and exposed four pairs of plates at two-minute intervals, each pair centered on the crater Alphonsus. This was the formation where Pickering had suspected that gases, seeping from craterlets punctuating a delicate network of rilles on the crater floor, supported the growth of vegetation that appeared as irregular dark patches under a high Sun.

Predictably, the rilles appeared well defined in the infrared images but not in the violet images. For that matter, the same was true of many of the rilles, craterlets, and hillocks in the adjacent formations of Arzachel and Ptolemaeus. Nonetheless, Alter suspected that the blurring of the rilles on the floor of Alphonsus was slightly more pronounced than the blurring of comparably fine details elsewhere.

If real, the effect was admittedly quite subtle, and Alter cautioned that "each observer must decide for himself whether there is a greater loss… along the northern part of the rille than there is in other places. Such a loss, of course, would suggest outgassing from this rille, which contains the famous black spots of Alphonsus with one or more craterlets in the center of each."[46]

Few who examined Alter's photographs found his suggestion at all convincing. Indeed, the images might have been quickly forgotten had they not been followed by dramatic news from behind the Iron Curtain. A Russian astronomer, Nikolai Alexandrovich Kozyrev (1908–1983) (Figure 17.14), announced that he had managed to capture on a photographic plate an actual spectral signature of an emission of gas in Alphonsus.

Intrigued by Alter's images, in the autumn of 1958 Kozyrev began to examine Alphonsus with the 50-inch (1.27 meter) Zeiss reflector of the Crimean Astrophysical Observatory, which was equipped with a prism spectrograph. On the night of November 3, 1958, the phase of the Moon was one day before Last Quarter, so Alphonsus stood well placed for observation not far from the terminator. Kozyrev placed the slit of the spectrograph across the central peak of the crater and opened the shutter of the camera to begin a 30-minute exposure. He kept his eye glued to the ocular of a six-inch (152-mm) refractor mounted atop the massive reflector to serve as an auxiliary guiding telescope. The instrument's drive mechanism, geared to compensate for the Earth's rotation, was designed to keep the image of a star motionless in the field of view. With such a "sidereal rate" drive, the Moon's motion against the stellar background appears as a drift of half a second of arc per second of time in right ascension, while its drift in declination can exceed a quarter of a second of arc per second of time. Consequently, Kozyrev had to make frequent manual corrections in both coordinates to keep the slit of the spectrograph centered over the crater's central peak, a tedious and exhausting exercise. While guiding the exposure, Kozyrev noticed that the central peak "appeared brighter and whiter than usual," until "suddenly, for a period of less than a minute, the brightness of the peak dropped to normal."[47] (It was late afternoon lighting in Alphonsus at the time, so these impressions are hardly startling.) He immediately halted the exposure and inserted a second plate to record the spectrum of the peak, now "in its normal state."[48] This second exposure lasted ten minutes.

On the first plate, exposed when the central peak had appeared bright, Kozyrev could make out a set of faint emission bands centered at 474 nm and 440 nm in the blue region of the spectrum, but these features were absent on the subsequent comparison plate. He attributed these delicate bands to ionized molecules of diatomic carbon in a rapidly expanding, rarefied cloud of gas released from the central peak and excited to fluorescence by hard solar ultraviolet radiation. Curiously, the chemical composition of the gas was not similar to terrestrial volcanic emissions, but seemed to resemble the materials found in the nuclei of comets.

Kozyrev's account appeared in the February 1959 issue of the popular journal *Sky & Telescope*, replete with reproductions of his spectrograms. The sensational story prompted scores of amateurs and a few professionals to carefully scrutinize Alphonsus. The talented Greco-French observer, Jean-Henri Focas (1909–1970), saw nothing unusual with the 24-inch refractor at the Pic du Midi Observatory high in the French Pyrenees. Despite using inferior telescopes, on that very night two American amateurs, H. F. Poppendiek and W. H. Bond, reported seeing a "diffuse cloud,"[49] while in Britain H. P. Wilkins, still active after the O'Neill's Bridge debacle, suspected a reddish-brown tint on the southern slopes of the central peak.[50]

Expert spectroscopists who examined Kozyrev's images suspected that his "emission bands" were simply artifacts of faulty guiding. Guiding errors would be far less pronounced in the second comparison spectrum, which was exposed with the benefit of the half hour of practice spent guiding the first spectrum and for only one-third the length of time, convincingly accounting for its dearth of "emission bands." Moreover, doubts were expressed that emission bands rather than absorption bands would be observed in any expanding cloud of rarefied gas illuminated by sunlight.

Nevertheless, many did embrace Kozyrev's report as unimpeachable, objective evidence of volcanic activity on the Moon, the long-awaited vindication by a highly trained astrophysicist equipped with modern instruments of so many amateur reports of obscuring mists and strange glows that had been received with skepticism and indifference by most professionals. Dinsmore Alter was especially effusive in his praise: "Kozyrev's spectrum is the most important single lunar observation ever made."[51] But he was hardly alone. In 1959 the American Astronomical Society awarded Kozyrev its Gold Medal.

One might expect that witnessing even the rather quiescent emission of gas from a lunar volcano would be a once-in-a-lifetime chance occurrence, rather like catching a glimpse of the Loch Ness Monster on a clear day. Accordingly, eyebrows were raised when Kozyrev announced that he had managed to record a second event in Alphonsus, and this time nothing less than a *bona fide* volcanic eruption. On the night of October 23, 1959, when conditions of illumination were very similar to those on November 3 of the previous year, he again trained his spectrograph on the central peak of Alphonsus and made a 15-minute photographic exposure. This time there were no "peculiarities in the appearance of the crater" noted with the guiding telescope, so no comparison spectrum was made.[52] Blue emission bands were absent on this plate, but Kozyrev's eye detected a very slight "uniform increase in contrast" between 530 nm in the yellow region of the spectrum and 660 nm in the orange. He interpreted this subtle contrast enhancement as the thermal blackbody radiation emitted by a flow of lava at a temperature of 1,200 K.[53]

This time reaction to Kozyrev's announcement was considerably more muted. Doubts were further compounded in 1963 when he reported that he had repeatedly recorded the emission lines of excited molecular hydrogen in spectra of the crater Aristarchus.[54] He surmised that the gas was escaping from the Moon's interior, which certainly seemed implausible for a small, rocky world so depleted in volatile elements.

Much of the credence that had been placed in Kozyrev's work seems to have been attributable to his inaccessibility and to ambiguities in the translations of his communications. Certainly many accounts of his observations have been stealthily improved in the retelling. By far the most common embellishment was that he had seen a discrete cloud or a reddish glow rather than a mere brightening of the central peak of Alphonsus on the fateful night of November 3, 1958.

With the Cold War at its height at this time, direct exchanges between Western scientists and their Soviet counterparts were limited. During a visit to the United States in 1959, one of Kozyrev's colleagues, the astronomer V. I. Krassovsky, confided to his hosts that not only were Kozyrev's spectra "defective," but that Kozyrev himself was "personally unstable."[55]

Few could have imagined the ordeal that lay in the background of this appraisal. The details are as follows. In 1936 the dreaded Secret Police, engaged in the arrest and deportation of a quarter of the population of Leningrad during Stalin's Great Terror, descended on nearby Pulkovo Observatory, heralding a purge of the scientific community. Kozyrev was among the many members of the Pulkovo staff who were arrested. At first he was held in solitary confinement in a freezing punishment cell for several months. Next he was locked up with an insane cellmate for more than a year. In 1938 he was sent to a labor camp at Norilsk in Siberia, where a fellow inmate denounced him for his belief in the expansion of the universe, then contrary to Soviet dogma. This brought a sentence of ten years inprisonment, which Kozyrev appealed. The tribunal admitted a mistake; instead of life, the sentence was amended to execution by firing squad. Kozyrev endured the horrific strain of expecting to be shot at any moment until a second appeal resulted in the revocation of the death sentence several months later. He was finally released in January of 1947 and set to work trying to rebuild his shattered career.[56]

Doubts about Kozyrev's lunar spectra are certainly deepened when they are considered in the context of some of his other spectrographic "discoveries." In 1954 he announced that he had obtained high-dispersion spectrograms of a glow emanating from the night side of Venus—the "ashen light" that had so exercised the imagination of Gruithuisen. These spectra, he reported, contained a host of emission lines and absorption bands; two of these features could be attributed to neutral and singly ionized molecular nitrogen.[57] While the reality of the ashen light continues to be debated to this day, even many of its proponents reacted with incredulity to Kozyrev's claim that the emission he recorded was 50 times brighter than the "airglow" that occurs at altitudes of 60 to 120 miles in the Earth's atmosphere, where atoms and molecules of rarefied gases excited by sunlight release radiant energy at night as they gradually revert to their ground state. It is notable that Kozyrev reported that the spectral signature of ionized oxygen was absent on his plates, because recent evidence suggests that it is excited oxygen that produces a feeble glow high in the atmosphere of Venus that fluctuates in intensity and shifts its position from day to day.[58] In 1955 Kozyrev published a bizarre claim that the characteristic ruddy color of Mars is an illusion caused by the optical properties of the planet's atmosphere.[59] In 1966 he announced the presence of absorption bands in spectra of Saturn's rings which suggested that they are enveloped in a tenuous atmosphere of ammonia.[60] Several attempts by other observers to verify this result met with failure, and data from the Voyager spaceprobes have now ruled out such a possibility. During a transit of Mercury across the face of the Sun in 1973, Kozyrev reported that he was able to detect the emission lines of hydrogen some 25 seconds before the planet's first contact with the solar limb. He attributed this luminescence to an equilibrium atmosphere about 1⁄100 as dense as the Earth's, continually replenished by the solar wind.[61] The ultraviolet spectrograph aboard the Mariner 10 spacecraft did detect a hydrogen halo during its flyby of Mercury the following year, but it proved to be ten trillion times more rarefied than the one postulated by Kozyrev, far beyond the threshold of his instrument.

Kozyrev's spectroscopy calls to mind the sad case of the sincere but self-deluded René-Prosper Blondlot (1849–1930), a reputable physicist at the University of Nancy who made valuable contributions to Maxwell's theory of electromagnetism. In 1903 Blondlot announced the discovery of a new kind of radiation that he called "N-rays." He was awarded a prize by the French Academy for this work the following year, but his experiments proved unreproducible. The *coup de grace* was administered by the brilliant Johns Hopkins University experimental physicist Robert W. Wood (1868–1955), who surreptitiously removed a vital prism from Blondlot's spectroscope during a visit to his laboratory. The absence of this prism did not diminish Blondlot's ability to "see" the spectral lines of his N-rays.[62]

This cheerless recitation of Kozyrev's failings as an observer and interpreter of data is necessary only because his Alphonsus spectra continue to be cited with annoying frequency as incontrovertible evidence that the Moon is not quite geologically dead. Although in retrospect Kozyrev's spectrograms hardly seem to be the unimpeachable evidence of lunar activity that they seemed at the time, they galvanized not only the amateur community but NASA as well, particularly after a report by two lunar cartographers at Lowell Observatory dispelled many lingering reservations and became, after Kozyrev's observation of the putative emission in Alphonsus, the second-most celebrated TLP case.

Early on the evening of October 23, 1963, James Greenacre and Edward Barr trained the 24-inch refractor on the craters Aristarchus, Herodotus, and nearby Schroeter's Valley. The region was under a high angle of solar illumination, but favorable libration afforded an opportunity to inspect the interior of the craters to check proof copies of the charts they had prepared under contract with the U.S. Air Force's lunar mapping program (Figure 17.15). Unfortunately, the rising Gibbous Moon hung only 25° above the eastern horizon and the atmosphere was very turbulent, producing a "boiling" image. Greenacre recounted: "At first the seeing quality was rated about 2 on a scale of 10. In the next few minutes it improved somewhat, with moments of 3 and 4 seeing during which I zoomed the eyepiece to about 500 power."[63] (Needless to say, recourse to such a high magnification despite poor seeing would be considered an objectionable practice to any experienced lunar or planetary observer.)

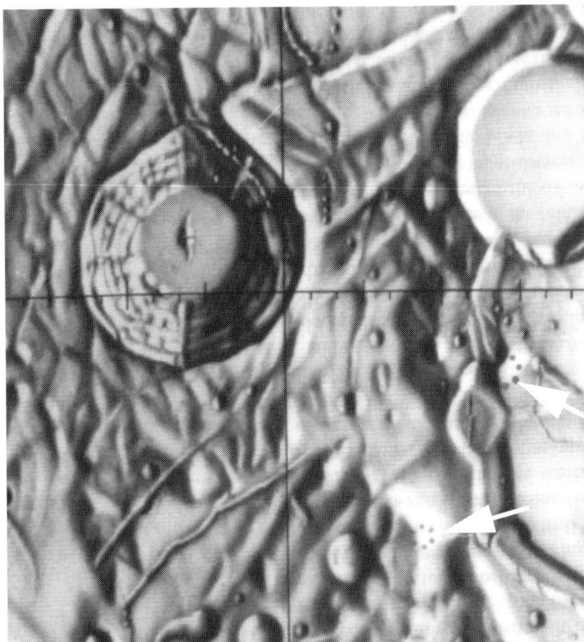

Figure 17.15 On this map of the Aristarchus region from the U.S. Air Force Aeronautical Chart and Information Center, Greenacre denoted the reddish glows that he saw through the 24-inch Clark refractor by inserting arrows and dotted lines. Courtesy *Sky & Telescope*.

Now Greenacre noticed a "reddish orange color over the domelike structure on the southwest side of the Cobra head. Almost simultaneously I saw a small spot of the same color on a hilltop across Schroeter's Valley [about 25 miles away, according to a sketch he later prepared]. Within about two minutes these colors had become quite brilliant and had considerable sparkle."[64]

Twenty-five minutes later an "elongated pink streak appeared along the interior rim of Aristarchus [some 40 miles from the farther of the two spots], which did not sparkle like the other two spots." During the next five minutes, all three features took on a "light ruby red" hue, giving the impression of "looking into a large polished gem ruby," but they remained invisible through the six-inch refractor that served as the large instrument's finder. Suddenly it became apparent that the colors were beginning to fade, and within ten minutes "everything seemed the same as before the color phenomena were first noticed."[65] Astonishing! It was as if Mount Shasta, Mount St. Helens, and Mount Hood in the Cascade Range had somehow become active and then subsided in unison.

The most likely explanation is simply that Greenacre's reddish glows (and the blue and violet counterparts reported by others) arise not on the Moon but in the Earth's atmosphere. When the Moon or a bright planet is viewed at a modest elevation above the horizon, its telescopic image exhibits blue and red fringes along its top and bottom edges. Similarly, bright stars are spread out into tiny vertical spectra. This phenomenon, known as atmospheric prismatic dispersion, is caused by the greater refractive index of air at shorter wavelengths, which causes the blue component of the image to be lifted to a greater extent than the red component. At an altitude of 30 degrees, the separation between blue and red light is usually about one second of arc, so visually the spurious colors are readily apparent only for objects at or below this altitude.[66] The blue component is often subdued or even absent because atmospheric scattering is so much more pronounced at shorter wavelengths; hence the Moon often appears golden or even squash-orange when it hangs low in the sky. Interestingly, large telescopes are more vulnerable to such effects than small ones, for the simple reason that one arc-second of dispersion cannot be detected with a 4-inch aperture, but represents twice the resolving power of a 9-inch instrument. Thus the large telescope is severely limited at large distances from the zenith.[67] The stratification of air into discrete layers of differing temperature and density during temperature inversions and winter anticyclones often markedly accentuates dispersion. Atmospheric turbulence convincingly explains the "considerable sparkle" of the two small spots that Greenacre described.

Spurious as it may have been, the Greenacre-Barr report of red glows gave a tremendous boost to NASA's "Moon-Blink" project, then in its infancy. The idea behind "Moon-Blink" was straightforward. If the temporary lunar colorations were beyond the grasp of moderate apertures, color filters promised to reveal them. A color filter is the optical analog of a semipermeable membrane, passing light of certain wavelengths and blocking others. If an object appears equally bright through red, green, and blue filters, one may suppose that it is in fact grey. But if it appears perceptibly darker through a blue filter than through a red one, one may infer that it has a warm color (i.e., a color from the red end of the spectrum). Observers were encouraged to view the Moon through red and blue filters mounted on a rotary disc so that they could be rapidly alternated.[68] With the aid of such a device, a reddish glow will appear to flicker or "blink." Positive results came in at first as a trickle, then grew to a maddening torrent.

Frustrated by uncorroborated TLP reports by solitary amateurs, in 1964 NASA scientist Gerald Guter organized a network of amateur lunar observers and ham radio operators across the United States. When a TLP was detected, the observer would telephone a ham radio operator, who would relay the report to a ham in a distant city who, in turn, could telephone another observer. Some hams arranged telephone "patches" that allowed observers to converse directly with one another. Christened the "Argus-Astronet" after the creature with a hundred eyes in Greek mythology, its organizers touted it as an astronomical analog of the Distant Early Warning (DEW) Line of Arctic radar stations poised to detect incoming Soviet bombers and missiles.[69] The University of Arizona's Lunar and Planetary Laboratory in Tucson, the Lowell Observatory in Flagstaff, Arizona, and the Corralitos Observatory near Las Cruces, New Mexico all agreed to participate in the program.

NASA had established the Corralitos Observatory for the express purpose of conducting systematic surveillance of the Moon in the hope of detecting and recording TLPs. Operated by personnel from Northwestern University at Evanston, Illinois, the centerpiece of the facility was a 24-inch (610-mm) Cassegrain reflector equipped with a sensitive, state-of-the-art image-orthicon television camera. A motorized three-color filter wheel permitted lunar features to be successively examined in violet (390 nm), green (550 nm), and red light (700 nm). The observer sat in a climate-controlled room beneath the telescope, where he watched a monochrome television monitor in seated comfort to minimize fatigue. The monitor displayed a detailed image measuring six arc-minutes square, about 1⁄25 of the lunar disk. The entire visible hemisphere of the Moon could be examined in three colors at a resolution of one second of arc every 15 minutes. As a check on the observer's visual inspection, the monitor was frequently photographed.

Piggybacked atop the reflector rode a 5-inch (127-mm) refractor equipped with its own television camera and monitor. This instrument employed a filter that excluded visible light and transmitted infrared wavelengths of 850 to 1,040nm in order to record any thermal anomalies like a volcanic eruption or impact.

The Corralitos telescopes operated every clear night and the staff monitored the Argus-Astronet radio network, fully prepared to immediately verify and record any TLP report. After three years of operation, during which over 3,000 hours of lunar surveillance were logged, J. Allen Hynek, Chairman of the Northwestern University Astronomy Department, reported that "no localized lunar events were detected" even though "color changes extending two or three seconds of arc, and even more compact brightenings, should have been detectable."[70] On several occasions the Corralitos staff failed to confirm events when they were alerted by the Argus-Astronet visual observers, "even when the changes were of a nature that ought to have been easily detectable by the image-orthicon system."[71] (As Hynek was learning from his role as scientific consultant to the U.S. Air Force's "Project Blue Book" investigation of UFOs, for true believers absence of evidence is never evidence of absence!)

The Corralitos results are far less widely known than another NASA-sponsored project. This was the *Chronological Catalog of Reported Lunar Events*, the bible of the TLP literature, issued in 1968. Barbara Middlehurst of the University of Arizona and Goddard Space Flight Center and Patrick Moore had independently started to collect reports of anomalous lunar events, but their lists proved so similar they decided to combine them into a single (and presumably definitive) collection of 579 "temporary changes on the Moon."[72] Published by NASA, their effort certainly did convey the impression of being the last word on the subject. Moreover, the authors suggested that rather than merely produce a compilation of raw, undigested reports, they had been selective, and "as far as possible... eliminated reports of events that, for one reason or another (e.g., possibly because of special lighting effects, multiple reflections, and changes of appearance caused by libration), are considered to be spurious."[73] There is no mention, however, of atmospheric turbulence with its inevitable effacing of fine detail; neither of atmospheric prismatic dispersion effects, nor of instrumental chromatic aberration, despite the fact that the vivid reds, blues, and violets so often reported were extremely suggestive of these phenomena. It is hard to credit the editors' claim to have culled unreliable data when, out of hundreds of reports perused, they saw fit to exclude precisely six.[74]

Some reports were accepted on the basis of "the high stature of the observer as a scientist." That being the case, it is perhaps not surprising to find in the list a report by Hevelius (the "red hill," Mons Porphyrites; the crater long since designated as Aristarchus), or another by Cassini of the well-known and hardly mysterious Cassini Bright Spot, which the editors chose to describe as a "small white cloud" near Pitatus.[75] William Herschel's 1787 account of Aristarchus, Copernicus, and Kepler illuminated by Earthshine as "volcanoes in eruption" was also included without critical comment.[76] However, the reader also encounters less familiar names. There are, for instance, the "observers at Worms" (who saw a starlike appearance on the dark side of the Moon in 1540), "anonymous" (a "sterre... in the bodie of the Moone" standing directly between the points of the horns), a "friend of Weidler" (an "appearance like lightning on the face of the Moon during a solar eclipse in 1738"), and "Beccaria's nephew and niece" (a "bright spot," not more explicitly described, on the disk of a fully eclipsed Moon in 1772).[77]

Furthermore, it is worth noting that no fewer than 108 of the reports in the catalog—almost one in five—are sightings by a Baltimore physician and amateur astronomer, James C. Bartlett, Jr., of a "blue radiance" or "violet glare" in and around the brilliant crater Aristarchus, observed during the years 1949 to 1967 with Newtonian reflectors of only 3.5 to 5 inches (89 to 127 mm) aperture. The brightest feature on the lunar surface, Aristarchus is particularly prone to exhibit spurious colors due to both atmospheric prismatic dispersion and instrumental chromatic aberration. That the secondary spectrum of a doublet refractor would introduce a "blue radiance" or "violet glare" is obvious, but even reflectors can hardly be regarded as immune: the simple oculars of Ramsden and Huygenian design which were widely employed during this period suffer from readily perceptible color errors at appreciable distances from the center of the field of view.

Bartlett's observations represented a veritable embarrassment of riches, the sheer quantity of the sightings only serving to deepen suspicions about their quality. According to Bartlett, the anomalies were "rather consistently visible." Why, then, were they not more commonly reported? To this legitimate question Bartlett could do no better than reply that "the phenomenon—which is a delicate one at best—cannot be seen at all by those whose vision is less sensitive in the blue end of the spectrum."[78] Unfortunately, attempts to record his recurrent glows photographically using blue-sensitive emulsions were no more successful.

The Middlehurst and Moore catalog seemed to show a clear tendency of TLP to avoid the rugged, crater-saturated highlands and an equally clear tendency to cluster along the edges of the maria where there are numerous rilles. This finding seemed to fit well with the notion that TLP represented emissions of gas from beneath the lunar crust, the explanation favored by Middlehurst and Moore.[79] The correlation was not quite as definite as might have appeared, however, since there would have been a strong tendency for "feedback" to skew the statistics from which this inference was drawn: when events were reported in Alphonsus and Aristarchus, these formations were immediately singled out for disproportionate scrutiny by other observers. Moreover, the alleged penchant of TLP to occur at the periphery of the maria might also be accounted for by effects of atmospheric prismatic dispersion. As the British lunar observer F. E. Fitton pointed out:

> Any bright or extended area of the lunar surface may be considered to consist of an infinite number of point sources, each of which produces its own spectrum. Since in the bright area all such spectra overlap, the colours re-mix to white and "edge-colours" cannot appear until the bright area interfaces with a dark area, a situation which may not occur until the lunar limb is reached if the area in question is in the "highlands" regions, where in high Sun conditions there are few such interfaces... Where adjacent point sources share similar brightness or darkness, as in the highlands or plains generally, the spectra produced must be appropriately bright or dim, and will therefore re-mix to low abedo white in the plains and to high albedo white in the the bright highlands. At a bright/dark interface [i.e., on the periphery of the maria] the edge-color produced by the "last alignment of point sources" representing the limit of the bright area cannot be neutralized by the far dimmer spectra of the adjacent dark area and hence the edge-colour persists.[80]

To skeptics, the fact that the TLP myth continued to grow despite the increasingly long odds was the result of a lapse of critical judgement distinctly reminiscent of the fairy tale of "the Emperor's New Clothes." Reports peaked in 1969, the year of the first manned landing, and even the Apollo astronauts were enlisted in the search. That year Kozyrev reported new spectra of Aristarchus with features that he attributed to the presence of ionized molecular nitrogen and hydrogen cyanide.[81] He speculated that this release of gas on the Moon might be related to a terrestrial earthquake that occurred the previous day, both events triggered by tidal stresses, but by now his pronouncements seemed blasé and elicited few comments.

A 1971 revision of the original catalog by Moore added another 134 reports. And by 1978, through the efforts of another NASA-funded lunarian, Winifred Sawtell Cameron, the stream of reports had swollen to 1,468. It looked increasingly like an exercise in pure Baconian empiricism, consistent with the philosophy that Wilkins and Moore had advocated years earlier in their book *The Moon*: "Selenography must be founded on observation, not on preconceived and often erroneous conceptions; let us be observers first and theorists afterwards."[82]

Observers first they may have been, but observers whose ideas were intertwined inextricably with an underlying belief in the existence of ongoing lunar volcanic activity. One would have expected the fortunes of the TLP to rise and fall with the viability of such notions—yet to a certain extent they managed to survive independently of them. They became a phenomenon in search of an explanation. Indeed, along with the time-honored explanation of gases being vented from the lunar crust, triggered perhaps by moonquakes or meteorite impacts, other explanations, sometimes vague or at best suggestive, were advanced—the electrostatic levitation of dust particles, localized thermoluminescence of surface materials or fluorescence of soils induced by solar ultraviolet or corpuscular radiation, and triboelectric or piezoelectric effects. Perhaps a combination of all of these. As re-

Figure 17.16 The eminent French planetary astronomer Audouin Dollfus, with his pioneering video-polarimeter installed at the focus of the 1-meter Cassegrain reflector at the Meudon Observatory. His 1992 observations of polarization anomalies in Langrenus with these instruments have sustained the hopes of TLP enthusiasts, despite the fact that the phenomena that he recorded would not have been discernible to a visual observer. Their meaning remains uncertain. Courtesy Audouin Dollfus.

cently as 1991, the experienced British lunar observer J. Hedley Robinson wrote: "I am of the opinion that tidal strains or thermal shocks causing outgassing and producing a piezoelectric effect might be the most plausible explanation."[83]

Even today, a handful of dedicated amateurs continue to stand watch with their color wheels in readiness, awaiting the next outbreak of violet glows at Aristarchus or the latest outgassing at Alphonsus. They bear tribute to the continuing allure of the old Moon, the nineteenth-century Moon—the Moon of Webb and Schmidt, of Birt and Pickering and H. G. Wells.

Theirs was certainly a more exciting Moon than the inert world offered up by modern geologists, where no changes occur apart from random meteorite strikes. However anachronistic the quest for changes on the Moon may be, the hope dies hard. Moreover, we ought not to be too hard on these dedicated observers. There can be no question of their honesty. They saw what they believed, and believed what they saw. They were inspired by a dream—like Coronado in search of the Seven Cities of Cibola or Quiros in pursuit of the Great Southern Continent. One can admire their patience and sympathize with the passion of their quest. One can even understand their reluctance to give up.

The authors were about to go to press when word reached us of two more TLP observations. The first was reported by one of the world's greatest planetary observers, Audouin Dollfus (b. 1924) (Figure 17.16) of the Meudon Observatory. Though it was not published for seven years, on December 29, 1992, Dollfus imaged the crater Langrenus (**Figure 17.17**) with a video camera at the focus of the

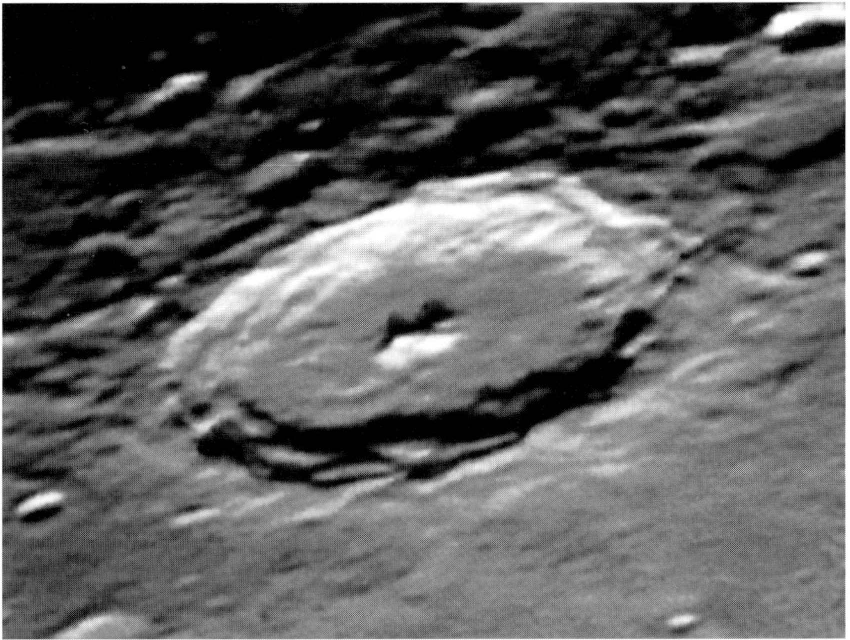

Figure 17.17 A high-resolution image of Langrenus by Steve Massey.

observatory's 1-meter Cassegrain reflector, in both ordinary and polarized light. That night nothing out of the ordinary was detected, but on the following night a diffuse brightening appeared in polarized light in the area surrounding the central peak. (Nothing unusual could be distinguished in ordinary light.) The brightening persisted and changed shape over the next few nights, suggesting to Dollfus the presence of a cloud of small particles levitated above the lunar surface by an emission of gas from the crater's central peak.

Was it a gaseous emission? A meteorite impact? At the moment it is impossible to say, but it is worth noting that the phenomenon, whatever its nature, would not have been perceptible at all to a visual observer.

The second TLP case attracted far more attention. Shortly after Dollfus communicated his report, Sascha B. Culkins of Caltech and Bonnie J. Buratti of the Jet Propulsion Laboratory announced with considerable fanfare that two amorphous bright spots located at one of the most celebrated TLP sites—the head of Schroeter's Valley near the crater Aristarchus (Figure 17.18)—had grown redder. This apparent alteration of surface features, based on a comparison of before and after images obtained by the Clementine spacecraft, seemed to furnish a striking vindication of the glows that had been reported by telescopic observers in that area in April of 1994. However, it proved to be yet another false alarm. Buratti later admitted that once the images were carefully corrected for lighting geometry and other effects, the purported color differences disappeared.[84]

Figure 17.18 The region surrounding the brilliant crater Aristarchus—the brightest spot on the Moon under high illumination—and the "Cobra Head" at the mouth of Schroeter's Valley near the crater Herodotus, has accounted for the majority of TLP reports. Though complex, the region certainly looks quiescent enough in this stunning image obtained from lunar orbit by the Apollo 14 astronauts in 1971. Courtesy NASA.

It is unlikely these will be the last reports produced to challenge the current consensus that the Moon is dead. Dollfus wrote in a fashion reminiscent of his countryman Camille Flammarion in the aftermath of the Linné "eruption": "The consequence is that the Moon should not be considered as a celestial body presently completely inert."[85]

The Moon may well be dead, but TLP resolutely refuse to die.

References

1. T. J. J. See, *Researches on the Evolution of Stellar Systems* (Lynn, Massachusetts, 1910), vol. 2, pp. 331–336. See had, according to Donald E. Osterbrock in *The Yerkes Observatory, 1892–1950: The Birth, Near Death, and Resurrection of a Scientific Research Institution* (Chicago: University of Chicago Press, 1997), p. 159, "one of the strangest careers on record in American astronomy." After completing his undergraduate work at the University of Missouri, See went to Europe for graduate studies, and was awarded a doctorate from the University of Berlin in 1892. He was hired by William Rainey Harper as the first member of the astronomy faculty at the University of Chicago. After a stormy tenure there, he went on to Lowell Observatory, where he soon had a bitter falling out with his coworkers. He went on to the U.S. Naval Observatory in Washington, but in 1903 was relegated for the rest of his career to the obscurity of the timekeeping station at the Mare Island Navy Yard near San Francisco.
2. Mott T. Greene, *Changing Views*, p. 13.
3. Ibid., p. 14.
4. Ibid., p. 290.
5. Stephen Brush, "Planetary Science: From Underground to Underdog," in *Transmuted Past: The Age of the Earth and the Evolution of the Elements from Lyell to Patterson* (Cambridge: Cambridge University Press, 1996), p. 54.
6. A. C. Gifford, "The Mountains of the Moon," *New Zealand Journal of Science and Technology*, 7 (1924), 129–142. In drawing up the brief account given here, the authors have also consulted the A. W. Bickerton-A. C. Gifford manuscripts in the Alexander Turnbull collection in the National Library of New Zealand, Wellington.
7. Wilfred Owen, "The Show"; MS version. *The Collected Poems of Wilfred Owen,* edited C. Day Lewis (New York: New Directions Books, 1965), p. 51.
8. W. Goodacre, "Fauth's New Moon Charts: a review," *Journal of the British Astronomical Association*, 43 (1933), 212.
9. To use Charles Augustin Sainte-Beuve's term in "What is a Classic?"; see W. J. Bate, *The Burden of the Past and the English Poet* (New York: W. W. Norton, 1972), p. 32.
10. Compare Sir William Dampier's comment on late 19th-century physics: "It came to be believed that the main lines of scientific theory had been laid down once for all, and that it only remained to carry measurements to the higher degree of accuracy represented by another decimal place." Sir William Dampier, *A Shorter History of Science* (London, 1945), p. 92.
11. Robert Jourdain, *Music, the Brain, and Ecstasy* (New York: William Morrow, 1997), p. 81.
12. Ibid.
13. T. W. Webb, *Celestial Objects for Common Telescopes*, vol. 1, p. 92.
14. W. Beer and J. H. Mädler, *Der Mond*, p. 250.
15. W. K. Hartmann, "Radial Structures Surrounding Lunar Basins, I: The Imbrium System," *Communications of the Lunar and Planetary Laboratory*, 2 (1963), 1–15:1.
16. J. E. Spurr, *Geology Applied to Selenology* (Lancaster, Pennsylvania: Science Press, 1945), p. ix.
17. P. Moore and P. J. Cattermole, *The Craters of the Moon: An Observational Approach* (New York: W. W. Norton, 1967), p. 56.
18. Ralph B. Baldwin, "An Overview of Impact Cratering," *Meteoritics*, 13 (1978), 364–379.

19. Ralph B. Baldwin, "The Meteoritic Origin of Lunar Craters," *Popular Astronomy*, 50 (1942), 365–369:365.
20. Ibid, p. 367.
21. Ibid, p. 369.
22. Walter H. Haas, "Does Anything Ever Happen on the Moon?" *Journal of the Royal Astronomical Society of Canada*, 36 (1942). The article was re-issued as a monogrpah by the Association of Lunar and Planetary Obseervers in 2000. In the introduction to the new edition, Haas wrote that "youthful enthusiasm may make better reading than the cautious conservatism of later yars, In truth, one's thinking is not likely to remain unchanged for 58 years."
23. Walter H. Haas, "Is the Moon Changeless?" *The Strolling Astronomer* 1, 5 (1947), 4–6.
24. Ibid.
25. Ibid.
26. K. P. Marshall and Martin Mobberly, "The Lunar Crater Plato," *Journal of the British Astronomical Association*, 96, 3 (1986), 161.
27. Ibid.
28. Wilhelms, *To a Rocky Moon*, p. 36.
29. Both, *A History of Lunar Studies*, p. 31.
30. Richard Baum to William Sheehan, June 14, 1993; personal communication.
31. Baldwin, *The Face of the Moon*, p. 152.
32. Moore and Cattermole, *The Craters of the Moon*, p. 50.
33. Wilhelms, *To a Rocky Moon*, p. 14.
34. H. P. Wilkins, *Our Moon* (London: Frederick Muller, 1954), p. 134.
35. Firsoff, *Strange World of the Moon*, p. 80; "…the psychological approach provides a simpler answer." Citing as an analogous case the canals of Mars, he added: "Such features may appear perfectly obvious to someone conditioned to expect them, but an unprejudiced eye might skip them."
36. H. P. Wilkins, "Recent Research on the Moon—2. Bubbles and Streaks," *Journal of the British Interplanetary Society*, 14,3 (1955), 133–134.
37. Wilkins, *Our Moon*, p. 134.
38. K. W. Abinieri and A. P. Lenham, "Banded Lunar Craters," *Journal of the British Astronomical Association,* 65, 4 (1955).
39. Wilkins, *Our Moon*, p. 130.
40. Ibid.
41. Ibid., p. 128.
42. Thomas A. Dobbins and Richard M. Baum, "O'Neill's Bridge Remembered," *Sky & Telescope*, 95, 1 (1998), 105–108.
43. Dinsmore Alter, *Introduction to the Moon* (Los Angeles: Griffith Observatory, 1958), p. 17.
44. Letter from Dinsmore Alter to R. M. Baum, November 19, 1954.
45. Dinsmore Alter, ed., *Lunar Atlas* (New York: Dover, 1964), p. 306.
46. Dinsmore Alter, *Pictorial Guide to the Moon* (New York: Thomas Crowell, 1967), p. 147.
47. N. A. Kozyrev, "Physical Observations of the Lunar Surface" in Zdenek Kopal, ed., *Physics and Astronomy of the Moon* (New York: Academic Press, 1962) p. 366–367.
48. Ibid.
49. H. F. Poppendiek and W. H. Bond, *Publications of the Astronomical Society of the Pacific,* 71 (1959), 233.
50. H. P. Wilkins, *Monthly Notices of the Royal Astronomical Society*, 119 (1959), 421.
51. Alter, *Pictorial Guide to the Moon*, p. 149.
52. Kozyrev, "Physical Observations of the Lunar Surface," pp. 375–379.
53. Ibid.
54. Barbara M. Middlehurst, "Lunar Transient Phenomena," *Icarus*, 6 (1967), 140.
55. Ronald Doel, "The Lunar Volcanism Controversy," *Sky & Telescope*, 92, 4 (1996), 26–30; 29.

56. Alexandr Solzhenitsyn, *The Gulag Archipelago* (New York: Harper & Row, 1973), pp. 480–484.
57. N. A. Kozyrev, *Isvestia Krymskoy Astrophysicheskoy Observatorii*, 12 (1954), 169–176.
58. "Greenhouses in Space: Unearthly Findings," *Science*, 140 (1991), 167.
59. N. A. Kozyrev, *Isvestia Krymskoy Astrophysicheskoy Observatorii*, 15 (1955), 169–181.
60. M. S. Bobrov, *The Rings of Saturn* (Moscow: Nauka Press, 1970): translation: NASA TT F–701 (Washington, D.C.: U.S. Government Printing Office, 1972).
61. V. A. Firsoff, *The Solar Planets* (London: David & Charles, 1977), p. 93.
62. Martin Gardner, *Fads and Fallacies in the Name of Science* (New York: Dover, 1957), pp. 302–303. Even more serious questions are raised by Kozyrev's forays into experimental physics. In 1951 he embarked on a prolonged series of experiments with gyroscopes, torsion balances, and pendulums in the laboratory of the Pulkovo Observatory, inspired by ruminations on the nature of time during his dreary years in captivity. Fifteen years later he published a number of utterly incredible claims: he had observed quantum effects on the macroscopic scale of Newtonian mechanics; time possesses a variable spatial density and can be shielded against by interposing chiral organic compounds; information can be propagated instantaneously through space, seemingly in violation of Special Relativity. The gyroscope experiments led him to infer that the distance from the equator to the north pole of a rapidly rotating planet should be ever so slightly less than the distance from the equator to its south pole, and he claimed to have confirmed this exceedingly subtle asymmetry by measuring photographs of Jupiter and Saturn. N. A. Kozyrev, *Possibility of Experimental Study of Time* (Washington, D.C.: U.S. Department of Commerce Joint Publication Research Service, 1968).
63. James A. Greenacre, "A Recent Observation of Lunar Color Phenomena," 26, 6 *Sky & Telescope* (1963), 316–317.
64. Ibid.
65. Ibid.
66. H. E. Dall, Atmospheric Dispersion," *Journal of the British Astronomical Association*, 71, 2 (1961), 75–78.
67. Ibid.
68. Winifred S. Cameron, "An Appeal for Observation of the Moon," *Journal of the Royal Astronomical Society of Canada*, 59, 5 (1965), 219f.
69. D. C. Kirkman, "Astronomical 'DEW' Line," *Popular Astronomy*, 59, 537 (1965), 10–11.
70. "Systematic Search for Lunar Events," *Sky & Telescope* (1968), 299–300.
71. Ibid.
72. Barbara M. Middlehurst et al, *Chronological Catalog of Reported Lunar Events* (Washington, D.C.: NASA, 1968), p. 1.
73. Ibid., p. 2.
74. Ibid., pp. 3–4.
75. Ibid., p. 5.
76. Ibid., pp. 6–7.
77. Ibid.
78. James C. Bartlett, "Aristarchus: The Violet Glare," *The Strolling Astronomer*, 20, 1–2 (1967) 23.
79. Barbara M. Middlehurst and Patrick Moore, "Lunar Transient Phenomena: Topographical Distribution," *Science*, 155 (1967), 404.
80. L. E. Fitton, "Transient Lunar Phenomena—A New Approach," *Journal of the British Astronomical Association*, 85, 6 (1975), 511–527; 523.
81. "Earth-Moon System?" *Nature*, 222 (1969), 404.
82. H. P. Wilkins and Patrick Moore, *The Moon* (London: Faber and Faber, 1955), p. 46.
83. Winifred S. Cameron, "Lunar Transient Phenomena," *Sky & Telescope* 81, 3 (1991) 265–268.
84. "Lunar Surface Change: A False Alarm," *Sky & Telescope*, 99, 3(2000), 22.
85. Audouin Dollfus, "Lueurs sporadiques sur la Lune," *Comptes Rendus de l'Academie des Sciences*

Paris, t. 327 (1999), Serie IIb, p. 709–714. It should be noted that Dollfus adds that "visual survey with small telescopes by non-professional observers appears not to be a relevant approach to firmly document this type of phenomenon. The implications are too important for the geophysics and the time evolution of the lunar globe." Dollfus, personal communication to William Sheehan, June 25, 1999.

Chapter 18:
Apollo's Moon

Prior to 1959 it could be said that "everything we knew about the Moon was based on low-resolution remote sensing, numerical models, and speculation, all carried out by a small number of largely-ignored scientists. We couldn't even see half of the Moon, we couldn't tell what it was made of, and we had no idea whether it was young or old. To explain any observed lunar phenomenon, there were a variety of theories, all of them untestable, and most of them wrong."[1]

In the 1950s, and even the early 1960s, the impact and volcanic theories continued to have strong adherents. The situation was complicated by the fact that *both* impact and volcanic processes are in evidence on the surface of the Moon. Even as late as 1963, F. E. Wright, of the long-disbanded "Committee on Study of the Surface Features of the Moon," still attempted to split the difference much as Nathaniel Southgate Shaler had done sixty years earlier:

> Both volcanic and meteor impact hypotheses are supported by observations. Some lunar craters closely resemble terrestrial volcanic craters. Appearance, non-random distribution over the lunar surface, and association with zones of apparent crustal weakness all point to a volcanic origin for these craters. On the other hand, an impact origin is plausible for most of the large ring craters and for other features, such as the radial and tangential fault patterns about Mare Imbrium.[2]

In retrospect, it is supremely ironic that it was not the program long favored by most amateur and professional students of the Moon—the careful and minute study of special objects such as domes, rilles, and tiny craterlets, with their apparent vicissitudes—but a broader, low-resolution view of the Moon that would point the way to further progress.

At the dawn of the Space Age, two leading professional scientists interested in the Moon were Harold Clayton Urey (1893–1981) (Figure 18.1) and Gerard Peter Kuiper (1905–1973) (Figure 18.2), both members of the faculty at the University of Chicago. A Nobel laureate in chemistry, Urey's interest in the Moon was stimulated when he happened to read Baldwin's *The Face of the Moon* on a train; soon afterwards, he pasted photographs together into a map of the Moon and hung it on the wall of his office. He became convinced that the Moon had coalesced cold from stony ("chondritic") meteorites and had not melted since. That being the case, it should have been unaffected by the process of differentiation into crust, mantle and core that had taken place on the Earth. Its rocks would then represent

Figure 18.1 Two of the twentieth century's leading cosmo-chemists, Harold Clayton Urey (left) and Cyril Ponnamperuma, during a visit to NASA's Ames Research Center. Urey was the leading proponent of a "cold Moon" that had changed little since its formation. Though nearly all of Urey's ideas have failed the test of time, he did provide an important impetus for the scientific study of the Moon. Courtesy *Sky & Telescope*.

Figure 18.2 A 1957 photograph of the Yerkes Observatory staff, during Gerard Kuiper's brief and stormy tenure as director. Kuiper is seated with folded arms at the center of the front row. The renowned British historian of selenography Ewen Whitaker is third from the right in the third row. Courtesy of the late Alan Lenham.

relatively pristine material, chemically unchanged since its formation. The Moon would be, then, of overwhelming scientific importance, a veritable Rosetta Stone of the early Solar System.

Urey was not particularly interested in the formation of any particular surface features; he did, however, accept the idea that Mare Imbrium had been formed by a giant impact—though, curiously, he identified its origin not with the center of Imbrium but with Sinus Iridum on its northern edge. He suggested, rather implausibly, that a large object had struck near Sinus Iridum at a low angle and partially rebounded before crashing into Mare Imbrium. Consistent with his view that the Moon had always been cold—and thus unaffected by volcanism—he maintained that the marial lavas were impact melts, despite the obvious fact that craters such as Archimedes had formed between the time that the basin floors had formed and the time that lavas had overflowed them.

Kuiper, one of few professional astronomers in the pre-Space Age with a strong interest in the Moon and planets, was well known for his discoveries of the atmosphere of Saturn's satellite Titan during World War II and of previously unknown satellites Miranda and Nereid just after the War. In the early 1950s he began to observe the Moon visually with large telescopes. Almost at once he realized the need for better maps of its surface features.

At first Urey and Kuiper stimulated one another's work, but "both were too successful, too confident, and too self-centered to remain long in that relationship."[3] Their views began to diverge. In contrast to Urey, Kuiper believed that within the first billion years of the Moon's formation its interior melted, from the heat of its own radioactivity. The maria were formed by lava from within the Moon rather than from impact melts. Moreover, Kuiper was convinced that the central peaks of craters could never be formed by rebound of rock; this meant they must be volcanic. Urey and Kuiper's professional differences became personal as well. In time they came to loathe one another, and eventually they both left Chicago. Urey departed for the San Diego (La Jolla) campus of the University of California, while Kuiper—after obtaining funding for a new lunar atlas, in which he was to be assisted by two British emigrees to the U.S., Ewen A. Whitaker (b. 1922) and David William Glyn ("Dai") Arthur (b. 1917)—left the University of Chicago's Yerkes Observatory, where he had served a brief and stormy reign as director, to found the Lunar and Planetary Laboratory of the University of Arizona in Tucson. Whitaker and Arthur followed Kuiper to Arizona to work on what developed into the famous U.S. Air Force lunar mapping project, in which a team of observers collaborated in making detailed observations with the 24-inch Clark refractor of the Lowell Observatory (Figure 18.3). In Arizona, Kuiper surrounded himself with a group of gifted students who went on to become leaders in the lunar and planetary astronomy of their era, including William K. Hartmann (b. 1939) (Figure 18.4) and Charles A. Wood (b. 1942) (Figure 18.5). During the 1950s and 1960s, *Sky & Telescope* magazine reproduced many memorable lunar drawings in a regular column written by another member of Kuiper's LPL staff, the optician Alika Herring.

Figure 18.3 Co-author William Sheehan and the 24-inch Clark refractor of the Lowell Observatory, Flagstaff, Arizona. The observing chair is the one used by Percival Lowell when he made his observations of Mars. The telescope saw extensive service in the U.S. Air Force lunar mapping project during the 1960s. 1982 photograph by Deborah Sheehan.

Figure 18.4 William K. Hartmann, one of Kuiper's most gifted students at the University of Arizona. In the early 1960s he coined the term "multi-ring basin" for the large impact formations on the Moon that were revealed to Kuiper, Hartmann and others on the plaster-of-Paris sphere Kuiper and Dai Arthur set up at the University of Arizona for the purpose of producing a "rectified" lunar atlas. Hartmann later proposed the now generally accepted large-impact theory of the formation of the Moon. Hartmann is an artist as well as an astronomer—in an inscription of one of his books, Carl Sagan wrote: "You have a good corpus callosum." Hartmann, who currently divides his time between Tucson and Hawaii, is shown here sketching volcanoes in Volcanoes National Park on the Big Island of Hawaii. Photograph by William Sheehan.

Figure 18.5 Charles A. Wood, a student under Kuiper at the University of Arizona who went on to become one of the leading lunar geologists of his generation. At the time of this 1998 photograph, he was chairman of the Space Sciences Department at the University of North Dakota. Photograph by William Sheehan.

Soon after Kuiper arrived in Arizona in 1960, he and Dai Arthur set up a 36-inch white sphere onto which they projected the best photographs of the Moon taken under the most favorable conditions of libration at various observatories. This effort led to a "rectified" atlas showing the lunar formations as they would appear if viewed from directly overhead, undistorted by foreshortening. The results were astonishing. Hartmann, who arrived at the University of Arizona in the summer of 1961, soon after the globe had been set up, recalls the "excitement in walking around to the side of the globe and viewing large expanses of hitherto 'unseen' regions in ordinary perspective with circular craters and unfamiliar mare outlines. On its black supports in the dark photo studio, brightly illuminated by its projector down the hall, the globe seemed a realistic new planet: we spent much time viewing this new Moon."[4]

Hartmann, indeed, was literally seeing a new Moon, or at least the old Moon from an entirely new perspective. "Walking around looking at it from different sides," Hartmann remembers, "I realized there were a lot of these giant concentric ring bulls-eye features that had not been properly recognized. Baldwin saw some of it; but not so much as a systematic repeated pattern, I think, as we recognized with the globe."[5] To describe these characteristic features Hartmann coined the term "basin" in order to distinguish them from the marial lava fields that covered some of them, such as Imbrium, and with which they had tended to be confused in earlier writings on the subject.

The most striking of the multi-ring basins proved to be Mare Orientale—the eastern sea. It had been first glimpsed in the 1930s by Wilkins, who dubbed it "Mare X" (Figure 18.6). It was next seen in 1946 by the young Patrick Moore, who first suggested the name Mare Orientale. At the time it was an entirely apt name, since the feature was then on what was officially the eastern limb of the Moon; later, when the astronautical convention was adopted, it ended up on the western limb. However, the name is too well established now to be changed. Al-

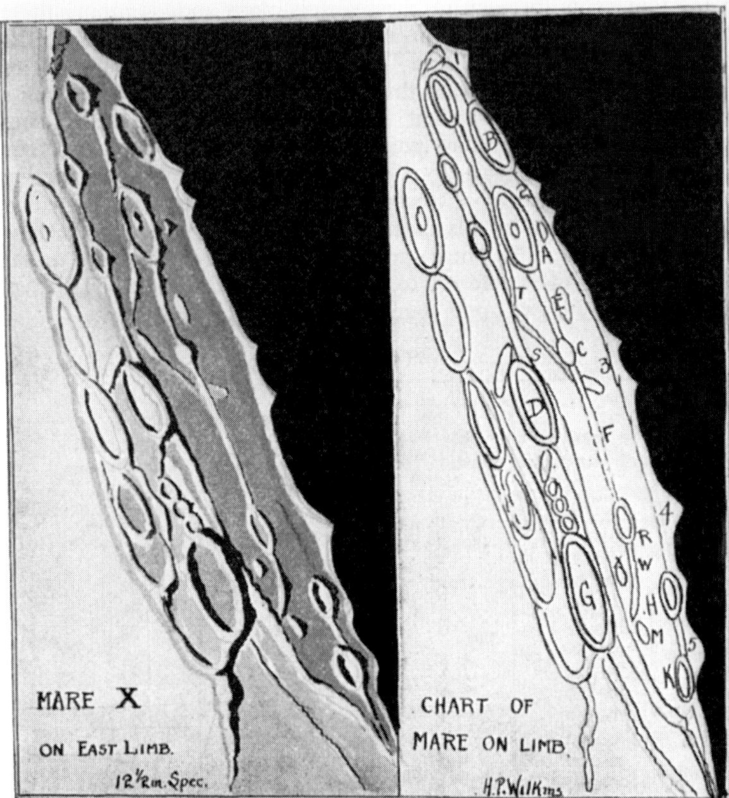

Figure 18.6 The discovery drawing of "Mare X" by H. P. Wilkins, from the *Journal of the British Astronomical Association*, Volume 48, No. 2 (1937). This formation is only visible under favorable conditions of libration. In 1946, it was independently discovered by Patrick Moore, who first proposed the name "Mare Orientale"—the Eastern Sea. Unfortunately, since then the astronautical convention for lunar directions supplanted the traditional astronomical one, so that the Eastern Sea now lies on the Moon's western limb. Courtesy Harold Hill and the British Astronomical Association.

ways severely foreshortened as seen from the Earth, neither Wilkins nor anyone else guessed that Mare Orientale would one day prove to be one of the most imposing and significant of all the formations on the Moon (Figure 18.7).[6] Hartmann no longer remembers who first saw it as a set of "ring-like cliffs," himself or his colleague Harold Spragley, "but I remember being struck by a 'eureka' experience" for several days after seeing it. What old maps showed only as an obscure "Mare Orientale" and some mountain ranges such as the "Rook Mountains" suddenly emerged as an "enormous system of elegant symmetry, with several cliff-like rings and a complex mass of radial valleys and striations."[7] Imbrium itself was another multi-ring basin which had subsequently filled with lava.[8] There were many others, some in much more battered condition, degraded by later impacts,

Figure 18.7 Mare Orientale, the archetypal multi-ring basin, imaged by Lunar Orbiter 4 in 1967. This basin forms a giant "bull's eye" on the Moon's western limb. Three concentric circular rings can be seen; the outermost is the Cordillera Mountain scarp, which measures almost 900 kilometers in diameter. Courtesy NASA.

lava-flooding, and tectonic disturbances such as fractures caused by lava-emplacement.[9] It became clear that radially-oriented lineaments such as those found around Orientale and Imbrium were characteristic of all of these basins.

The recognition of the multi-ring basins as the basic gestalt organizing the large-scale structures of the Moon's surface was decisive. The fact that this recognition came so late is remarkable given the conspicuousness of these features—they include some of the prominent patches of the naked-eye face of the Moon that have stared back at man since time immemorial. In part, the failure of earlier researchers to recognize the multi-ring basins as an important *leitmotif* can be explained by their long preoccupation with ever-smaller details. Ralph Baldwin once remarked, "too many people were enamored with learning more and more about less and less and hence did not see the big picture."[10]

The larger features of the Moon had been seen for so long that they were taken for granted. Familiarity breeds contempt; they had become, in a manner of speaking, invisible. They had long since retreated cognitively into the background. Hartmann—an artist as well as a scientist by training—notes that the new rectified view of the Moon provided a glimpse of the "big picture," unmasked the archetypal features of the pattern in Mare Orientale and allowed the same pattern to be recognized in other less defined examples. He adds: "It is intriguing to realize that we could see no details of Orientale or other systems that had not long

Figure 18.8 Impact theorist Eugene Shoemaker, right, the most brilliant astrogeologist of his generation, and his wife Carolyn, the most prolific comet discoverer of all time, at the Texas Star Party in 1996, shortly before Shoemaker's untimely death in an automobile accident in the Australian outback. Shoemaker had long dreamed of becoming the first geologist on the Moon; a goal he probably would have accomplished had he not been diagnosed with Addison's disease. Some of his ashes were carried to the Moon by the Lunar Prospector spacecraft, which impacted in the neighborhood of the south pole. Courtesy Stephen James O'Meara.

been visible for three centuries to Earth-centered observers... Many researchers searched their way through succeeding layers of finer and finer detail, all duly confirmed and analyzed, but missed fundamental patterns of structure in the first, coarsest layer of details."[11] In a further comment, Hartmann pointed out that sometimes "God isn't in the details. God is in the first order relationships."[12]

At the same time Kuiper and his associates were pursuing brilliant new lunar mapping projects, Eugene Merle Shoemaker (1928–1997) (Figure 18.8) of the U.S. Geological Survey, later to become famous as a co-discoverer of the Shoemaker-Levy 9 comet, which broke into pieces that impacted on Jupiter in 1994, was publishing a series of classic papers destined to tip the balance decisively in favor of the impact theory of crater origin once and for all. The first, appearing in 1959, effectively completed Gilbert's pioneering survey of Meteor Crater and provided new insights into the physics of impacts.[13] Shoemaker built on work by astronomer-geodosist John A. O'Keefe (b. 1925) and the geologist Edward Ching-Te Chao (b. 1919) in which they had identified, in the quartz-rich sandstones found near the crater, crystals of coesite—a mineral formed by subjecting quartz to enormous pressures. No volcanic steam explosion could ever generate pressures on such a scale, and thus any lingering doubts that Meteor Crater had been blasted out by a cosmic impact were dispelled. Shoemaker's calcuations showed how the meteorite, striking the target rock in the Arizona desert (an event which occurred less than 50,000 years ago), had formed two interacting shock waves. The first engulfed the meteorite, vaporized it, and melted rock at the immediate point of impact but absorbed only a relatively small fraction of its energy. Most of

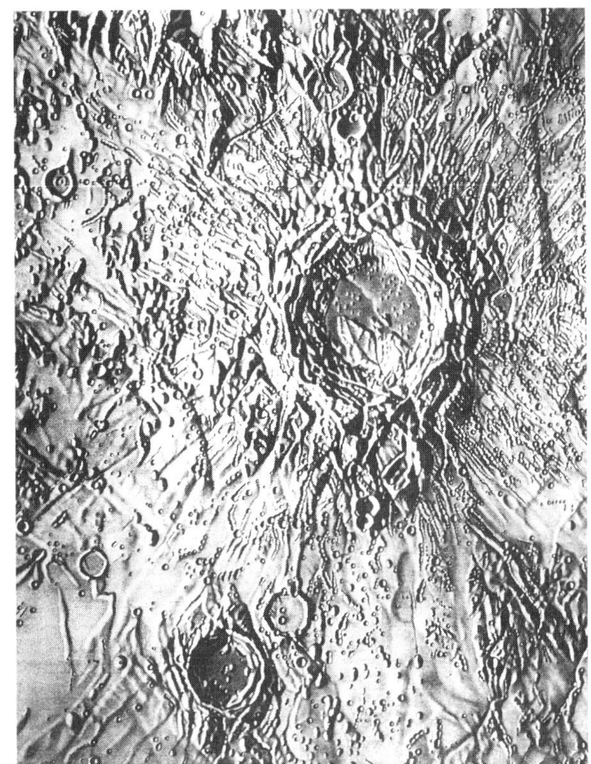

Figure 18.9 "The monarch of the Moon," the crater Copernicus was the subject of one of Eugene Shoemaker's classic papers in 1960. Fresh from his definitive study of the impact mechanics at Meteor Crater in northern Arizona, Shoemaker demonstrated that the crater could not possibly have been formed by volcanic activity. The loops and chains of "satellite" craters—abundantly represented in this shaded relief map of Copernicus issued by the U.S. Air Force Aeronautical Chart and Information Center—were secondary impact features formed by debris thrown out at high energies during the impact. Courtesy U.S. Air Force.

the energy went to produce a second shock wave traveling radially away from the point of impact, excavating the crater and throwing a rim of disintegrated material—an "ejecta blanket"—around it.

In 1960, Shoemaker applied the insights gleaned from his analysis of the impact process at Meteor Crater to Copernicus, the 97-kilometer wide "monarch of the Moon" (Figure 18.9).[14] Telescopic observation had revealed, scattered throughout its ray system, countless tiny "satellitic" craters, each on the order of one kilometer across. Some of these features had been perceived by the keensighted Gruithuisen, but each increase of aperture had revealed more. Beer and Mädler had counted sixty-one; Schmidt, describing this part of the lunar surface as "pierced like a sieve," had estimated their number at about 300; Pease, using the 100-inch Hooker reflector at Mt. Wilson in 1936, found "the smoothed surface about Copernicus becomes a billion crater pits, each showing typical crater shadows at low Sun."[15] The loops and chains formed by these satellitic craters were long regarded by volcanists as among the most compelling points of evidence of endogenic cratering processes, on the grounds that it was hard to imagine how they could possibly be accounted for by random impacts.

Fresh from his landmark study of Meteor Crater, Shoemaker immediately grasped (as New Zealand's Charles Gifford had also grasped in the 1920s, although no one had paid much attention at the time) that no volcanic process could ever have produced an explosion capable of throwing out masses of millions of tons of material across distances of 200 kilometers and more. Instead, he realized, the loops and chains of satellitic craters near Copernicus must be secondary impacts, caused by a peppering of debris thrown out at much higher energies than the thick and rugged ejecta material that had sloshed over the crater's outer flank. Other features of crater morphology were also convincingly explained in terms of Shoemaker's impact mechanics: central peaks were formed by violent rebound from the shock, while the terraced walls of many of the larger craters were due to landslip—slumping in of rim materials toward the center of the crater. The characteristic shallowness of lunar craters was due to their being partly filled with rim debris.

Shoemaker was instrumental in forming the astrogeology branch of the United States Geological Survey at Menlo Park, California (later moved to Flagstaff, Arizona). In 1962, he published the first geologic map of part of the Moon—the Kepler region. Among the questions not resolved up to that time was that which had intrigued Clerk and Hutton long before: the precise stratigraphic sequence of the formations of the Moon. It was obvious that the heavily cratered uplands were "pre-Imbrian" and had formed earlier and in a much shorter time than the relatively few craters superimposed on the mare (which defined the "Eratosthenian" and "Copernican" systems), but in an important 1962 paper, Shoemaker and his collaborator Robert Hackman (1923–1980) emphasized the extent to which ejecta from the Imbrium impact made up a stratigraphic unit overlying previous lunar landscapes. In doing this, they added another dimension to the study of the Moon's landscapes—the dimension of time.[16] Instead of being a mere map, the Moon unfolded as a narrative—a narrative in which, it became increasingly evident, the first chapter was an episode of massive bombardment.

It would be pleasant to be able to say that finding answers to such scientific questions was the principal motivation for the space program. Not so. The Race to the Moon was a vigorous but peculiarly ephemeral phenomenon that played out in the aftermath of World War II, when the United States and the Soviet Union, allies during the war, faced one another as antagonists across a vast ideological gulf. On October 4, 1957, the Soviets shocked the world when they launched the first artificial satellite, Sputnik, atop an ICBM known as the R-7, which looked "like a mechanical Cossack in billowing pantaloons."[17] Even then their "Chief Designer" Sergei Pavlovich Korolev (1907–1966)—his name long afterwards forgotten in the West—and his American rival, the German-born Wernher von Braun (1912–1977), were ardently dreaming of moonships.

The Moon became the quintessential goal for Soviets and Americans in order to demonstrate the awesome range and accuracy of their ICBMs. It was the ideal target—"close enough to reach, audacious enough to capture the imagination."[18] The Moon was the mountain in whose shadow we lived in awe; it was, as mountaineer Edward Whymper described the Matterhorn, "an affront to man's conquest of nature." The Matterhorn-moon was sharp, angular, steep-sided, inac-

Figure 18.10 The eastern limb and far side of the Moon, photographed by the Apollo 16 astronauts. The dark dark feature at the 8 o'clock position is Mare Crisium. The paucity of mare features on the Moon's averted hemisphere had been anticipated by a number of prescient earthbound observers, notably Nathaniel Shaler and H. P. Wilkins. Courtesy NASA.

cessible, and it became the grandest goal ever aspired to by restless humanity. But it was not primarily a scientific goal.

There can be little doubt that rockets of power sufficient to reach the Moon would never have been developed had it not been for the Cold War and the need of the Soviet Union and the United States to demonstrate to one another, in sublimated form, their ability to hurl multi-megaton warheads around the globe and onto one another's soil. It was almost as an afterthought that these weapons were yoked to the purpose of launching satellites and, by 1958, the first lunar probes.

The Russians had the early successes—on September 14, 1959, at precisely two minutes, 24 seconds after midnight, Moscow time, Lunik 2 became the first manmade object to reach the lunar surface, impacting on the rim of the crater Au-

Figure 18.11 An oblique view of the 185 km-diameter crater Tsiolkovsky, photographed from lunar orbit by the Apollo 15 astronauts. One of the most prominent features on the lunar far side, Tsiolkovsky has a smooth, lava-flooded floor and complex central peak. Courtesy NASA.

tolycus. It was soon overshadowed, however, by Lunik 3, the first spacecraft ever to circle around to the Moon's far side, obtaining blurry photographs of that hitherto unknown region of long-forgotten speculation. The far side proved to have few "seas," as earlier students like Nathaniel Shaler and H. P. Wilkins had correctly surmised from their careful observations of limb regions (Figure 18.10). However, there was at least one lava-flooded sea, Mare Moscoviense, and the most prominent feature of all, a magnificent dark-floored crater (Figure 18.11), was fittingly named for Konstantin Tsiolkovsky (1857–1935), the deaf schoolteacher and rocket visionary from the town of Kaluga whose monumental work had inspired Korolev and through him the whole Russian space program.

In those days Russian plans were closely guarded secrets, and even the location of the launch facility, the Baikonur Cosmodrone, was never revealed (it was not actually located at Baikonyr but 300 kilometers to the southwest, near the town of Tyuratam in Kazakhstan). But America's intentions were clearly stated in a stirring speech given on May 25, 1961 by President John F. Kennedy. Hoping to restore America's prestige and his own after the botched invasion of Cuba at the Bay of Pigs only a month earlier, Kennedy tossed American caps for the horns of the Moon with the announcement: "I believe that this nation should commit itself to achieving the goal, before this decade is out, of landing a man on the Moon."[19] The attempt to reach this imposing goal thus became "part of the battle along the fluid front of the Cold War."[20]

Kennedy unleashed an era of "budgetless financing" for the American Moon program, which at its peak employed 417,000 people in NASA and its supporting industries. In all, Project Apollo would require an investment of $24 billion in 1969 dollars, which works out to $2 billion a ticket for each man who eventually walked on the surface of the Moon. Again, though the Soviet Moon program was always shrouded in secrecy, there can be no doubt that it involved a comparable commitment of manpower and rubles. Despite later attempts by the Soviets to rewrite history and to deny that they had ever planned to send men to the Moon at all—hence that the "Race to the Moon" had been a myth—they were still testing their Zond spacecraft, and thus very much in the race, in the months before the Americans succeeded in sending Apollo 8 around the Moon in December 1968.[21] Zond was essentially a stripped-down, unmanned version of the Soyuz vehicle that was intended to ferry cosmonauts to the Moon.

The massive build-up for the "Moon Race" led to a golden era of lunar exploration and of Solar System studies more generally. There were the Luna, Ranger, Surveyor, and Lunar Orbiter automated probes, followed by the Apollo circumlunar flights and manned landings. After the Americans circled the Moon with Apollo 8, then landed in the Mare Tranquillitatis with Apollo 11, the point had been proved. Kennedy's Moon had been reached, and the Americans had won.[22] That the other Apollo missions took place at all was largely a matter of sheer momentum. Apollo 12 landed at Oceanus Procellarum, another mare site and the location at which the Surveyor 3 had been resting since it softlanded in 1967, in November 1969 (Figure 18.12). It was followed by the aborted Apollo 13 mission, then by Apollo 14, which made a pinpoint landing at a site on the Imbrium ejecta blanket at Fra Mauro. The later Apollos carried lunar rovers, complete with computers for navigating on the surface. Indeed, they were—in terms of scientific data returned and the grandeur of the landscapes explored—"great explorations," comparable to the voyages of discovery of Columbus and Magellan in the fifteenth and sixteenth centuries.[23] Apollo 15 reached Hadley Rille in the foothills of the Apennines (Figure 18.13); Apollo 16 explored the crater Descartes in the southern highlands; and Apollo 17 was aimed at the Taurus-Littrow valley near the edge of Mare Serenitatis.

By the time Apollo 17 returned to the Earth, public interest in the Moon program had largely evaporated. Kennedy was long gone, and so was his successor, Lyndon Johnson. Richard Nixon occupied the White House. The United States,

Figure 18.12 This image shows the Surveyor 3 spacecraft, which made a soft landing on April 20, 1967. In the background is the Apollo 12 Lunar Module, which landed less than 200 yards away in November of 1969. Courtesy NASA.

Figure 18.13 A highlight of the Apollo 15 mission was the traverse to the verge of Hadley Rille by astronauts David Scott and James Irwin. This photograph shows Irwin and the Lunar Rover. Courtesy NASA.

still carrying on its struggle along the extended front of the Cold War, had become bogged down in the quagmire of Southeast Asia. It had reached the Moon, but was losing a war with a Third World Soviet client. Instead of setting out for bolder goals in space—Mars, for instance, as recommended by Nixon's own Space Task Group—the mood had changed. Space no longer enjoyed the priority it had only a decade earlier. Apollos 18 and 19 were canceled, and so was production of Wernher von Braun's awesome launch vehicle, the Saturn V. A number of the great Moon rockets were left to rust. Some of them can still be seen, brooding in outdoor museums at Cape Canaveral, Huntsville, and Houston. Meanwhile, Soviet manned lunar efforts had ended even more ignominiously with explosions of their still unperfected giant booster rocket N-1 in 1969 and 1970. Although the Soviets later sent a series of unmanned spacecraft, including automated rovers (or *Lunakhods*) and sample-return vehicles, their interest was hardly more sustained than the Americans. By 1976 they too had abandoned their program of lunar exploration.

The moment for manned lunar exploration had come and passed. As had happened in the aftermath of the Beer and Mädler map and in the inter-war years of the twentieth century, the fire of humanity's desire for the Moon had once more, at least temporarily, flickered and gone out.

In retrospect it seems rather strange that throughout the Apollo era, the volcanic theory—and the TLPs—still had a profound influence on lunar studies. Beginning with the robot probe Ranger 9, which impacted in Alphonsus in March of 1965, many of the destinations on the Moon, including most of the Apollo landing sites, were selected for their supposedly volcanic features. Apollo 16, for instance, visited the highland crater Descartes where earlier images had shown what seemed to be lava flows, cinder cones, and explosion craters; but the astronauts, John Young and Charlie Duke, brought back not the expected volcanic rocks but impact-shocked breccias. Apollo 17's destination in the Taurus-Littrow hills on the shore of Mare Serenitatis had similarly been chosen in part because Alfred Worden, pilot of the Apollo 15 command module, had reported seeing "a whole field of cinder cones" there. Apollo 17 astronauts Eugene Cernan and Harrison Schmitt located one of these features—an ordinary fresh impact crater surrounded by an ejecta blanket comprised of dark soils rich in iron and titanium.

Only by the end of the Apollo era, the fires in the Moon that earthbound observers had so long and so zealously sought had grown nearly dark. Though there were volcanic features on the Moon—the lava-fields covering the great basins, for instance—their fires died out well over a billion years ago. The long pursuit of eruptions and gaseous emanations, the passionate search for evidence of change that had been the Holy Grail of so many observers, proved in the end to be a chimera, a latter-day quest for the Philosopher's Stone. "Whatever the real or psychological cause of transient phenomena," concludes Don Wilhelms, "it is not volcanism."[24]

Searching for volcanic eruptions on the Moon, the theme of so much of this history, turns out to have been the equivalent of awaiting a Krakatoa-like explosion in the Appalachians of the Eastern United States or the Cumbrian hills of Wales.

After Apollo, the Moon no longer wore the mask of the volcanic analogy; it no longer masqueraded as another Earth. Instead it had become "itself, itself alone"—a small planetary object with a unique history and destiny, a world which had been fashioned and shaped and modified over vast lapses of time. It was no longer a cryptogram, a text written in inscrutable characters—no longer a kind of cosmic Balshezzar's writing or Dighton Writing Rock.[25] Its face had yielded to partial decipherment.

Above all else, the face of the Moon proved to be incredibly ancient. Containing a record of events from an almost inconceivable past, it is a palimpsest of violent episodes, all trace of which had long since been erased by erosion from the Earth. Whether it had been ejected from the Earth, formed simultaneously and then been captured, or come about through some other agency entirely, there could at least be no doubt about its great antiquity.

In fact, the Moon formed at almost the same time as the Earth. According to the most recent analysis, its age is 4.51 billion years as against the Earth's 4.55 billion years.[26] Early on, the Solar System was still cluttered with planetesimals—whirling debris left over from the formation of the planets. The Moon and the other planets, including the Earth, were subjected to a hailstorm of debris and massively battered. Impacts occurred at rates so intense that the whole surface of the Moon became saturated, one impact crater forming only by piling on and obliterating the one beneath it. This is the era of lunar history recorded in the rugged southern highlands.

"The beauty of nature lies in its details, the message in generality."[27] The message here is that the phase of planetary bombardment was universal. The Earth too was once shell-shocked and battered beyond belief. But the Earth has long since forgotten that part of its history. The Moon, by contrast, is a Mnemosyne-world, a world of incredibly ancient memory, retaining the impressions of its infancy in its almost inconceivably old rocks.

The planetary shards that formed these ancient scars were high-velocity projectiles, traveling at speeds of tens of kilometers per second. When they were suddenly stopped at the lunar surface, the kinetic energy of their motion was instantaneously transformed according to the expression $e = \frac{1}{2}mv^2$, i.e., energy is proportional to mass multiplied by the square of velocity. In a nutshell, this formula shows how relatively small objects are able to produce tremendous explosions. A ten-kilometer object, travelling at cosmic speeds, carries the energy equivalent to millions of hydrogen bombs, enough to blast out a crater on the Moon hundreds of kilometers wide. And there are scores of such craters on the surface of the Moon.

Toward the end of the torrential bombardment, some particularly large bodies began smashing into the Moon, forming the multi-ring basins. An object the size of the state of Rhode Island formed the great grey eye of mighty Imbrium. Its main ring, some 1,140 kilometers wide, is defined by the Caucasus, Apennine, and

Figure 18.14 The South Pole-Aitken basin, which extends well over onto the lunar far side. It was missed by the Lunar Orbiter spaceprobes of the mid-1960s, whose trajectories always passed over it when it was poorly illuminated, thus becoming the last remaining bastion for amateur Moon-mappers. This mosaic of 1500 images obtained by the Clementine spacecraft in 1994 shows the south pole at center and extends to 70 degrees S latitude at the edge. After the Hellas basin on Mars, the South Pole-Aitken basin is the largest formation of its kind known in the entire Solar System. Part of its ejecta blanket forms the lofty Leibnitz mountain range near the south limb of the Moon. In the last few years, radio astronomers have discovered evidence that water ice may lurk in some of the shadowed craters near the pole itself. Courtesy NASA.

Carpathian ranges, which are actually huge ejecta blankets. There is also a fairly well-defined but fragmentary inner ring that contains the Spitzbergen, Teneriffe, and Straight Ranges, along with such isolated peaks as Pico and Piton. Hurtling debris from the impact gouged out bordering highland terrain to produce radial grooves and blanketed many of the pre-Imbrium walled plains in the widespread hummocky deposit known as the Fra Mauro formation, the landing site of Apollo 14. The dark-floored crater Julius Caesar, located not far from the center of the Moon's near side, has been filled to a depth of one kilometer by Imbrium ejecta even though it lies at a distance of 600 kilometers from the Imbrium basin.

Imbrium itself is certainly the largest and most spectacular impact feature on the Moon's near side, but it is not the only example of a multi-ring basin. Another is Nectaris; its inner ring lies within the boundary of the mare, its outer ring is marked on the west by the prominent Altai Scarp and on the east by the Pyrenees

Figure 18.15 Harold Hill at the eyepiece of his 10-inch f/10 Newtonian reflector, located at his private observatory at Dean Brook House, near Wigan, Lancashire. Without a doubt the most skillful draughtsman of the lunar surface to have appeared in the last 30 years, Hill depicts lunar formations using a very pleasing pen-and-ink stippling technique that is now widely imitated. He has been especially interested in observing some of the less heralded lunar formations. Since the late 1940s he has made a special study of the formations of the south polar region under favorable conditions of libration. Much of this work is yet to be published. Photograph by William Sheehan, 1993.

mountains. Other basins on the near side of the Moon include Serenitatis, Crisium, Humorum, Foecunditatis. The best-preserved lunar basin, and almost surely the youngest, is Mare Orientale. Its outer ring, the Cordilleras mountains, measures some 900 kilometers across, and two of the inner rings are known as the outer and inner Rooks. But the largest basin on the whole Moon—and after the Hellas basin on Mars, the largest known anywhere in the Solar System—is the 2,500 kilometer-wide South Pole-Aitken basin (Figure 18.14), centered at 56° S, 180° W. Incidentally, this region was hidden in shadow during the Lunar Orbiter mapping missions of the 1960s, so detailed maps remained a last bastion of amateurs like John Westfall and his associates of the Association of Lunar and Planetary Observers and, working independently, the talented British lunar mapper Harold Hill (b 1920) (Figure 18.15). It was finally imaged in detail by the Clementine spacecraft early in the 1990s.

The ages of several of the basins have been conjectured from samples brought back by the Apollo astronauts and the Soviet Lunikhods, the automated sample-return missions of the early 1970s. That of Nectaris is thought to be around 3.92 billion years; Serenitatis 3.87 billion; Crisium and Imbrium 3.84 billion, and Orientale 3.80 billion years.

In contrast to the maria, which are predominantly a near-side phenomenon, the basins are well-represented on both the near and far sides of the Moon. They seem to have a preferential polar concentration which may be accounted for by the fact that the Moon formed quite close to the Earth.[28] Baldwin attempted to explain the formation of the ring structures in terms of radiating concentric shock waves frozen at some distance from the inner basin rim, and even compared the impact-produced shock to a tsunami.[29] This seems to be more or less correct. Another of his insights was that the inner rings were somehow basin equivalents of the central peaks of lesser formations.[30] The details of how the ring structures formed are even now not entirely clear, however. One thing is certain: they are characteristic only of the largest craters of the Moon—those exceeding 140 kilometers in diameter.

Originally, heat from all the impacts that formed the Moon melted it throughout. Even the surface was molten, and covered by what has been called the magma ocean. Later as the heat leaked out and was lost by radiation into space, the outer layers of the magma ocean cooled first. Hence the lighter highland rocks represent the most primitive crust—they are anorthosites, rocks rich in feldspar, a mineral whose crystals are first to form when molten rock cools and which therefore tends to float to the surface. The heavier basalts tended to settle at the base of the magma ocean.

At the end of the basin-forming era, the basalts were partially remelted by the heat produced by the decay of radioactive materials trapped within the Moon. The ensuing extrusion of lavas onto the lunar surface was assisted by the presence of an extensive system of cracks and fissures produced by impacts and subsequent crustal adjustments. After being extravasated onto the surface from depths of more than 100 kilometers, the heavy basalts attempted to sink back again, but by then the outer crust was firm enough to support them. The basalts thus contributed to the mass concentrations ("mascons") detected by their anomalous gravitational attraction that perturbed the orbits of the Lunar Orbiter and Apollo spacecraft. The mascons are the dense, solidified lava lakes and upwarped mantle rocks that fill these basins.

Eventually the lavas filled all the near-side basins and disguised their identities as *maria*—Imbrium, Foecunditatis, Nectaris, Tranquillitatis, Crisium. Orientale is an intermediate case; it remained partially "dry." These marial basalts are actually rather thin veneers, comprising less than one percent of the volume of the lunar crust. Most of the far-side basins never did fill with lava, in part because the crust there was thicker and presented too great a barrier for the lavas to penetrate. Of course, all of the lava-flows did not happen at once. The basalts in Mare Tranquillitatis sampled by Apollo 11 are rich in titanium (explaining their slightly bluish hue) and date to 3.7 billion years, while those at the Apollo 12 site in Oceanus Procellarum are only 3.1 billion years old. Clearly the Moon is not, as Urey hoped, made up of primordial material; it has melted and differentiated, and different regions of the interior have melted to produce different kinds of lavas.

At last extremely fluid lavas were extruded into the existing lava fields to form the meandering, sinuous rilles found chiefly along the margins of the maria. These features are actually collapsed lava tubes. Other characteristic features of the maria formed when the solidified lava lakes slumped inward; they cracked at their outer edges to form arcuate rilles or grabens and buckled in their interiors to produce wrinkle ridges. Still other features associated with what can be defined, broadly speaking, as the epoch of lunar volcanism include the domes. Some do have summit craters or networks of fissures, proving that they are indeed volcanic. They seem to have been produced by eruption of more viscous basaltic lavas onto the surface. There are even, evidently, a few *bona fide* cinder cones, notably the long-studied dusky spots in Alphonsus that had suggested patches of vegetation to Gruithuisen and Pickering. However, most of the dark-haloed craters are nothing more than ordinary impact craters that excavated darker basalts from beneath a superfical layer of lighter material. Although volcanic processes as well as impacts have played a role in shaping the surface of the Moon, the dominant forms of the lunar surface—the walled plains and craters—are clearly impact features.

The oldest lavas began to flow onto the surface just after the major basin-forming era ended, about 700 million years after the formation of the Moon itself. By then, the rate of impacts was also falling off. As the Moon continued to cool, its heat continued to retreat further and further inwards, until all signs of volcanism ceased (for that matter, the interior of the Moon is still warm and even seems to be molten, which means the temperature must be more than 1300° C). By 3 billion years ago, when the lava flows had mostly ended, the earthward face of the Moon presented nearly the same features it shows today, with only the occasional impact registering the continuing passage of time and serving as scribe to the Moon's ongoing history of planetary evolution. The youngest craters are clearly identifiable by their brilliant ray systems—they are indeed splash features as G. K. Gilbert realized, consisting of pulverized debris shot from impact-centers and still fresh enough not to have been weathered by the steady infall of micrometeorites. The age of mighty Copernicus, tentatively dated from an Apollo 12 sample, seems to be 810 million years, while Tycho, dated from a farflung ray which crossed the Apollo 17 landing site, seems to be 109 million years old—a mere heartbeat in selenological time. Some of the dinosaurs might even have glanced up and seen the spark of its formation.

Whence came the Moon in the first place? This has long been a riddle wrapped in a mystery inside an enigma. Wind the Moon's motion backward to the beginning; reverse the effects of the tides; follow it as it spirals back ever closer to the Earth and as the Earth's rotation speeds up. By extrapolating back from current rates of tidal evolution, one finds that our satellite ought to have been in actual contact with the Earth 1.6 billion years ago, or only a third of the actual age of the Earth and the Moon. This cannot have been the actual time of the Moon's origin, however. In fact things are rather more complicated—the so-called "tidal dissipation function" has not been constant throughout the history of the Earth-Moon system. As a result, one cannot reconstruct its history in detail—much less come to a theory of its origin—from considerations of celestial mechanics alone.[31]

The fission theory has long since had to be abandoned, since it is impossible for the Earth to have rotated sufficiently fast to have cast part of itself off into space. Gone too is the theory that the Moon emerged out of the Pacific—a theory at odds, in any case, with the ideas of plate tectonics, according to which the age of the Pacific Basin must be considerably less than 500 million years. Capture theories that envision a Moon that formed as a separate object and was then reeled in by gravitational forces also face an array of difficulties, the most insurmountable of which is the problem of angular momentum. In simple terms, it is nearly impossible to get the Moon to travel at just the right rate of speed for capture to occur.

The old theories were clearly hopelessly inadequate. On the other hand, there was no way around the embarrassing fact that the Moon actually exists!

By the early 1970s, a new paradigm for the origin of the Moon began to emerge, drawing on accretion theories of planetary origin worked out largely by the Russian astronomer Viktor S. Safronov. As far back as 1966, Safronov suggested that the axial tilts of the planets—especially Uranus, which appears to have been reoriented onto its side—had been determined by the last large impact to which they had been subjected during the accretion era.[33] Spacecraft images of Mars and Mercury in the early 1970s showed that multi-ring basins are prevalent on other bodies in the Solar System, meaning that these planets, like the Moon, had at one time been walloped by large bodies up to 100 kilometers across. Finally, it was observed that the composition of the Moon, at least to a rough approximation, resembled the material of the Earth's upper mantle.[34]

William K. Hartmann recalls,

> With lunar origin, I kept thinking the Moon to first order seems a lot like mantle material—same density, etc. But all the detail scientists—the geochemists with parts per billion accuracy—told me, "No, there's such and such a difference with such and such isotope ratios." I thought, "Well, they know what they are talking about so I guess my idea is wrong."[35]

But Hartmann persisted. He realized that following the formation of a planet by accretion in a certain zone in the Solar System, there must be a second-largest body in that zone, a third-largest body, and so on. With Don Davis he carried out a series of numerical calculations of the size distributions of planetesimals by the end of the accretion process, and found that the distribution must include several largish bodies which, arriving at the Earth's surface at 13 kilometers per second, would be sufficient to eject two lunar masses of material to near-escape velocity. Assuming that a collision of such a body with the Earth took place after the Earth's iron core had already formed, the material ejected by the impact would be depleted in iron, just as in the fission theory. The ejected material would form a cloud of hot dust that would be rapidly depleted in volatiles. The particles in such a swarm would rapidly collapse into the equatorial plane of the Earth where a satellite could form.[36]

Hartmann briefly presented his idea of a big-impact origin of the Moon at the Conference on Satellites of the Solar System at Cornell University in 1974. Meanwhile, a similar idea had occurred independently to A. G. W. Cameron of Harvard, who was present at the same meeting, and his colleague W. R. Ward.[37] As later recounted by Cameron, they

> were led to the suggestion of a collisional origin of the Moon through the following consideration. The angular momentum of the Earth-Moon system is less than sufficient to spin the Earth to rotational instability; we were nevertheless interested to determine the mass of the body which, striking a tangential blow to the protoearth, could impart the angular momentum of the Earth-Moon system to the protoearth. The required projectile turned out to be about the mass of Mars.[38]

The tangentially-directed whack gave the Earth a rapid prograde paleo-rotation of about five hours, a rotation which the Moon has been braking ever since. More recent calculations seem to indicate that the impactor that formed the Moon may have been somewhat larger than at first thought, perhaps two or three times as massive as Mars. The ejecta coalesced only 22,500 kilometers from the Earth. After the outer ring of material accreted to form the Moon—possibly in a period as short as one year—its gravitational forces began scattering the inner disk material back onto the Earth.[39] Then the Moon began its long and tortuous tide-driven drift outward from the Earth. It will continue its drift outward until, in the remote future, the Earth and the Moon will be locked in a frozen embrace like Pluto and its satellite Charon. Then the Earth will turn on its axis once in the same period the Moon revolves around it, about once every 50 days.

References

1. G. A. Briggs, "What Good is a New Moon?" in *Lunar Geoscience Working Group, Status and Future of Lunar Geoscience*, NASA SP–484 (1986), pp. iii–iv.
2. F. E. Wright, "The Lunar Surface: An Introduction," in *The Moon, Meteorites, and Comets,* eds. B. M. Middlehurst and G. P. Kuiper (Chicago: University of Chicago Press, 1963), p. 47.
3. Donald E. Osterbrock, personal communication to William Sheehan.
4. W. K. Hartmann, "Discovery of Multi-ring Basins: Gestalt Perception in Planetary Science," *Proceedings of the Conference on Multi-ring Basins: Formation and Evolution*, Houston, Texas, 10–12 November 1980. Proceedings of the Lunar and Planetary Science Conference 12A, (New York: Pergamon), 79–90.
5. W. K. Hartmann to William Sheehan, December 10, 1997; personal communication.
6. Wilkins wrote: "Under conditions of extreme libration a large, very foreshortened, dark plain appears on the south-east [southwest according to the current astronautical convention] limb of the Moon… Generally but a portion is visible, the whole being seen only under the most favourable circumstances when it appears as a narrow dark streak requiring considerable telescopic power for the detection of its rampart and surface details." H. P. Wilkins, "The Lunar 'Mare X,'" *Journal of the British Astronomical Association*, 48 (1937), 80–82. Our appreciation to Harold Hill for calling this publication to our attention.
7. Hartmann, "Discovery of Multi-ring basins," pp. 84–86.
8. W. K. Hartmann, "Radial Structures Surrounding Lunar Basins I: The Imbrium System," *Communications of the Lunar and Planetary Laboratory*, 2 (1963), 1–15.
9. W. K. Hartmann and G. P. Kuiper, "Concentric Structures Surrounding Lunar Basins," *Communications of the Lunar and Planetary Laboratory*, 1 (1962), 51–66.
10. Baldwin quoted in Hartmann, "Discovery of Multi-ring Basins," p. 89.
11. Ibid.
12. W. K. Hartmann, personal communication to William Sheehan, December 10, 1997.
13. Eugene M. Shoemaker, "Impact Mechanics at Meteor Crater, Arizona," *U.S. Geological Survey Open File Report*, 1959; "Penetration mechanics of high velocity meteorites, illustrated by Meteor Crater, Arizona," *International Geological Congress Report, XII Session (Norden),* part 18 (1960), 418–435. Shoemaker's contributions are placed in context of the wider debate over Coon Mountain, now better known as Meteor Crater, in Hoyt, *Coon Mountain Controversies.*
14. Eugene M. Shoemaker, "Interpretation of Lunar Craters," in *Physics and Astronomy of the Moon*, ed. Z. Kopal (New York, Academic Press, 1962), 283–359.
15. Quoted in F. E. Wright, F. H. Wright, and Helen Wright, "The Lunar Surface: Introduction," in *The Moon, Meteorites and Comets*, eds. Barbara M. Middlehurst and Gerard P. Kuiper (Chicago: University of Chicago Press, 1963), p. 31.

16. Eugene M. Shoemaker and Robert J. Hackman, "Stratigraphic Basis for a Lunar Time Scale," in *The Moon*, ed. Z. Kopal and Z. K. Mikhailov (London: Academic Press, 1962), pp. 289–300.
17. Walter A. McDougall, *The Heavens and the Earth: A Political History of the Space Age* (New York: Basic Books, 1985), pp. 60–61.
18. Andrew Chaikin, *A Man on the Moon* (New York: Viking Press, 1994), p. 578.
19. "President Speaks," *The New York Times,* May 26, 1961.
20. John M. Logsdon, *The Decision to Go to the Moon* (Cambridge, Massachusetts: MIT Press, 1970), p. 128.
21. James E. Oberg, *Red Star in Orbit* (New York: Random House, 1981).
22. The literature on Apollo is vast. For a bibliography through 1994, see Chaikin, *A Man on the Moon.*
23. Paul D. Spudis, *The Once and Future Moon* (Washington: Smithsonian Institution Press, 1996), p. 70.
24. Wilhelms, *To a Rocky Moon*, p. 341.
25. A rock with figures and pictographs near the Taunton river in southeastern Massachusetts, whose marks, long a subject of controversy and wild speculation, are now believed to be native Algonquin. See Edward R. Tufte, *Envisioning Information* (Cheshire, Connecticut: Graphics Press, 1990), pp. 72–73.
26. "Dating a Mixed-up Moon," *Sky & Telescope*, 95, 3 (1998), p. 26.
27. Stephen Jay Gould, *Wonderful Life: The Burgess Shale and the Nature of History* (New York: W. W. Norton, 1989), p. 13.
28. N. Barricelli and R. Metcalfe, "The Lunar Surface and Early History of the Earth's Satellite System," *Icarus*, 10 (1969), 144. This paper showed that families of circumterrestrial planetesimals exterior to the Moon's orbit can be perturbed into lunar collisions with non-random latitude distributions.
29. Ralph Baldwin quoted in W. K. Hartmann and C. A. Wood, "Moon: Origin and Evolution of Multi-ring Basins," *The Moon*, 3 (1971), 3–78.
30. Baldwin, "Ancient Giant Craters and the Age of the Lunar Surface," *The Astronomical Journal*, 74 (1970), 570. Baldwin was extrapolating from work by Charles A. Wood, another student of Kuiper, who had observed that noticeable central peaks begin to appear in lunar craters of about 10 kilometers and are present with a frequency of 100% in craters more than 60 kilometers across. C. A. Wood, "Statistics of Central Peaks in Lunar Craters, *Communications of the Lunar and Planetary Laboratory,* 7 (1968), 157.
31. A. P. Boss and S. J. Peale, "Dynamical Constraints on the Origin of the Moon," in *The Origin of the Moon*, eds. W. K. Hartmann, R. J. Phillips, and G. J. Taylor (Houston: Lunar and Planetary Institute, 1986), 59–101.
32. For a comprehensive treatment of the various theories of the origin of the Moon, see Stephen G. Brush, *Fruitful Encounters: The Origin of the Solar System and of the Moon from Chamberlin to Apollo* (Cambridge: Cambridge University Press, 1996).
33. William Hartmann, personal communication to William Sheehan.
34. V. S. Safronov, "Sizes of the Largest Bodies Falling onto the Planets during their Formation," *Sov. Astron. AJ*, 9 (1966), 987–991.
35. Alan B. Binder, "On the Origin of the Moon by Rotational Fission," *The Moon*, 11 (1974), 53–76 and "On the Petrology and Structure of a Gravitationally Differentiated Moon of Fission Origin," *The Moon*, 14 (1975), 431–473.
36. William K. Hartmann and Donald R. Davis, "Satellite-sized Planetesimals and Lunar Origin," *Icarus*, 24 (1975), 504–515.
37. A. G. W. Cameron and W. R. Ward, "The Origin of the Moon," *Lunar Science*, 7 (1976), 120–122.
38. A. G. W. Cameron, "Formation of the Prelunar Accretion Disk," *Icarus*, 62 (1985), 319–327:319. Since it was first proposed, the giant impact hypothesis has been the subject of numerous papers, some of which are included in the proceedings of the conference on the origin of the Moon, orga-

nized by Hartmann in Hawaii in October 1984: *Origin of the Moon*, ed. Hartmann, Phillips, and Taylor.

39. "How to Make Earth's Moon," *Astronomy*, 26 (1998), 24.

Epilogue

It may be that the presence of our anomalously large Moon in the vicinity of the Earth has played a critical role in the evolutionary history of life. The Moon's most obvious effect has been the tides. We can already trace their presence in terrestrial sedimentary rocks three billion years old. In South Africa's Pongola supergroup of rocks of that era, there is a regressive sequence of sediments in which a tidal flat prograded seaweed over a subtidal basin. Some of the ripples are double-crested, reflecting the ebb and flow of the tides from a time when the Moon was less than 300,000 kilometers away and the Earth's rate of rotation was between 17 and 18 hours.[1] As that time, of course, all life still remained in the sea. When animals finally did advance from water to land, they were able to do so only very gradually. They required adaptations which would allow them to resist dehydration, to use atmospheric instead of water-dissolved oxygen, to move about on land, and to reproduce on land instead of water. All of these adaptations could never have occurred at once, and likely would not have occurred at all but for the existence of coastal areas and marshes where intermittent exposure of organisms to dry conditions produced by the tides allowed their gradual emergence. The Moon, in these intertidal basins, first summoned life from the sea. Even today, cycles of life still reflect lunar influence; horseshoe crabs emerge from the ocean to lay their eggs, an activity triggered by the Full Moon.

The Moon has also had a moderating effect on the Earth's climate. Its presence has acted like a massive counterweight to stablize the spin-axis of the Earth, thereby preventing the wild gyrations of climatic conditions found, for instance, on Mars.[2] We have only just begun to fathom how much we may owe to our old and battered companion.

There may be even deeper connections. In their first paper, Hartmann and Davis commented on the philosophically satisfying aspects of the big-impact theory:

> There has always been difficulty in accounting for all properties of all satellite systems by a simple evolutionary theory. Jupiter and Saturn have "miniature" solar systems with retrograde outriders. Uranus has its spin and satellites' angular momentum vector radically altered. Earth is a "dual" planet with a relatively large satellite. Mars has only two tiny moons. Venus and Mercury have none. This heterogeneity becomes more satisfyingly accountable if it is viewed as the product of events involving statistics of small numbers. Does the second-largest planetesimal in each system hit the planet after 107 or 108 years? Is it large or small? Does it hit the planet dead center? Retrograde? A glancing blow prograde? Or is it captured? Or is it destroyed by a

> planetesimal-planetesimal collision so that it has no appreciable effect on the planet other than to produce many small craters? Or does it hit a preexisting satellite of the planet, perhaps converting it to several small satellites? Only one of these kinds of fates can befall the second-largest planetesimal. And this fate, the product of small-number statistical chance encounters, may determine whether the planet acquires a tilted axis, a massive circumferential swarm of dust, a captured satellite, or perhaps loses a larger satellite, gaining small fragmentary satellites.[3]

The Moon is, therefore, the product of a chance event. The residue of an ancient and haphazard collision, it records the glancing blow of a large planetary body which gave the Earth not only its Moon, but also its axial tilt and its rapid prograde rotation.

Statistically, the Earth might with equal chance have ended up like its "sister" planet Venus. The starting conditions are determined largely by chance, but they lead in rapidly divergent directions—to one planet which is hospitable, and to a second which is hostile and forbidding beyond belief. The giant impact imparted to the Earth its relatively fast prograde rotation, whereas Venus's slow—and now retrograde—spin was assembled from numerous small impacts whose effects largely canceled one another.[4] The energy imparted by the giant impact may have made the Earth hot enough for it to support the development of plate tectonics, which never arose on Venus. It may also have blasted away enough of the Earth's early volcanically-enriched atmosphere to allow the Earth's surface temperature to drop below the boiling point of water.[5] Thus, instead of a hell-hole like Venus, where the only liquids able to flow on the surface are molten metals and lava, the Earth evolved into a comparative Eden among the planets. It has developed oceans of liquid water, a breathable atmosphere, and a moderate environment for the development of life. All thanks to the Moon.

Humble shard of debris it may be, a mere fragment of planetary bric-a-brac. But the Moon has clearly been potter as well as pot. The Earth and Moon have developed together. Only because of the Moon—a mere husk of rock drifting through space, unaware of its own existence or of ours—the Earth has developed in a way that has been favorable to life and has allowed complicated creatures to emerge such as ourselves, with brains capable of pondering the silent and enigmatic features of its face.

> The mild moon,
> Who comforts those it sees not, who knows not
> What eyes are upward cast...[6]

Why, then, did we go to the Moon? Now we can at least do better than the usual answer—"Because it's there."

Three thousand years ago Homer sang of the moonlit scene of watchtowers and fires before Troy, scarcely dreaming what epic was inscribed above him in the Moon's inscrutably blank and impassive mask. There lay legible but in cipher the drama of the birth of worlds, of giant wars, of Olympian and Titanic struggles, tales of the robust and infant Earth and the wan and care-lined Moon, "majestic though in ruin."

Is there not something in this more than Homeric, Virgilian, Dantean?

Generations of monomoniacal Ahabs, peering through telescopes, failed to decipher this great white whale of the Moon. For that we had to moor by its scaly rind for ourselves.

> To the Moon!
>
> There is much in that sound to inspire proud feelings; but whereto does all that circumnavigation conduct? Only through numberless perils to the very point whence we started, where those that we left behind secure, were all the time before us.[7]

In the end we have gone to the Moon and discovered the Earth. The destiny of this small neighbor planet has been inextricably intertwined with our own. No story can be more important to us. Its story is our story. It may even be: no Moon, no man.

We are now expert in lunar phrenology, having learned to read the characters in its bumps and scars. "O head! thou hast seen enough to split the planets and make an infidel of Abraham, and not one syllable is thine!"[8] In the Moon we have found the impassive mask of a primitive planet, which wears the expression the Earth once wore in the earliest chapter of its life. We have looked long into the face of the man in the Moon, and have found it a reflection of ourselves.

References

1. "Thank the Moon You're Here," *Sky & Telescope*, 95 (1998), 20–21.
2. W. R. Ward, "Comments of the Long-term Stability of the Earth's Obliquity," *Icarus*, 50 (1982), 444–448.
3. William K. Hartmann and Donald R. Davis, "Satellite-sized Planetesimals and Lunar Origin," *Icarus*, 24 (1975), 504–515.
4. John A. Wood, "Moon over Mauna Loa: A Review of Hypotheses of Formation of the Earth's Moon," in *The Origin of the Moon*, eds. W. K. Hartmann, R. J. Phillips, and G. J. Taylor (Houston: Lunar and Planetary Institute, 1986), pp. 17–55.
5. Stephen J. Brush, *Fruitful Encounters: The Origin of the Solar System and of the Moon from Chamberlin to Apollo* (Cambridge: Cambridge University Press, 1996), p. 246.
6. John Keats, *The Fall of Hyperion*, Book 1, lines 269–271.
7. Herman Melville, *Moby Dick*, in Great Books of the Western World (Chicago: Encyclopaedia Britannica, 1952), p. 176.
8. Ibid., p. 230.

Index